JOHN I. JOHNSON, JR.
BIOPHYSICS DEPARTMENT
MICHIGAN STATE UNIVERSITY
EAST LANSING, MICHIGAN 48823

S0-EPR-343

TASTE AND SMELL
IN VERTEBRATES

TASTE AND SMELL
IN VERTEBRATES

A Ciba Foundation Symposium

Edited by
G. E. W. WOLSTENHOLME
and
JULIE KNIGHT

J. & A. CHURCHILL
104 GLOUCESTER PLACE, LONDON
1970

First published 1970

Containing 163 illustrations

I.S.B.N. 0 7000 1454 3

© Longman Group Ltd. 1970

All rights reserved. No part of this publication may be reproduced, stored in a retrieval system, or transmitted, in any form or by any means, electronic, mechanical, photocopying, recording or otherwise, without the prior permission of the copyright owner.

Printed in Great Britain

Contents

O. E. Lowenstein	Chairman's introduction	1
R. G. Murray A. Murray	The anatomy and ultrastructure of taste endings	3
Discussion	Andres, Beidler, de Lorenzo, Duncan, Eayrs, Lowenstein, Murray, Pfaffmann, Reese, Zotterman	25
C. Pfaffmann	Physiological and behavioural processes of the sense of taste	31
Discussion	Adey, Anderson, Beidler, Døving, Pfaffmann, Wright, Zotterman	45
L. M. Beidler	Physiological properties of mammalian taste receptors	51
Discussion	Andres, Beidler, Duncan, Hellekant, Lowenstein, Murray, Ottoson, Reese, Zotterman	68
H. T. Andersen	Problems of taste specificity	71
Discussion	Andersen, Beidler, Hellekant, Lowenstein, Martin, Pfaffmann, Wright, Zotterman	79
G. Hellekant	The influence of the circulation on taste receptors as shown by the summated chorda tympani nerve responses in the rat	83
Discussion	Beidler, Hellekant, Lowenstein, Mac Leod, Zotterman	96
G. Borg H. Diamant Y. Zotterman	Neural and perceptual responses to taste stimuli	99
Discussion	de Lorenzo, Hellekant, Lowenstein, Zotterman	112
T. S. Reese M. W. Brightman	Olfactory surface and central olfactory connexions in some vertebrates	115
Discussion	Andres, Beidler, Døving, Lowenstein, Mac Leod, Moulton, Ottoson, Reese, Wright, Zotterman	143
A. J. D. de Lorenzo	The olfactory neuron and the blood-brain barrier	151
Discussion	Andres, de Lorenzo, Lowenstein, Reese, Wright	173
K. H. Andres	Anatomy and ultrastructure of the olfactory bulb in fish, amphibia, reptiles, birds and mammals	177
Discussion	Andres, Mac Leod, Murray, Reese	194
K. B. Døving	Experiments in olfaction	197
Discussion	Andres, Beets, Beidler, Davies, Døving, Mac Leod, Pfaffmann, Reese, Wright	221
D. G. Moulton G. Çelebi R. P. Fink	Olfaction in mammals—two aspects: proliferation of cells in the olfactory epithelium and sensitivity to odours	227
Discussion	Amoore, Andres, Beets, Beidler, Davies, Døving, Moulton, Murray, Ottoson, Reese	246

E. H. Ashton J. T. Eayrs	Detection of hidden objects by dogs	251
Discussion	Amoore, Ashton, Duncan, Hellekant, Lowenstein, Wright, Zotterman	261
J. T. Davies	Recent developments in the "penetration and puncturing" theory of odour	265
Discussion	Adey, Beets, Beidler, Davies, Døving, Duncan, Lowenstein, Mac Leod, Martin, Moulton, Wright, Zotterman	281
J. E. Amoore	Computer correlation of molecular shape with odour: a model for structure-activity relationships	293
Discussion	Amoore, Beets, Beidler, Davies, Duncan, Martin, Ottoson, Wright, Zotterman	306
M. G. J. Beets E. T. Theimer	Odour similarity between structurally unrelated odorants	313
Discussion	Amoore, Beets, Davies, Lowenstein, Moulton, Wright	321
R. H. Wright R. E. Burgess	Specific physicochemical mechanisms of olfactory stimulation	325
Discussion	Andersen, Andres, Beets, Beidler, Davies, Lowenstein, Mac Leod, Martin, Pfaffmann, Wright	337
D. Ottoson	Electrical signs of olfactory transducer action	343
Discussion	Beidler, Davies, de Lorenzo, Lowenstein, Murray, Ottoson, Zotterman	354
W. R. Adey	Higher olfactory centres	357
Discussion	Adey, Andersen, Døving, Pfaffmann, Zotterman	376
General discussion	Amoore, Beets, Beidler, Davies, Døving, Lowenstein, Moulton, Murray, Pfaffmann, Reese, Wright, Zotterman	379
O. E. Lowenstein	Chairman's closing remarks	389
Author index		393
Subject index		394

Membership

Symposium on Taste and Smell in Vertebrates, held 23rd–25th September, 1969

O. E. Lowenstein *(Chairman)*	Department of Zoology and Comparative Physiology, University of Birmingham, Birmingham 15, England
W. R. Adey	Department of Anatomy and Physiology, Center for the Health Sciences, University of California, Los Angeles, California 90024, U.S.A.
J. E. Amoore	Composition Investigations Vegetable Laboratory, U.S. Department of Agriculture, 800 Buchanan Street, Albany, California 94710, U.S.A.
H. T. Andersen	Nutrition Institute, University of Oslo, Blindern, Oslo 3, Norway
K. H. Andres	Anatomisches Institut der Ruhr-Universität Bochum, Germany
E. H. Ashton	Department of Anatomy, The Medical School, University of Birmingham, Birmingham 15, England
M. G. J. Beets	Vice-President, International Flavors and Fragrances (Europe), Hilversum, Holland
L. M. Beidler	Department of Biological Science, The Florida State University, Tallahassee, Florida 32306, U.S.A.
J. T. Davies	Department of Chemical Engineering, University of Birmingham, Birmingham 15, England
A. J. D. de Lorenzo	Division of Neurocytology, Johns Hopkins University School of Medicine, Johns Hopkins Hospital, Baltimore, Maryland, U.S.A.
K. B. Døving	Zoofysiologisk Institutt, Universitetet i Oslo, Oslo 3, Norway
C. J. Duncan	Department of Zoology, University of Durham, South Road, Durham, England
J. T. Eayrs	Department of Anatomy, The Medical School, University of Birmingham, Birmingham 15, England
G. Hellekant	Kungl. Veterinärhögskolan, Fysiologiska Institutionen, Stockholm 50, Sweden
P. Mac Leod	Laboratoire de Physiologie des Sensibilités Chimiques et Régulations Alimentaires, Collège de France, 11 Place Marcelin Berthelot, Paris Ve, France
A. J. P. Martin	The Technological University, Eindhoven, Holland, and Abbotsbury, Barnet Lane, Elstree, Hertfordshire, England

MEMBERSHIP

D. G. Moulton	Monell Chemical Senses Center, 529 Lippincott Buildings, 25th and Locust Streets, University of Pennsylvania, Philadelphia, Pennsylvania 19103, U.S.A.
R. G. Murray	Department of Anatomy and Physiology, Indiana University, Bloomington, Indiana 47401, U.S.A.
D. Ottoson	Fysiologiska Institutionen, Kungl. Veterinärhögskolan, Stockholm 50, Sweden
C. Pfaffmann	The Rockefeller University, New York, N.Y. 10021, U.S.A.
T. M. Poynder	Bush Boake Allen Ltd., 20 Wharf Road, London, N.1, and Physiology Department, University College, Gower Street, London, W.C.1, England
T. S. Reese	Laboratory of Neuropathology and Neuroanatomical Sciences, National Institutes of Health, Bethesda, Maryland 20014, U.S.A.
R. H. Wright	Olfactory Responses Investigation, British Columbia Research Council, Vancouver 8, B.C., Canada
Y. Zotterman	Symposium Secretariat of the Swedish Research Councils, Wenner Gren Center, Pylone Building, 23rd Floor, Stockholm, Sweden

The Ciba Foundation

The Ciba Foundation was opened in 1949 to promote international cooperation in medical and chemical research. It owes its existence to the generosity of CIBA Ltd, Basle, who, recognizing the obstacles to scientific communication created by war, man's natural secretiveness, disciplinary divisions, academic prejudices, distance, and differences of language, decided to set up a philanthropic institution whose aim would be to overcome such barriers. London was chosen as its site for reasons dictated by the special advantages of English charitable trust law (ensuring the independence of its actions), as well as those of language and geography.

The Foundation's house at 41 Portland Place, London, has become well known to workers in many fields of science. Every year the Foundation organizes six to ten three-day symposia and three or four shorter study groups, all of which are published in book form. Many other scientific meetings are held, organized either by the Foundation or by other groups in need of a meeting place. Accommodation is also provided for scientists visiting London, whether or not they are attending a meeting in the house.

The Foundation's many activities are controlled by a small group of distinguished trustees. Within the general framework of biological science, interpreted in its broadest sense, these activities are well summed up by the motto of the Ciba Foundation: *Consocient Gentes*—let the peoples come together.

Preface

THIS volume contains the proceedings of the last of five symposia on sensory mechanisms in vertebrates, and the Foundation would like to record its gratitude to Professor Otto Lowenstein who has throughout worked closely with us on the planning of the meetings and has taken the chair on all five occasions.

This final volume brings together a large amount of recent material on the two chemical senses, taste and olfaction, and we hope it will stimulate further interest in these relatively neglected areas of sensory function.

CHAIRMAN'S INTRODUCTION

O. E. LOWENSTEIN

THIS is the fifth of a series of symposia on sensory function in vertebrates which I have had the pleasure to organize with the Ciba Foundation. The first of the series was on colour vision; we went on to touch, heat and pain; the third symposium was on myotatic, kinesthetic and vestibular mechanisms and the fourth one on hearing mechanisms. We have now arrived at the point where what are sometimes considered to be the Cinderellas of the senses, taste and smell, are to be discussed. You may know only too well how anyone teaching sensory physiology usually enjoys himself until he comes to taste and smell, and then he tends to run out of teachable material. But I have a feeling that we are nearing a breakthrough in this field, and therefore it is very timely that we have this symposium now, as the last of the series. There has been an interruption by one year in the series, because the Third International Symposium on Olfaction and Taste, held at the Rockefeller University in New York in August 1968, beat us to this symposium last year and we are now assembled here to do what we originally intended to do then. But I think we are none the worse for this; a year has gone by and some very new information is available.

You may ask, of course, what—apart from having been asked by the Ciba Foundation to collaborate on a series of meetings on sensory function—is my legitimation for being chairman at a symposium on smell and taste, as I have never worked in this field. Apart from the general fact that I am a comparative sensory physiologist I have some reason for enjoying chairing this meeting. I was present on the stage, 35 years ago or more, when early work in this field was carried out at Munich under Karl von Frisch. I myself was a pupil of von Frisch, and I remember very vividly the efforts of many people (such as Klenk, Krinner, Minnich, Ritter, Schaller, Scharrer, Weis, Wrede and Vogl) who worked under him on various tasting and smelling substances. The very fact that many of their names may be unknown to some of you shows just how old my memories are and how far they go back in this field. And so I shall enjoy after 35 years seeing the scene unfolded again before my eyes. And I hope that from our papers and discussions we shall produce as useful a book on our topic as those covering the four preceding symposia.

THE ANATOMY AND ULTRASTRUCTURE
OF TASTE ENDINGS

R. G. Murray and Assia Murray

*Department of Anatomy and Physiology,
Indiana University, Bloomington, Indiana*

Beginning with the more widely accepted descriptions of taste buds as viewed with the light microscope (Bargmann, 1967; Bloom and Fawcett, 1968), in this paper we shall first discuss those modifications and clarifications to these descriptions which can be reliably supported with evidence from the electron microscope. Several considerations of a more speculative nature will then be discussed. The answers to these latter problems may be of great interest to this group, but unfortunately are of the type that yield slowly to the tedious methods of electron microscopic investigation.

The three primary areas of the tongue in mammals where the taste buds are found are those of the foliate, circumvallate and fungiform papillae. Past descriptions have tended to ignore possible differences in the structure of taste buds between these regions. One reason for this may be that discernible differences are largely those which have been demonstrated with the electron microscope. We have concentrated on the buds of the foliate area in the rabbit and will use these data as a baseline in the discussion to follow. This may be unfortunate, for most physiological experimentation on taste buds has been done on the fungiform papillae, especially of rats. However, an analysis of the primary similarities and differences between these areas will show that the differences may not in fact be of too fundamental a nature. Nevertheless, examination of these differences may eventually prove enlightening, especially in so far as differences in the physiological mechanisms between these areas can be demonstrated.

Our understanding of the taste buds from use of the light microscope has been carefully reviewed by Kolmer (1927). In spite of some difference on particulars, there is general agreement that the taste bud cells are modified epithelial elements, grouped in a barrel-shaped aggregate beneath a small opening in the epithelial sheet. The apical ends of these elongated cells show specializations in the region of this pore, and their middle and lower portions are in contact with nerves which supply the buds from the underlying connective tissue. Above the cells and in the pore is a homogeneous acidophilic substance. The elongated cells are generally assumed to be of two types, one of which is primarily supportive and the other the gustatory receptive cell. In addition there are basal cells which are confined to the lower third of the bud. From the upper surface of certain of the elongated cells, presumed to be gustatory cells, a slender process, usually called a

FIG. 1. Diagram summarizing the general features of the taste buds of the foliate papillae as viewed with the light microscope. The two principal cell types are symbolized by the central slender cell and the broader cell to its right, the former presumably a gustatory cell and the latter a supporting cell. A basal cell lies to the right and below the broader cell. Nerves enter from the underlying connective tissue and contact both of the principal cell types. At the pore many of the cells support a slender hair-like process and the bottom of the pore is filled with a dense, homogeneous substance.

hair, extends into the pore region and is considered to be the sensitive element. Numerous nerve fibres ramify within the bud making contact with the gustatory cells. At these points of contact the chemical stimulus, having been transduced by action of the gustatory cell, presumably, results in an excitation of the gustatory nerve fibres. These features have been summarized in Fig. 1.

This basic outline is consistent with what has been reported from the examination of taste buds with the electron microscope (Engström and Rytzner, 1956; Trujillo-Cenóz, 1957; de Lorenzo, 1958; Iriki, 1960; Nemetschek-Gansler and Ferner, 1964; Farbman, 1967; Murray and Murray, 1967; Scalzi, 1967). Moreover, use of this instrument has clarified certain confusions, and contributed significant new information which might be summarized under the following headings: (*a*) Clarification of the morphology and functional significance of previously recognized cell types; (*b*) The demonstration of an additional cell type which may be the principal gustatory receptor cell; (*c*) Focusing of attention on the taste pit as the region where the primary events of taste appreciation occur; (*d*) A more precise understanding of the relations of nerve processes to the taste bud cells; (*e*) Variations and similarities between buds in different locations. These various points will now be discussed in more detail.

RECOGNIZED CELL TYPES

With a low power electron micrograph of an entire taste bud we can demonstrate the general features (Fig. 2). Light and dark cells are evident, as are the pore with dense substance, the depression in the tips of the cells termed the pit, and the basal portion in contact with the connective tissue. The distinction between perigemmal and gemmal cells is based primarily on the presence or absence of the dense bundles of fibrils characteristic of stratified squamous epithelia here and elsewhere. The nerve fibres stand out as pale profiles between the cells. A somewhat higher magnification shows the region of the pit in greater detail (Figs. 3 and 12). Virtually all the light microscopic descriptions of the taste bud refer to taste hairs or slender processes which extend from the apical parts of the taste cells into the pore. Strictly speaking no hairs are discernible with the electron microscope. However, certain of the cells do extend in tapering form well out into the pit and pore region. These narrow portions ending in microvilli are particularly dense and might be visualized at the light microscopic level as hairs. Nearly all the cells which have these processes are in the category we have termed dark cells. Early electron microscopists (de Lorenzo, 1958; Iriki, 1960; Nemetschek-Gansler and Ferner, 1964) have equated these with

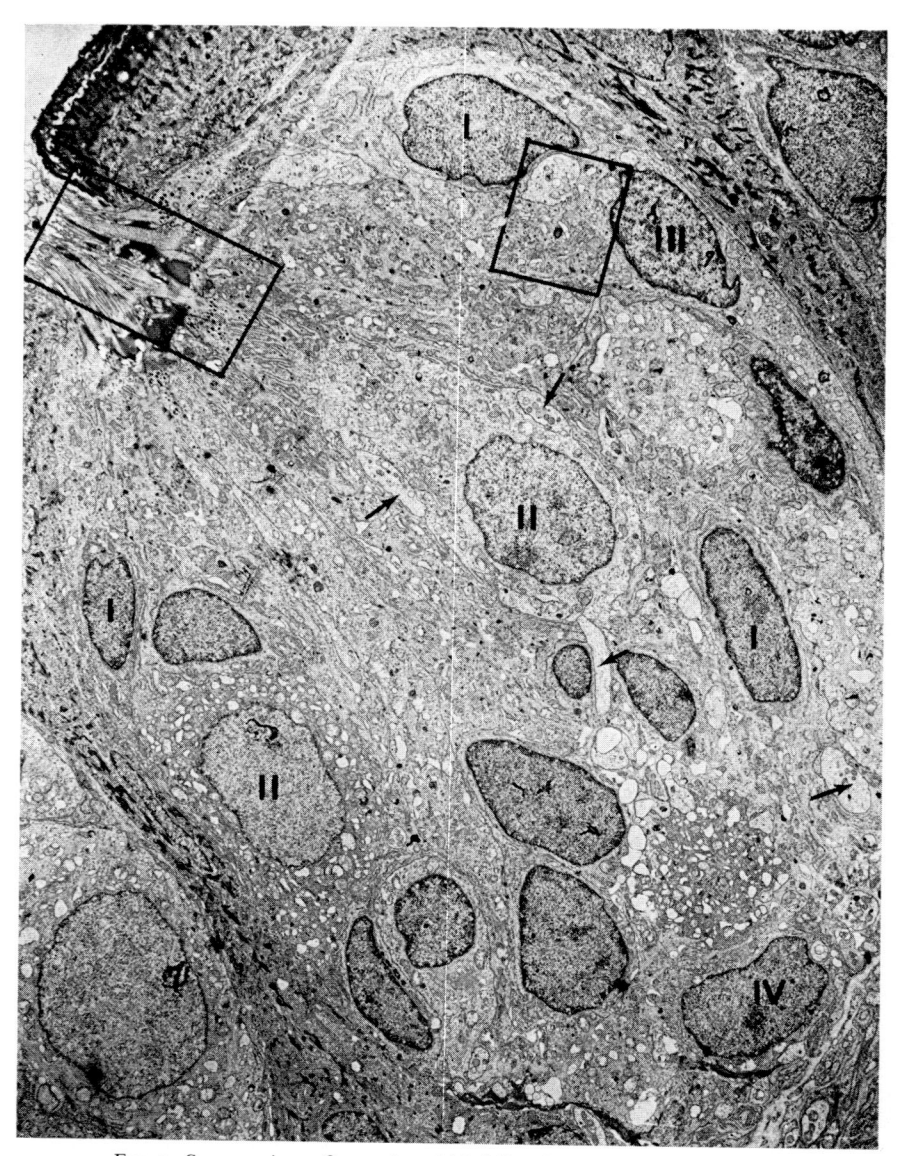

FIG. 2. Survey view of an entire rabbit foliate bud. Both vacuolated and non-vacuolated type II cells can be seen, surrounded by several type I cells. The pore is visible at the upper left and a portion of the underlying connective tissue lower right. Several nerve profiles are indicated by arrows. The area of the pore and of a portion of a type III cell, as marked by rectangles, are shown at greater magnification in Figs. 3 and 4. × 2900

Explanation of plates. All tissues are from rabbits, and except as otherwise indicated have been fixed in osmium tetroxide after aldehyde perfusion, dehydrated in grades of ethyl alcohol or acetone, embedded in Epon 812, sectioned on an LKB microtome with a diamond knife, and stained with uranyl acetate. In some cases, as noted, lead citrate stain was also used. The four cell types are designated by number throughout. The direction toward the pore is always upward.

FIG. 3. Detail of pore region from Fig. 2. Five type I cells, one type II cell and one type III cell border on the taste pit. This specimen illustrates the variability of the apical parts, for the type III cells usually end bluntly without the long neck pictured here, and the type II cells usually support several short microvilli. ×18 500

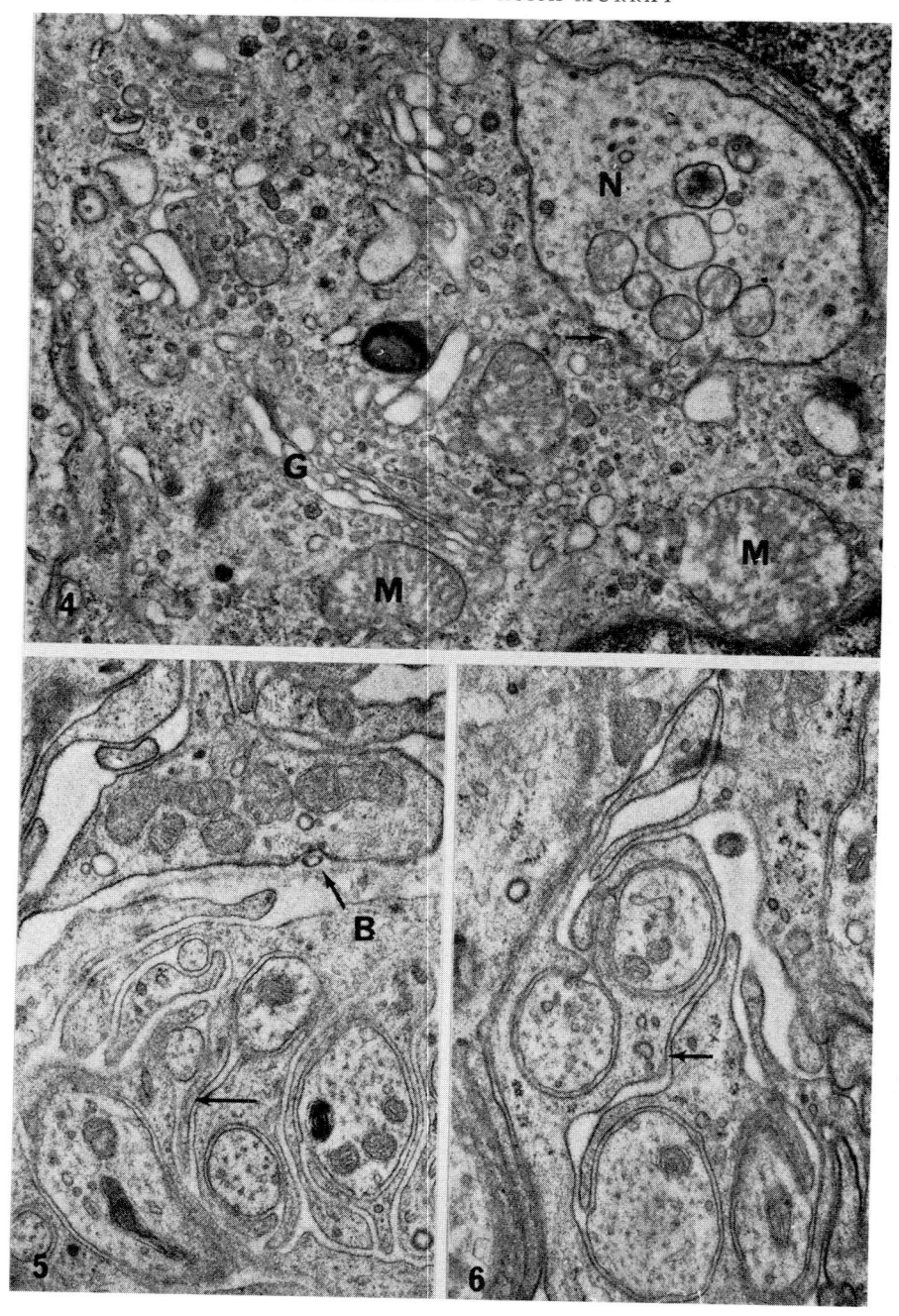

Fig. 4. Detail of part of a type III cell from Fig. 2. The cell is in synaptic contact at the arrow with a nerve (N). Many small synaptic vesicles are seen near the synapse and scattered through the cytoplasm, intermingling with occasional larger, dark-cored vesicles. A Golgi complex is present (G) and the large mitochondria (M) are typical of synaptic regions in these cells. × 14 500

the sensory or gustatory cells of the light microscopist. One obvious reason for this interpretation is that the slender end could be equated with the hair described by the light microscopist. However, these cells have other features which suggest that they are more appropriately thought of as primarily supporting elements, rather than the definitive gustatory receptors. They are most numerous, constituting more than two-thirds of the cells. In their apical cytoplasm numerous granules appear which are apparently the precursors of the very dense substance which occupies the pore region. These cells, although intimately related to nerve fibres, have a relation which is primarily that of envelopment in the manner of Schwann cells elsewhere in the nervous system (Figs. 5 and 6), rather than in the manner of synaptic relationship. Moreover, these cells may closely resemble the peripheral undifferentiated cells (Fig. 2), and appear to be the first stage in the progression of differentiation of epithelial cells as they replace the cells in the taste bud. Because of the considerable variation in the density of the cytoplasm of this cell type, we now believe that the term type I cell is most appropriate, and will henceforth refer to it as such.

The cells which others have considered supporting cells are the so-called light cells, which are distinguished from the type I cells by a paler, more voluminous cytoplasm, and a nucleus more rounded in outline (Figs. 2 and 7). They are second in frequency, constituting approximately 15–30 per cent of the cells in the bud. By logical extension from the reason stated above, this cell will be designated cell type II. They usually do not possess narrow apical extensions but end at the pit in short irregular microvilli. There are no obvious secretory granules and their relation to the nerve fibres is extensive and direct, but not apparently highly specialized and not in the enveloping manner characteristic of the type I cells. Since these cells possess the primary features which would be required of the gustatory cell (i.e. apparent contact with the stimulus and an intimate relationship to the nerve supply) they would appear to be appropriate candidates for the primary gustatory receptor cell, and we originally considered them as such (Murray and Murray, 1967). This reverses the more common interpretation that the dark slender cells are the gustatory cells and the larger light cells

FIG. 5. Nerve fibres ensheathed by Schwann cells in the connective tissue beneath a foliate bud. The basement membrane (B) lies against a process of a type I cell in the bud above. Arrow indicates a pair of infolded membranes of the Schwann cell (mesaxon). Not perfused, but directly immersed in osmium tetroxide. × 9000

FIG. 6. Nerve fibres ensheathed by the cytoplasm of a type I cell within a foliate bud. Note similarity to Fig. 5. Arrow indicates a mesaxon. Not perfused, but directly immersed in osmium tetroxide. × 9000

the supporting cells. One other important piece of evidence in this regard is that the type I cells not only surround the nerve fibres in a protective way, but surround the light cells in this fashion also (Fig. 7). Most of the type II cells are completely isolated or have very limited contact with each other.

In addition to the two types already discussed, a few "basal cells" (type IV) are always present, which presumably have recently entered the taste bud from the surrounding undifferentiated epithelium (Fig. 2). These cells closely resemble the dark cells, but it can be shown by serial reconstructions that they do not reach the pore nor have any apical specialization.

A UNIQUE NEW CELL TYPE

The new cell type which we have termed in our descriptions the type III cell (Murray, Murray and Fujimoto, 1969) is distinguished by the presence within its cytoplasm, in the deeper regions of the bud, of dark-cored vesicles (Fig. 8). These are not the same as the granules of the type I cells, but are smaller vesicles with a more granular content, more nearly resembling those vesicles which have been described as containing adrenaline (epinephrin) or related materials in other organs. These type III cells reach the pore, where they usually end in a relatively unspecialized manner although they may show a neck and microvilli similar to those of the type I cells (Fig. 3). Their most significant characteristic is their relationship to the nerve fibres. This consists of a classic synaptic arrangement (Fig. 4) with presynaptic densities within the cell along the membrane, aggregations of synaptic vesicles against the membrane, and the usual accumulation of mitochondria both within the cell and in the nerve ending on the postsynaptic side. The mitochondria in the cell near the synapse are unusually large, with numerous, well developed internal membranes. The type III cells are the only ones which have such classic synaptic relation to the nerves, and all cells of this type which have been examined sufficiently by serial sections have been shown to have such connexions. It seems clear, therefore, that these cells certainly should be classified as gustatory receptor cells. Their number within the taste bud is small, accounting for about 10 or 15 per cent of the cells within the bud. This value was arrived at by counting cells in serially sectioned buds, which is the only reliable means, we believe, by which such estimates can be made.

THE TASTE PIT

Focus on the taste pit as the site of primary reaction in taste appreciation has been justified by the demonstration (de Lorenzo, 1958; Iriki, 1960; Farbman, 1965) that the ends of the taste bud cells are tightly joined to one another, sealing off the underlying intercellular spaces (Figs. 12 and 13). At this symposium, Reese demonstrates in olfactory epithelium that such junctions are a barrier to the small molecules of horseradish peroxidase (see p. 124). Thus the taste substances cannot penetrate easily below the base of the pit, so their first significant contact would be with the irregular projections of the cells into this area. Although nerve fibres can be identified quite high in the bud, there can be no direct effect of taste substances on the nerves. Moreover the dense substance of the pit and pore covers the ends of all but the most extended cells (Figs. 2 and 3), and may play an important part in modifying the response of the cells to the taste stimuli.

NERVE PROCESSES

The nature of the ramification of nerves within the bud has been clarified by electron microscopic examination. There are both slender, compact segments and expanded, more empty portions which seem, in many cases, to alternate along the course of fibres. We have mentioned the encasement of the fibres by the type I cells and their special relationship to the type III cells (Figs. 4, 5 and 6). The pattern of branching within the bud has been studied in serial sections by Beidler (1968, personal communication), using buds in rat fungiform papillae. He found that one cell may have great numbers of contact points (as many as 30) with fibres. He did not relate these contacts to different cell types. Our serial sections have not been complete enough to trace all the nerves in a given bud. We can show, however, that each type III cell receives a number of fibres which synapse on the lower parts of the cytoplasm (Fig. 8). Those fibres which penetrate higher in the bud make extensive contact with the surfaces of type II cells, and may coil about them as they extend upward (Murray, 1970). Fibres which synapse on type III cells probably do not continue upward in the bud. However, we are unable to separate the fibres into distinct classes on morphological grounds. Whether the innervation of type II cells is by a completely distinct system of fibres, or whether a single complexly branching system supplies both cell types, is of course a distinction of great importance in the analysis of taste mechanisms. We do not, however, find much evidence of branching of fibres in the rabbit foliate buds. We hope, by continuing and refining our serial section investigations, to describe the pattern of distribution of nerves more accurately.

One feature of the nerve fibres within the taste bud which was first pointed out by Gray and Watkins (1965) is that the axolemma of these fibres apparently never shows the trilaminar or "unit membrane" type of structure (Fig. 9). This can be demonstrated by many techniques and is

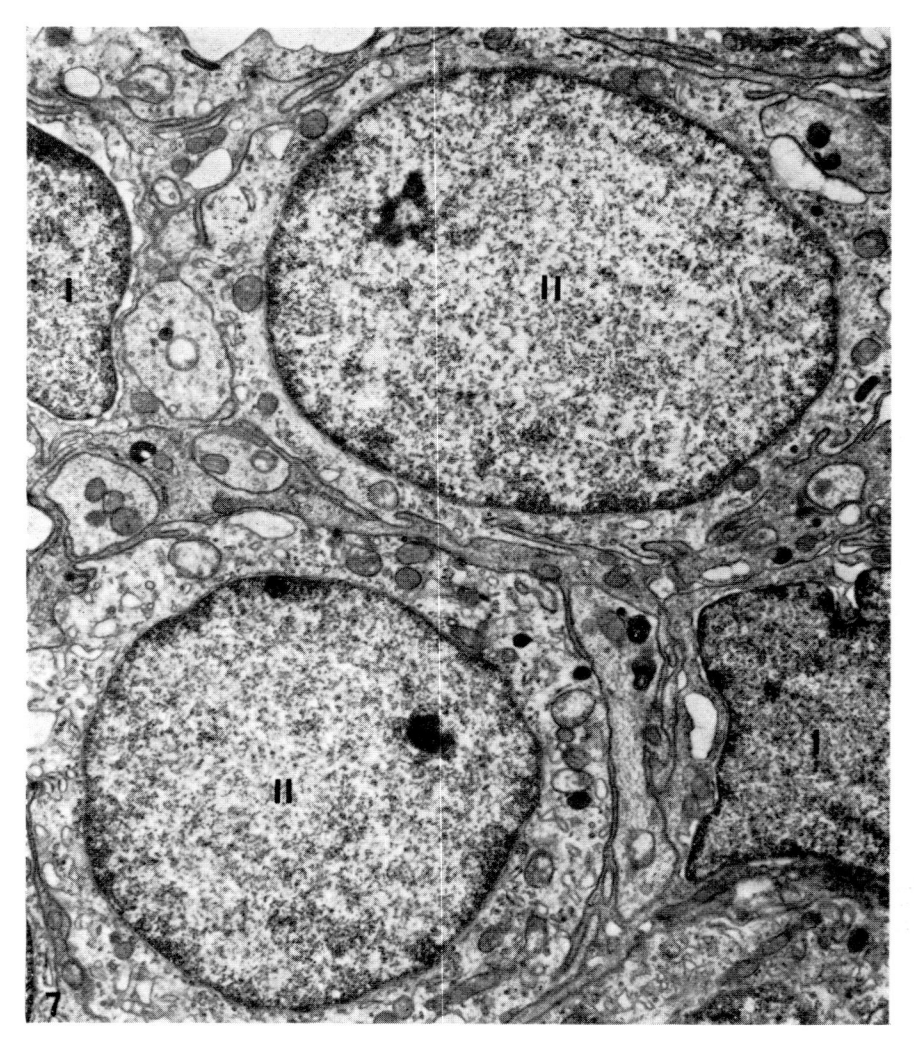

Fig. 7. Two type II cells in a foliate bud which has been sectioned across the vertical axis. The manner in which these cells are completely isolated by the darker cytoplasm of the type I cells is clearly shown. ×9000

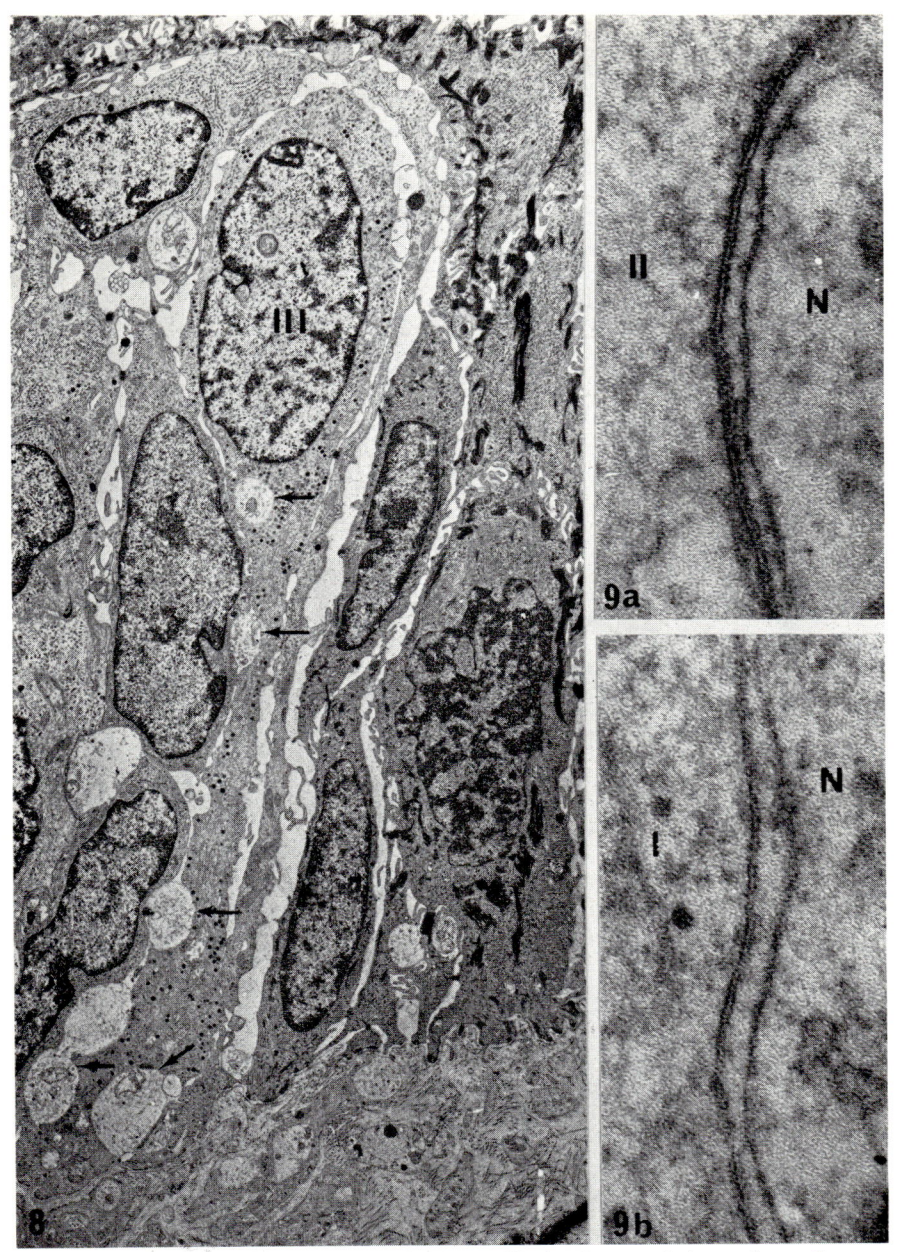

Fig. 8. Part of the basal zone of an obliquely sectioned bud. The extended cytoplasm of a type III cell, with its numerous dark-cored vesicles, is in synaptic contact with several nerves (at arrows). A mitotic perigemmal cell is at the right margin of the figure, and the two slender cells between it and the bud may be about to enter the bud as type IV cells. Stained also with lead citrate. ×4500

Fig. 9. Detail of the cell membranes of a type II cell (9a) and a type I cell (9b) adjoining nerve endings (N). Note that the triple-layered nature is well defined on the type II cell, less prominent on the type I cell, and not present on the axon membrane. Both figures are from adjacent areas of the same micrograph. Directly immersed in osmium tetroxide, and stained also with lead citrate. ×175000

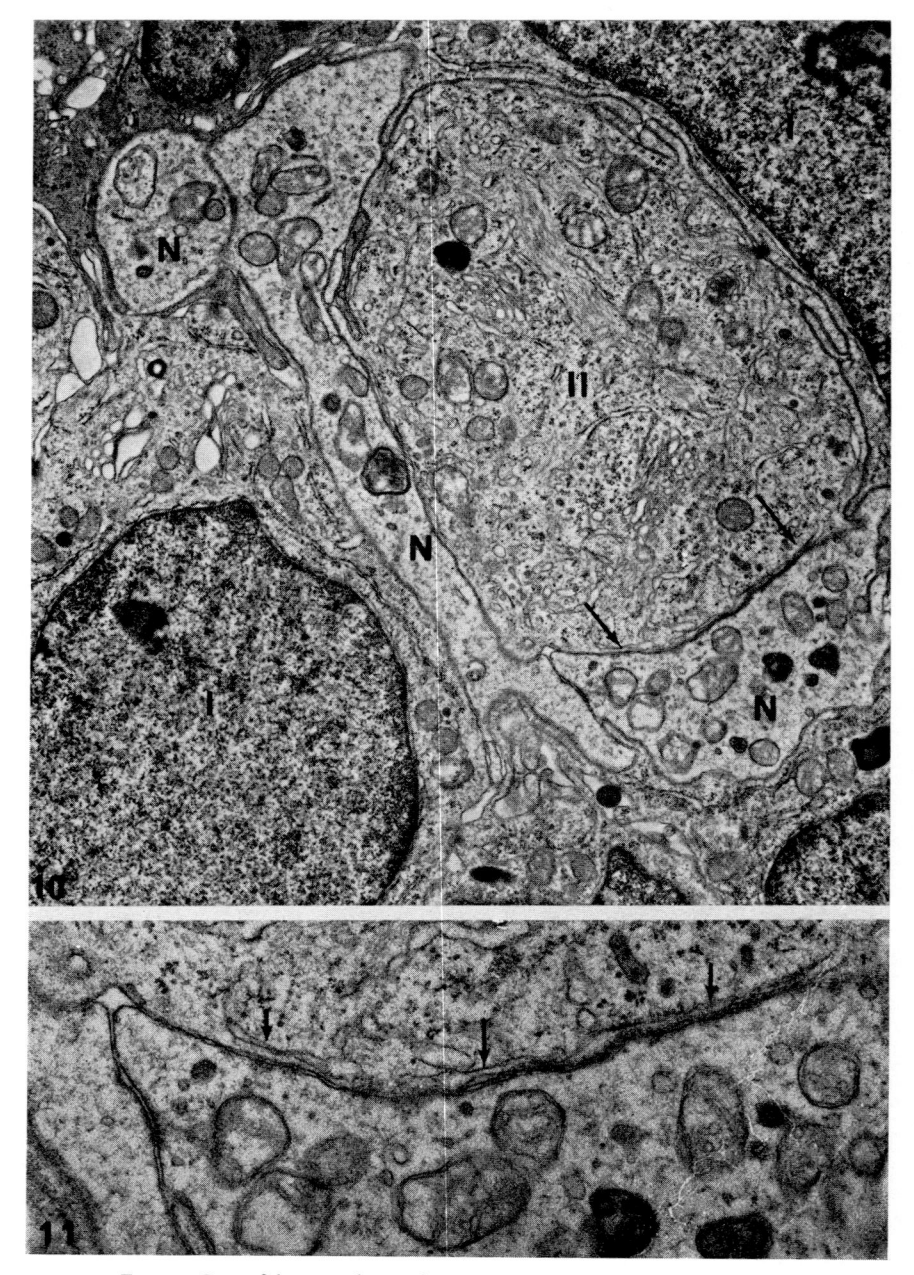

FIG. 10. Part of the cytoplasm of a type II cell surrounded by three nerve processes (N). The nerves are also in intimate contact with each other. The possible synaptic region marked by arrows is enlarged in Fig. 11. × 14 500

FIG. 11. Detail from Fig. 10. Beneath the cell membrane (arrows) is a flattened cisterna of endoplasmic reticulum. × 35 000

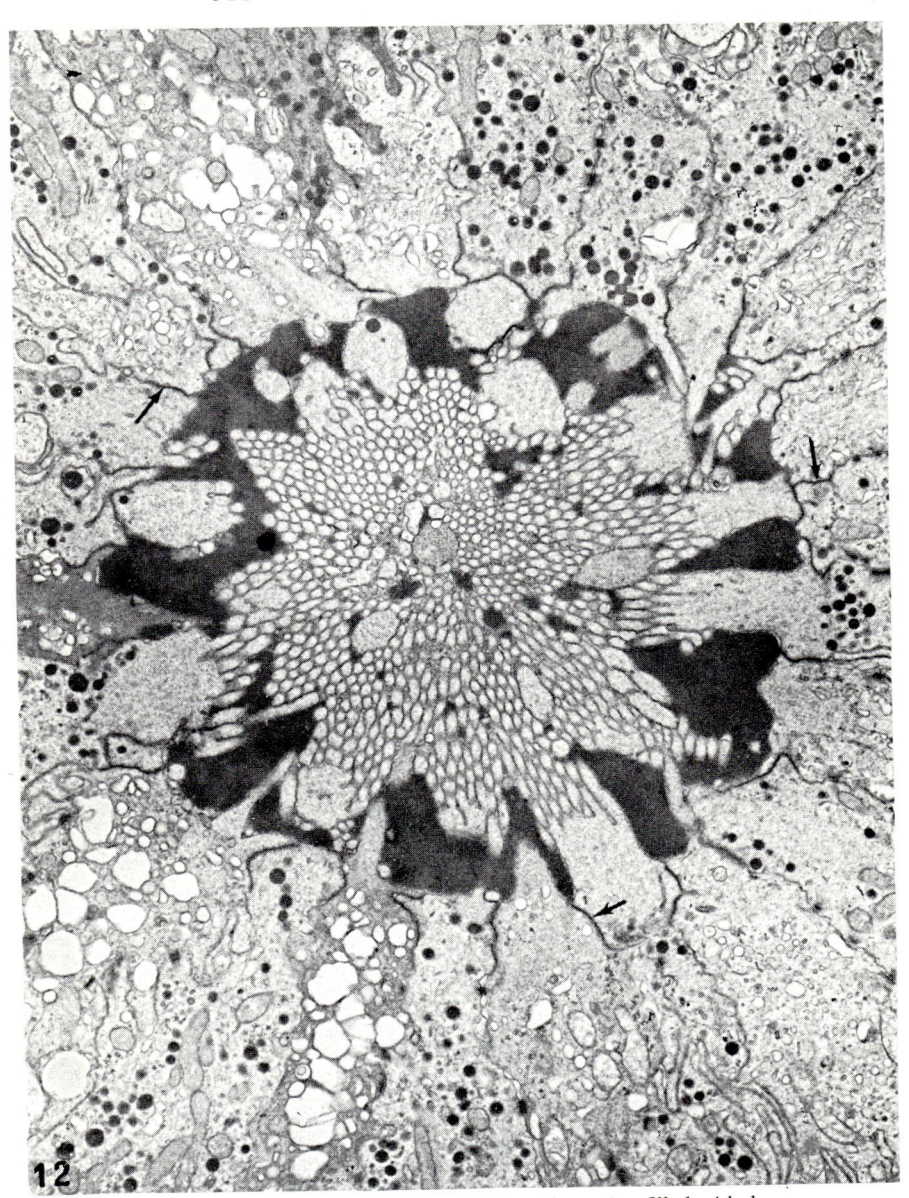

FIG. 12. Transverse section through a pore in the region filled with dense substance. Type I cells can be identified by their apical dark granules. Several of these show an extended neck portion ending in microvilli, numerous cross-sections of which fill the centre of the field. The highly vacuolated cells are of type II. Other cells without apical granules may be either type II cells or type III cells. The darkly stained contact regions adjacent to the pore (arrows) are the tight junctions. × 11 000

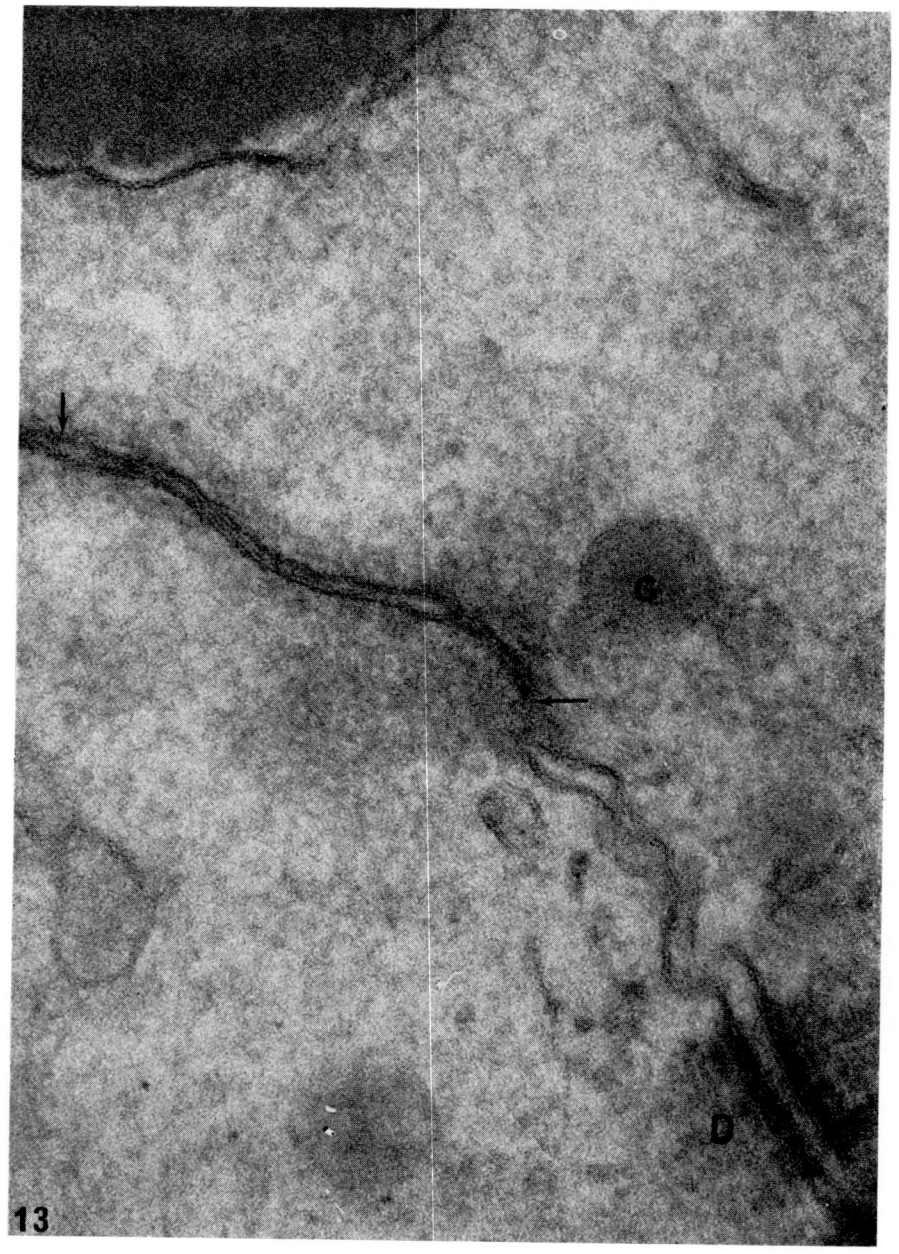

FIG. 13. Detail of a tight junction (between arrows) from Fig. 12. A trilaminar structure can be seen on the cell surface at the floor of the pit (upper left) as well as on the type I cell granule (G). In the tight junction the outer laminae of the two cells are closely adjacent. A *macula adhaerans*, or desmosome (D), is seen at lower right. × 150000

apparent in micrographs where membranes of the adjacent cells and of the vesicles within cells show a clear unit membrane structure. This feature has been useful diagnostically in determining for certain that a process under examination was indeed a nerve and not a slender extension of a taste bud cell.

COMPARISONS BETWEEN TASTE BUDS IN DIFFERENT REGIONS

Our descriptions of foliate buds apply almost verbatim to those of the vallate papillae, while the buds on fungiform papillae differ consistently in certain respects. These may relate either to their different innervation (VIIth nerve rather than IXth) or to the more exposed location of the buds at the surface of the papilla, rather than deep in a crypt. On the fungiform papilla a long narrow channel penetrates the surface epithelium (Fig. 14). Processes of taste cells reach part of the way along this channel, but never to the surface. Although type I and type II cells can be distinguished, the former are not marked by the dense granules seen in type I cells of the foliate buds. They do contain irregular vesicles with a dense content (Fig. 15). This difference is presumably related to the different character of the substance filling the pit and pore region, which in fungiform buds rather than being dense and homogeneous is more lightly staining and contains many vacuoles and vesicular elements. The type III cell can be identified in fungiform buds, with fairly typical synapses (Fig. 16), with both dark-cored and synaptic-type vesicles present (Fig. 17). Most of the cells of this type have been located in the lower part of the bud, and we have not yet traced such a cell to the pore. The numerous slender upward projections of the cells are compressed into a small region, so that even in our "near-serial" sections we cannot trace each cell reliably.

This tendency for the most typically synaptic contacts to occur in the basal part of the bud, and upon cells which may not reach the surface, is similar to the situation existing in fishes (Desgranges, 1965; Hirata, 1966; Uga and Hama, 1967). It may also be significant that the taste buds so far studied in fishes are those supplied by the VIIth nerve, as are the buds of mammalian fungiform papillae. The extent of this difference from the buds of the foliate area cannot be determined without much more detailed serial examinations.

IS THERE AN EFFERENT INNERVATION?

This brings us to the more speculative considerations. There are demonstrable efferent endings on receptor cells in other sensory epithelia, for example in the auditory and vestibular sense organs. Although there is no

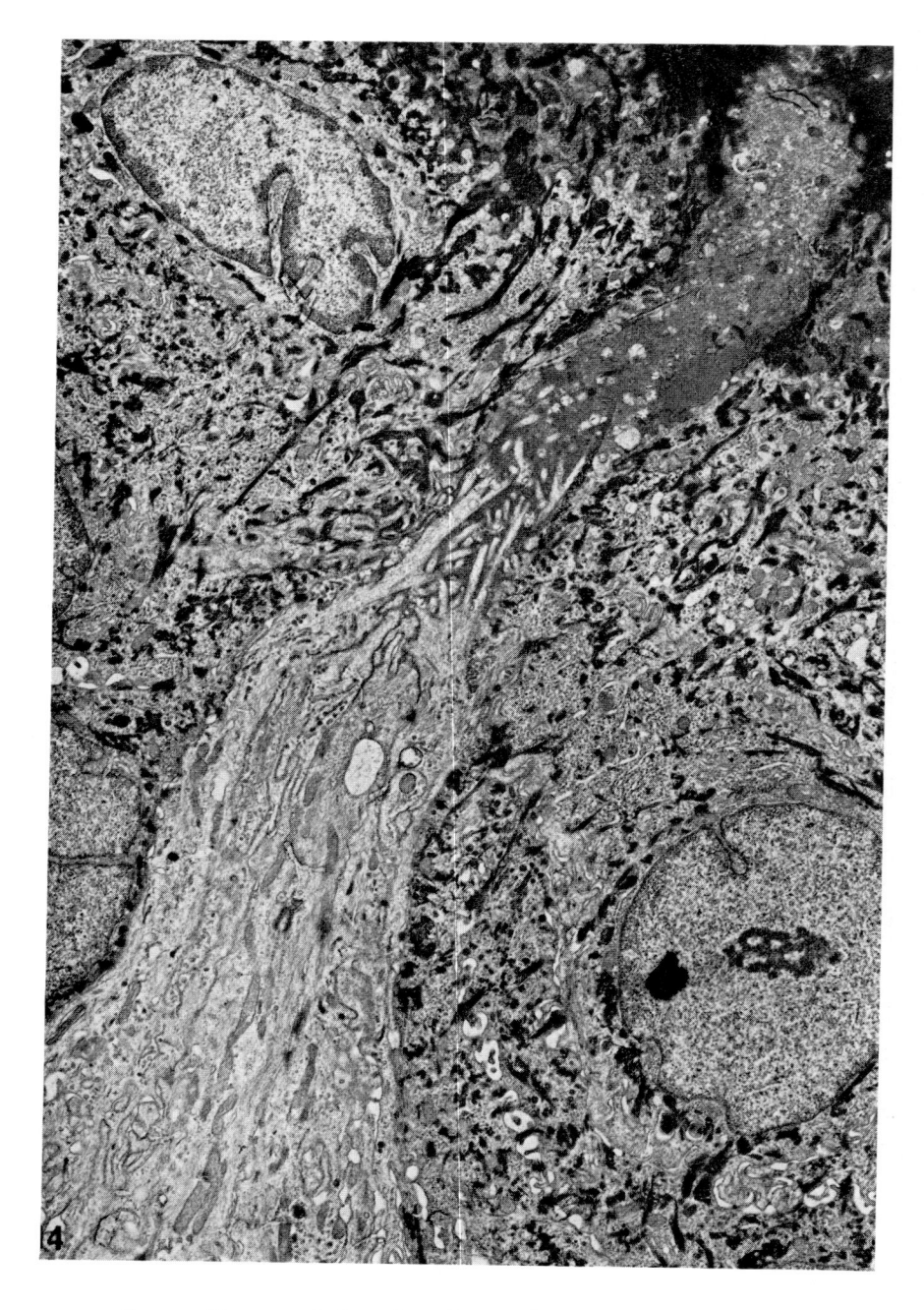

FIG. 14. Narrow apical portion of a taste bud in a rabbit fungiform papilla. Type I cells extend narrow necks with microvilli part of the way into the long narrow pore (upper right). The substance in the pore is vacuolated and less dense than that in foliate buds. ×8500

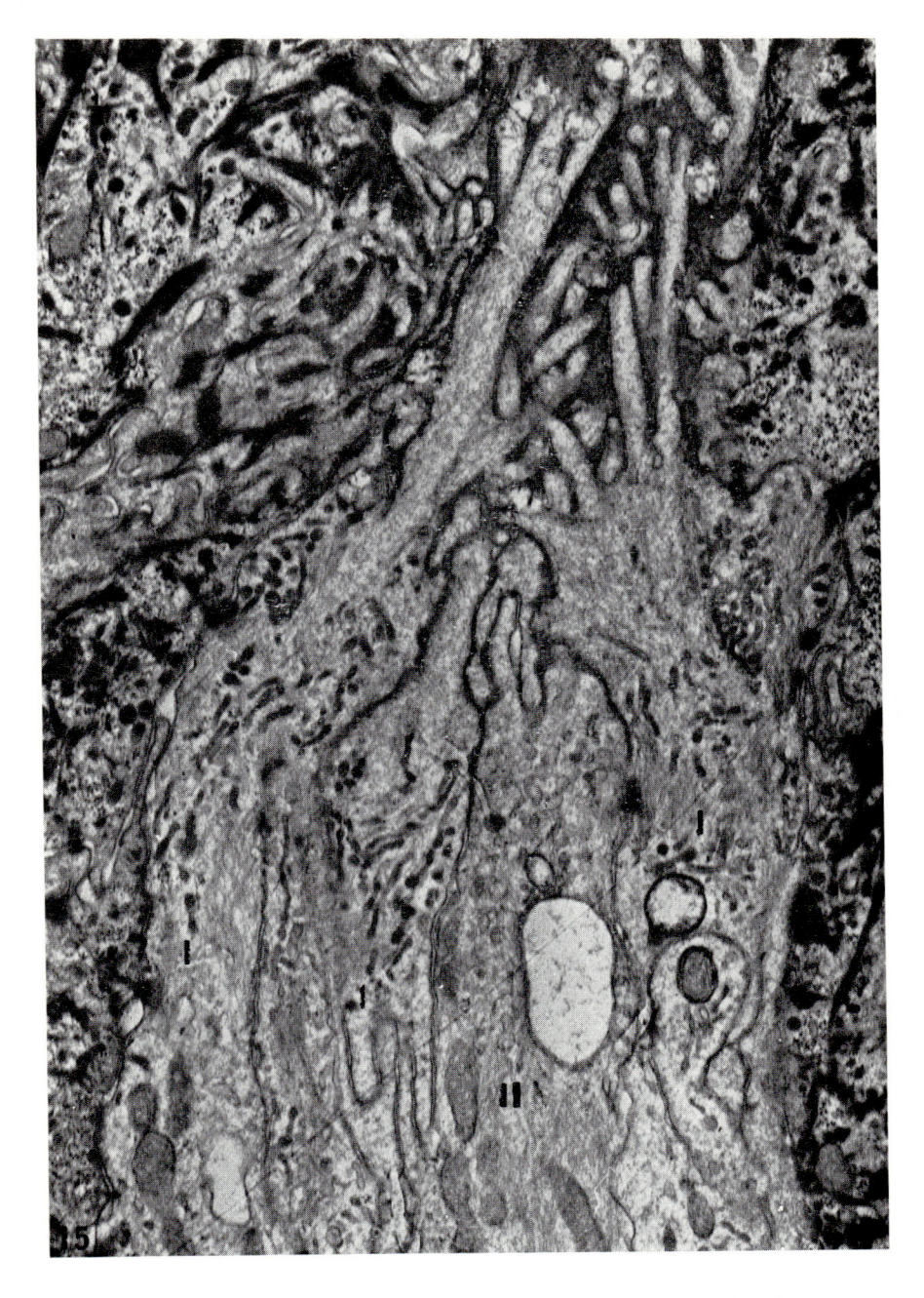

Fig. 15. Enlargement of a portion of Fig. 14. The worm-like densely stained bodies in the type I cells may be analogous to the dark granules seen in foliate buds, and the many vesicles in the type II cell are not unlike those seen in the apical parts of type II cells of foliate buds. × 20 000

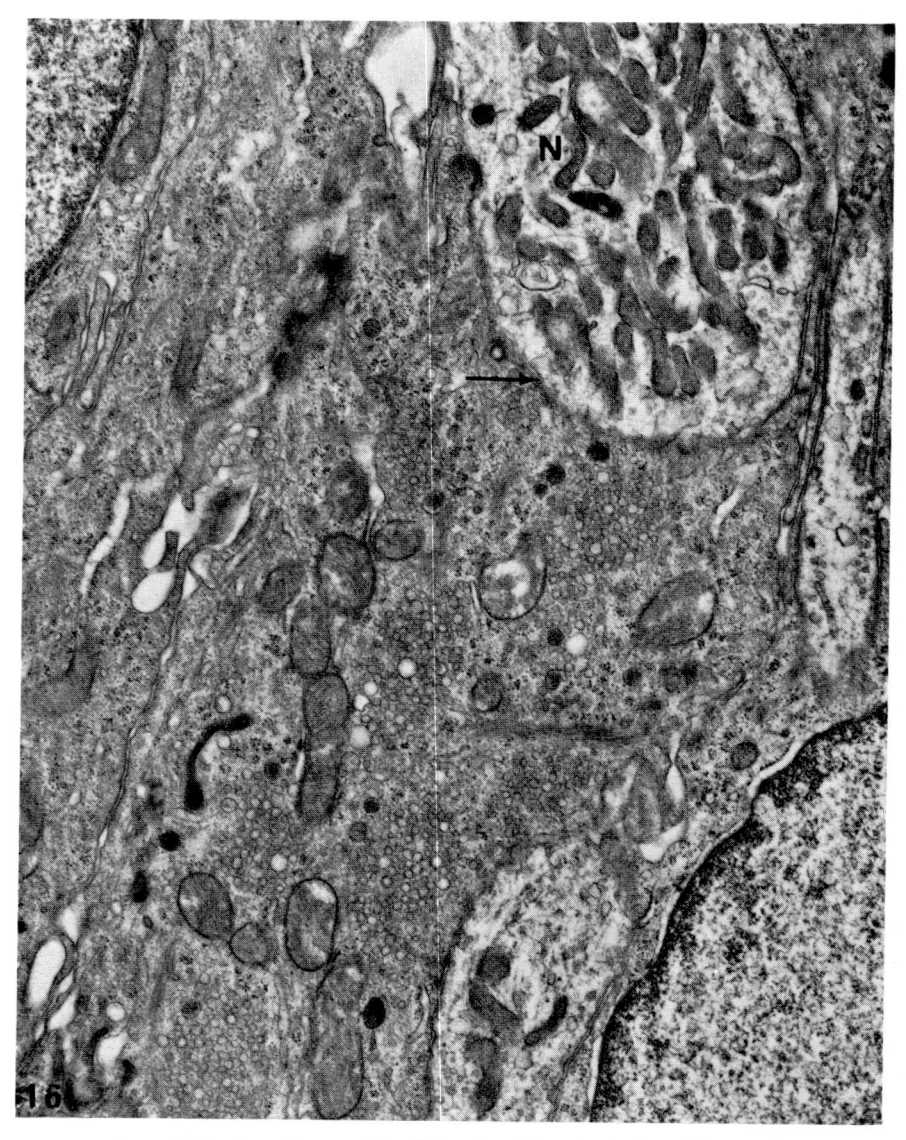

FIG. 16. Portion of a type III cell in a fungiform bud of a rabbit. Both dark-cored vesicles and dense clusters of synaptic vesicles are seen, some of the latter in relation to the synaptic membrane (arrow). The nerve ending (N) is crowded with small mitochondria and a very few small vesicles.
× 18 000

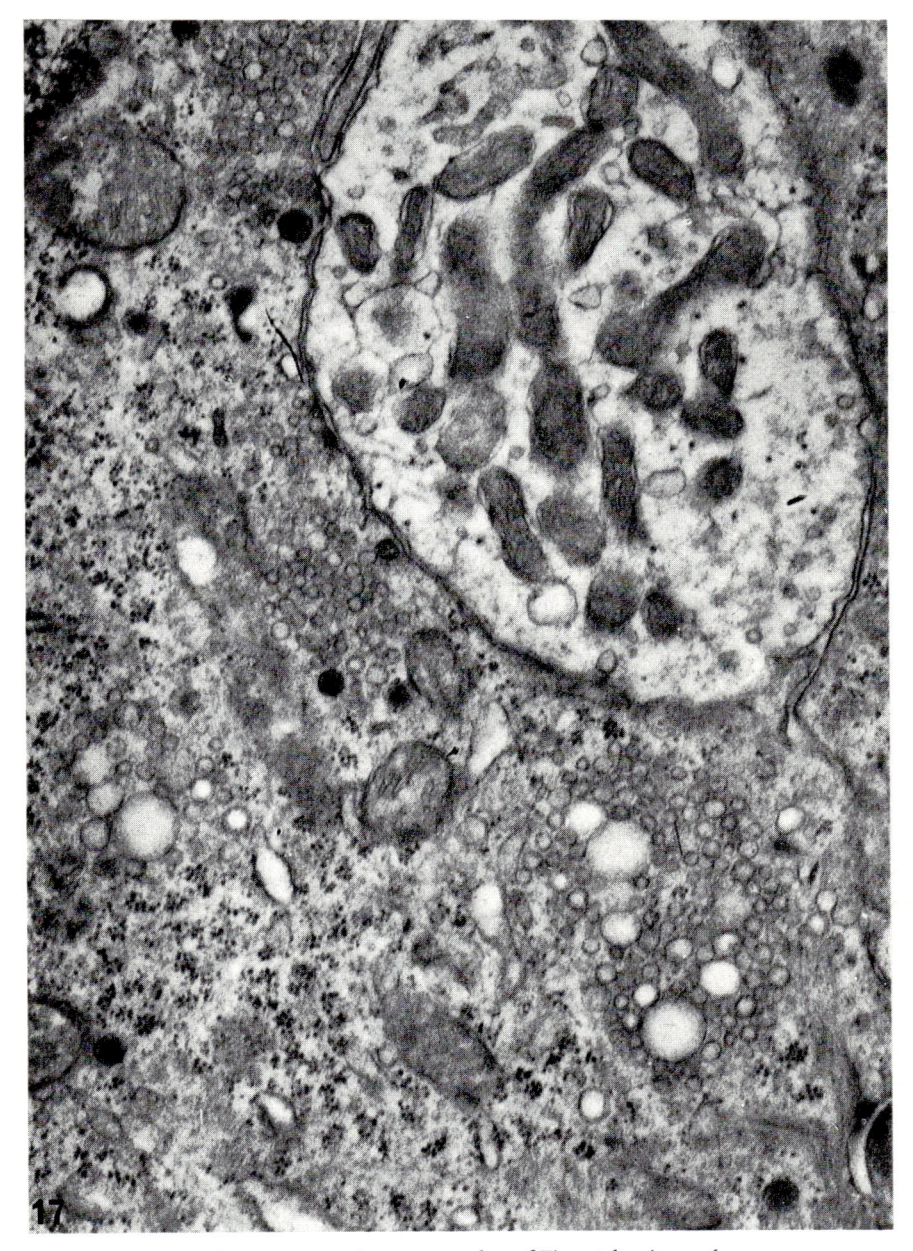

FIG. 17. The same synaptic contact as that of Fig. 16, but in another section, and at higher magnification. × 30 000

evidence as in these other areas of a special tract or bundle of efferent fibres leading to the taste buds, nerve endings within the bud should be analysed for characteristics suggestive of efferent function. The relatively unspecialized contacts of nerves with type II cells are of particular interest. There are no densities or groups of synaptic vesicles in the type II cells at these contacts, and the nerves may contain many mitochondria and a few vesicles of synaptic size. Frequently, but not in all cases, cisternae of endoplasmic reticulum (Figs. 10 and 11) lie against the plasmalemma, as is seen in the efferent endings of the auditory epithelium (Smith, 1967). These add up to interesting suggestive evidence but certainly do not constitute a demonstration of efferent endings. In particular, an efferent path leading into the buds is a prerequisite for adequate proof of this class of innervation. Efferent endings have been described in the taste buds of sub-mammalian species (Desgranges, 1966) and suggested in several mammals as well (P. Graziadei, 1968, personal communication).

CONTACTS BETWEEN NERVES

In all types of bud, most frequently in those of fungiform papillae, nerve fibres make contacts with each other (Fig. 10). Similar contacts elsewhere are thought, in some cases, to be synaptic in function (Gray and Guillery, 1966). Many of the contacts within buds have a highly structured appearance which may differ significantly from ordinary desmosomal junctions. However, all contacts so far observed have been symmetrical in form, and largely for this reason we have assumed that they are structural rather than synaptic in function.

CELL TURNOVER

The turnover in cells within the bud, most clearly demonstrated by Beidler and Smallman (1965), poses problems regarding the nerve supply to taste cells, which would need to be shifting constantly from one cell to another. Analysis of the rate of turnover was based on the assumption that all the cells in the bud were turning over at the same rate. If, however, the gustatory cells proper are few in number and very slow in turnover, they might remain unlabelled throughout the experiments in question. The type III cells may, indeed, represent such a class, and, in support of this speculation, there is no good evidence of transitional forms leading to this cell type. The well-organized synaptic contacts with these cells may remain nearly indefinitely, and the cell turnover could involve only the basal cells (type IV) and the type I and type II cells. Experiments with [^{3}H]-thymidine at the electron microscope level are in progress in our laboratory which may clarify this problem.

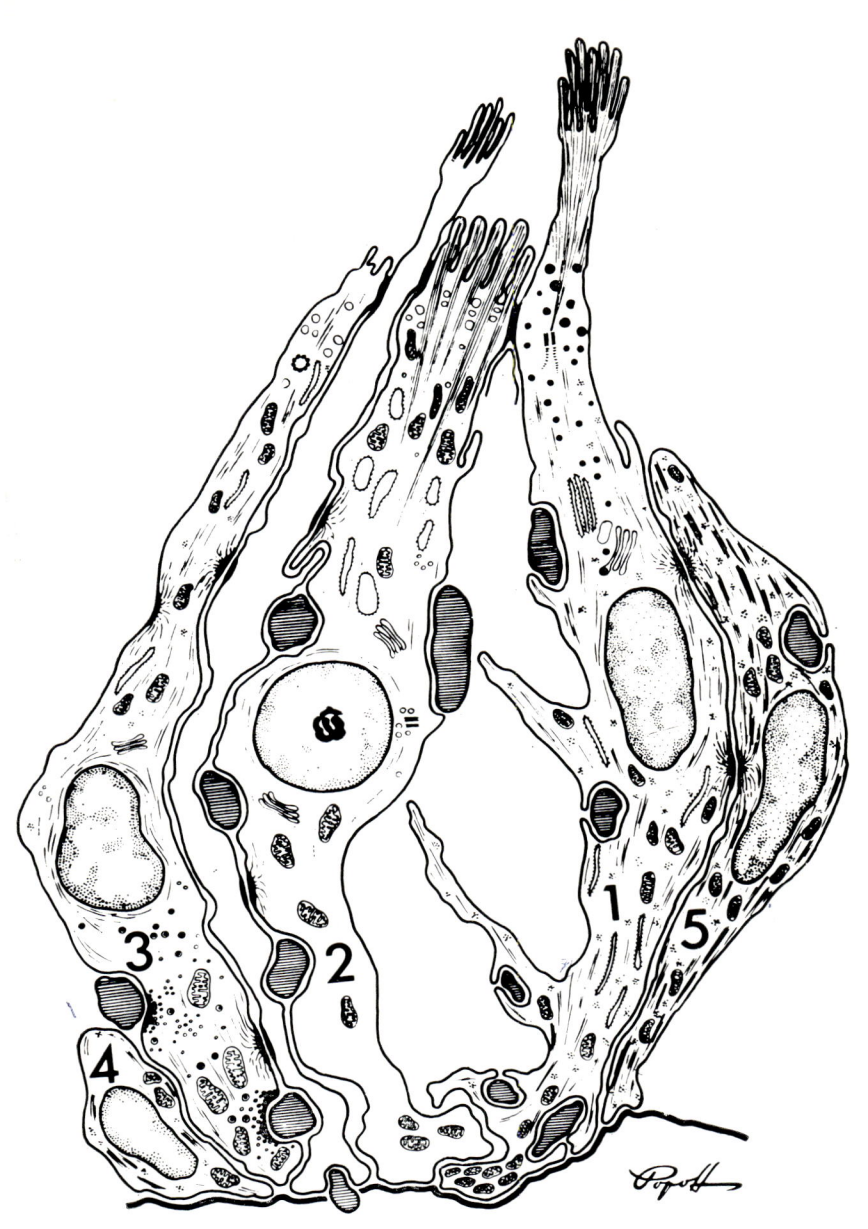

FIG. 18. Diagram of the principal features of the cell types in a rabbit foliate taste bud. (1) A dark cell or type I cell with characteristic apical granules, neck with slender villi, and numerous processes which surround both nerves and other cells; (2) A light cell or type II cell, relatively empty, showing blunt microvilli at the pit with deeply projecting cores, and numerous contacts with nerves (horizontal shading); (3) A type III cell with characteristic dark-cored and synaptic vesicles, ending irregularly above at the pit; (4) A basal cell of simple, compact structure; (5) A perigemmal cell with prominent fibril bundles and ribosomes, enclosing a perigemmal nerve process. For further explanation, see text. Magnification approximately $\times 3500$

CONCLUSIONS AND SUMMARY

In Fig. 18 we have indicated diagrammatically the principal cell types. Type I cells are most numerous and probably supportive in function, while type III cells are surely sensory elements, responding to taste substances in the pit region and initiating nervous activity at "typical" chemical synapses toward the base of the bud. Whether the type II cells are sensory cells of a different type, exert some local modifying effect controlled by incoming nervous impulses, or represent only a deteriorating form of another cell type, is not clearly indicated by present data. The basal cells (type IV) arise from adjacent perigemmal cells, and are less differentiated elements which serve to replace the other three types as they are lost. These conclusions can be most extensively documented for the taste buds of the rabbit foliate papillae, but apply equally to the vallate papillae of this and other mammals, and probably to buds of the fungiform papillae as well.

Acknowledgements

The authors wish to acknowledge the valuable consultation of Dr Sunao Fujimoto and the technical assistance of Mrs Rada Abrams, Mrs Nell Davis, Mr Herschel Lentz and Mr William White. This work was supported in part by Grant NB 07472 from the National Institutes of Health, United States Public Health Service.

REFERENCES

BARGMANN, W. (1967) *Histologie und mikroskopische Anatomie des Menschen*, 6th edn., pp. 654–655. Stuttgart: Thieme.
BEIDLER, L. M., and SMALLMAN, R. L. (1965) *J. Cell Biol.*, **27**, 263–272.
BLOOM, W., and FAWCETT, D. W. (1968). *A Textbook of Histology*, 9th edn., pp. 513–515. Philadelphia: Saunders.
DE LORENZO, A. J. D. (1958). *J. biophys. biochem. Cytol.*, **4**, 143–150.
DESGRANGES, J. C. (1965). *C.r. hebd. Séanc. Acad. Sci., Paris*, **261**, 1095–1098.
DESGRANGES, J. C. (1966). *C.r. hebd. Séanc. Acad. Sci., Paris*, **263**, 1103–1106.
ENGSTRÖM, H., and RYTZNER, C. (1956). *Ann. Otol. Rhinol. Lar.*, **65**, 361–375.
FARBMAN, A. I. (1965). *J. Ultrastruct. Res.*, **12**, 328–350.
FARBMAN, A. I. (1967). In *Symposium on Foods: Chemistry and Physiology of Flavors*, pp. 25–51, ed. Schultz, H. W., Day, E. A., and Libbey, L. M. Westport: Avi.
GRAY, E. G., and GUILLERY, R. W. (1966). *Int. Rev. Cytol.*, **19**, 111–182.
GRAY, E. G. and WATKINS, K. C. (1965). *Z. Zellforsch. mikrosk. Anat.*, **63**, 583–595.
HIRATA, Y. (1966). *Archvm. histol. Jap.*, **26**, 507–523.
IRIKI, T. (1960). *Acta med. Univ. Kagoshima*, **2**, 78–94.
KOLMER, W. (1927). In *Handbuch der mikroskopischen Anatomie des Menschen, Haut und Sinnesorgane*, vol. III, part 1, pp. 154–191, ed. Moellendorff, W. von, and Bargmann, W. Berlin: Springer.
MURRAY, R. G. (1970). In *Ultrastructure of Animal Tissues and Organs*, ed. Friedmann, I. Amsterdam: North-Holland. In press.
MURRAY, R. G., and MURRAY, A. (1967). *J. Ultrastruct. Res.*, **19**, 327–353.
MURRAY, R. G., MURRAY, A., and FUJIMOTO, S. (1969). *J. Ultrastruct. Res.*, **27**, 444–461.

Nemetschek-Gansler, H., and Ferner, H. (1964). *Z. Zellforsch. mikrosk. Anat.*, **63,** 155–178.
Scalzi, H. A. (1967). *Z. Zellforsch. mikrosk. Anat.*, **80,** 413–435.
Smith, C. A. (1967). In *Submicroscopic Structure of the Inner Ear*, pp. 106–131, ed. Iurato, S. Oxford: Pergamon Press.
Trujillo-Cenóz, O. (1957). *Z. Zellforsch. mikrosk. Anat.*, **46,** 272–280.
Uga, S., and Hama, K. (1967). *J. Electron Microsc.*, Chiba Cy, **16,** 269–277.

DISCUSSION

Pfaffmann: What is the evidence that leads you to separate these different cell types and conclude that there is no transition from the other types to the type III cell?

Murray: Sampling problems make it difficult to establish dynamic relations of this kind. Where we have examined type III cells in serial sections we can nearly always trace them to the pore, and demonstrate synaptic contacts with nerves. We find no cells that can be clearly identified as transitional to the type III cell. Perhaps it is a relatively long-lived cell as opposed to some of the other types in the rabbit foliate that are turning over more rapidly.

Beidler: The chance of finding a transitional cell for type III would be very small if these cells constitute only 10–15 per cent of the taste bud cells; if there are 50 cells in a taste bud, there would only be 5–8 type III cells in all.

Murray: This is why we haven't been too surprised that we have not found transitional cells.

Pfaffmann: I was really asking whether there is a transition from I to II and then to III.

Murray: We find cells which have necks like type I cells but no granules, and cells with granules but poorly developed necks. Since type II cells are frequently vacuolated and show other signs of degeneration while type I cells do not, we have favoured the interpretation that type II cells develop from type I. But each of the three main types could come from the type IV cells (basal cells). The absence of degenerating stages among the most numerous type I cells is difficult to explain in this case.

de Lorenzo: In our earlier studies we described free nerve endings in the apical pore. Have you seen nerve endings that are not ensheathed by either of your three cell types but seem to end as bare nerve endings, near the pore?

Murray: All nerve fibres end in close proximity to cells of the bud. Many such endings appear essentially unspecialized and thus might be in some sense considered "bare endings". Some nerves lie close to the base of the taste pit, but they never breach the tight-junction barrier to enter the pore.

Zotterman: Have you anything to tell us about the blood supply to the taste buds?

Murray: The blood supply is quite rich in the connective tissue beneath the buds, but there is never contact between the endothelium and the basement membrane of the taste bud; a space of several micrometres filled with connective tissue always intervenes.

Zotterman: Is there any accumulation of mitochondria in the bases of the taste bud cells, closer to the capillaries? There is a rather considerable distance between the tips of the cells and their bases; yet these cells must have a very high metabolic rate.

Murray: From the pore to the base is of the order of 30–50 μm. The basal parts of many of the cells, and most strikingly of the type I cells, are filled with mitochondria in a foot-like process against the basement membrane.

de Lorenzo: The blood supply to taste buds is really no different from that of most neuroepithelial cells. In the retina, for example, the receptor cells derive their blood supply from vessels literally miles away.

Lowenstein: Professor Murray, are there no nerve endings making synaptic contact with your type I cell, which appears to be the cell which was thought by the older microscopists to be the sensory cell?

Murray: There are no typically synaptic structures, but frequently the nucleus of a type I cell is in special relationship to a nerve ending filled with mitochondria. We know that it is a nerve ending and not just a nerve passing through because we have followed these in serial sections. This suggests that the nerves may be stimulating some change in the type I cells, which would fit with the assumption that they are transforming into other types.

Lowenstein: Although these cells are all peripherally located, is there any suggestion of anything like glia cells?

Murray: I would say that the type I cell has many of the features of glia cells.

Eayrs: Surely the type I cell has more the characteristics of a sensory cell, from your description, and the type II cell is more like a glial cell?

Murray: The fact that the type II cell is isolated by the type I cell argues against that. The type II cell doesn't surround anything, as glia cells do. The nerve fibre is intimately related to the type II cell, but at the same time the type II cell is enveloped by the type I cell.

DISCUSSION

Lowenstein: Are there now to your mind two types of sensory cell, or is *one* of the two remaining types (III or IV) *the* sensory cell?

Murray: There seems no question that the type III cell is a sensory cell, because of the synapses on it, and I am fairly certain that the type I cell is a supporting cell. The type II cell rests in limbo as far as I am concerned; I am not prepared to say what its function is. The type IV cell is a replacement cell, not yet differentiated as to type.

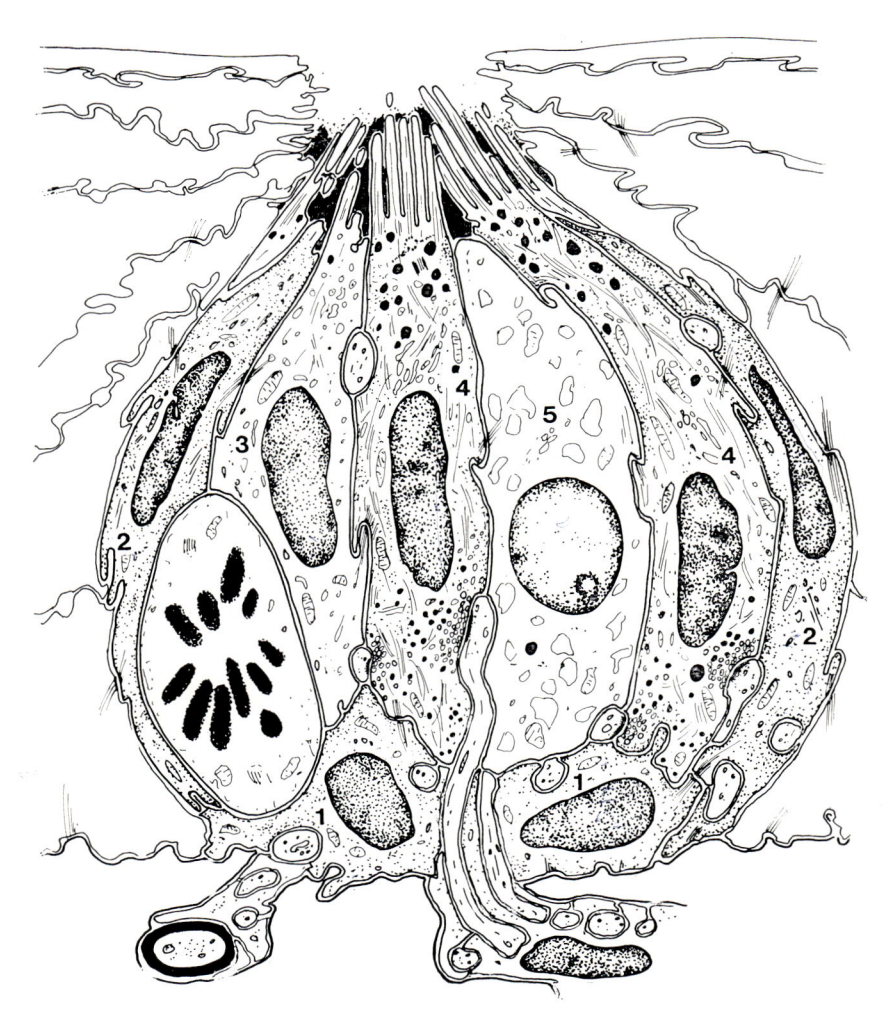

FIG. 1 (Andres). Drawing of a vertical section through a taste bud from the foliate papilla of a rabbit. See text for details.

Andres: I have studied the foliate papilla of the rabbit and the dog. From my results it seems clear that taste bud cells proliferate and later die. Five phases can be defined (see Fig. 1). Type I is the basal cell from which the proliferation seems to come, by mitosis. Type II is morphologically very similar to type I, but it is more extended and already reaches the pore with its apical pole. Type I and II surround the nerve fibres in a similar way to a Schwann cell. During further differentiation a light transitional phase occurs as type III which has a pale cytoplasm containing vesicles and endoplasmic reticulum. At the apical pole a few taste microvilli already project into the taste pore. Type IV is the mature sensory cell phase, which contains high-contrast secretory drops at the apical pole. Basally there are numerous granular vesicles near the presynaptic complex. The intracytoplasmic net of filaments is particularly striking in type IV cells. Type V is the dying phase in which the cells, blown up by autolysis, are shed.

Murray: I am keeping an open mind on such a sequence, because we have no conclusive evidence yet. The designations we have used are best interpreted as indicating the order of frequency. Type I is most frequent, then types II, III and IV in decreasing order of frequency.

Andres: In my studies, the cells that you call types II and III have a cilium extending from the centriole. It is a typical cilium with $9+2$ fibrils. We sometimes see two cilia. This is in both rabbit foliate and dog vallate taste buds.

Murray: I have never seen a cilium at the apex of any cell of the bud. Type I cells have two apical centrioles, one of which is a typical basal body, with a downward-projecting striated tailpiece, but we have never found anything projecting upward from the basal body.

Andres: Perhaps the cilium grows out only from time to time, at a certain stage?

Murray: We did find a rather typical cilium on a type II cell in the depth of the bud, which did not project into the pore; we found several other cases of "pseudo" cilia, without the internal tubular structure, also on type II cells.

de Lorenzo: I want to comment on taste bud innervation. In our studies we have never seen branching fibres with the electron microscope. With the Bielschowsky stain, when it works well, one sees an extensive amount of branching in the nerve plexus beneath the foliate papillae of the rabbit tongue.

Secondly, electron microscopists have a built-in prejudice about what constitutes a synapse! We usually look at classical synapses like those on anterior horn cells in the spinal cord. We should proceed with caution in looking at unfamiliar regions. In the last couple of years, J. B. Nadol and I have examined two other sensory systems in an attempt to study, in a physiologically well-documented area, fine dendritic terminals. We have examined stretch receptor neurons and muscle spindles in *Homarus*, and find that few dendrites, either in the sense receptors or in the motor systems, show any membrane specializations. Interestingly, when we use histochemical techniques, staining for acetylcholinesterase and choline acetylase, the endings show enzyme activity.

Zotterman: How far do the microvilli of the type I (dark) cells project in the pore, Professor Murray?

Murray: The longest ones extend to the surface and occasionally show bulbous projections on the end; we suspect from their rather empty look that these bulbous projections may be artifactual, but again they may not be.

Zotterman: Have you any histochemical or other information on the composition of the dark substance in the outer taste pore which covers the microvilli of the type II cells? This will be important in order to know whether sapid solutions can penetrate, and of what kind. This substance might exert an enormous influence on the penetration of water.

Murray: There is very little histochemical evidence, and that is contradictory. Scalzi (1967) thought that it was mucopolysaccharide. Erbengi and Ferner (1964) failed to find PAS-positive substance, but found ascorbic acid. ATPase has been demonstrated (Iwayama and Nada, 1967) and acid phosphatase both demonstrated (Scalzi, 1967) and denied (Iwayama and Nada, 1967). Dr S. Fujimoto, in my laboratory, stained with ruthenium red in block. This stained the pit substance intensely, but surprisingly failed to stain the granules in the apical parts of the type I cells, or the borders of any cells below the pit. This is perhaps another evidence of the barrier to diffusion presented by the tight junctions. One can't assume that the tight junctions are completely impenetrable, but they must make penetration much more difficult.

Reese: It would be important to know whether these are true tight junctions and whether they are continuous belts around the apices of taste cells. As I shall discuss in my paper (pp. 115–143), tight junctions might act as a barrier to even low molecular weight solutes. Therefore, it is possible that tasted substances may have ready access to only the exposed apices of taste cells. Also, the return flow of depolarizing current resulting from stimulation of a taste cell might be shunted through neighbouring taste cells.

Murray: In a section cut transverse to the long axis of the taste bud you can follow every contact almost all the way round and you can satisfy yourself that it is a continuous barrier. As to whether it is absolutely impervious, this seems to be a type of junction where the outer lines of the trilaminar structure come very close together but do not actually fuse.

Reese: That is the critical point and needs to be tested with electron microscopical tracers, injected intravenously as well as applied to the surface of the tongue.

Duncan: Is the dense substance found in the taste pit typical of the chemoreceptors of lower vertebrates?

Murray: No. Lower vertebrates so far reported (Murray, 1970) do not have a taste pit. The supporting cells may be secretory in nature, but the secretion evidently does not accumulate over the tips of the taste cells.

REFERENCES

ERBENGI, T., and FERNER, H. (1964). *Z. Zellforsch. mikrosk. Anat.*, **61,** 673–687.
IWAYAMA, T., and NADA, O. (1967). *Archvm histol. jap.*, **28,** 151–163.
MURRAY, R. G. (1970). In *Ultrastructure of Animal Tissues and Organs*, ed. Friedmann, I. Amsterdam: North-Holland. In press.
SCALZI, H. A. (1967). *Z. Zellforsch. mikrosk. Anat.*, **80,** 413–435.

PHYSIOLOGICAL AND BEHAVIOURAL PROCESSES OF THE SENSE OF TASTE

CARL PFAFFMANN

The Rockefeller University, New York

ELECTROPHYSIOLOGICAL studies of single primary afferent taste neurons uniformly agree that individual fibres very often have multiple sensitivities (Pfaffmann, 1941, 1955; Cohen, Hagiwara and Zotterman, 1955; Fishman, 1957; Ogawa, Sato and Yamashita, 1968; Sato, Yamashita and Ogawa, 1969). That is, when one of the four basic taste stimuli—sodium chloride, quinine, hydrochloric acid or sucrose—is placed on the tongue surface, most fibres respond to several of these stimuli. Most work on single afferent fibres in the mammal has utilized the chorda tympani nerve which responds well to sodium chloride or hydrochloric acid but is not very sensitive to sucrose or quinine in the rat. The posterior region of the tongue, innervated by the glossopharyngeal nerve, is more sensitive to quinine and sucrose. Thus, sensitivity to single units from the glossopharyngeal nerve was studied by Frank (1968) with the expectation that more units sensitive to all four basic stimuli would be encountered. As in most such studies, the criterion for a single fibre was uniformity of recorded spike height. In addition, since the papillae (foliate or circumvallate) of the posterior tongue are buried deeply in trenches, stimulus solutions were pumped in the lumen of the papillae through a small pipette.

Test stimuli were $0\cdot3$ M-sodium chloride, $0\cdot1$ M-hydrochloric acid, $0\cdot001$ M-quinine hydrochloride and $0\cdot3$ M-sucrose, at intensities which produce integrator responses of about 50 per cent of the maximal total nerve responses to these substances. This is an important consideration since if the stimuli are too strong, more fibres may be stimulated as high-threshold fibres are brought into action and low-threshold fibres would respond to a wider range of stimuli. Fibres were classified as having responded to a stimulus if there were an increase in response rate of at least 50 per cent during the first five seconds of stimulation over the resting rate. When a fibre responded to more than one stimulus, it usually did not respond equally well to all stimuli. The response criterion did affect the number of fibres so classified but even with an increase in the criterion to 500

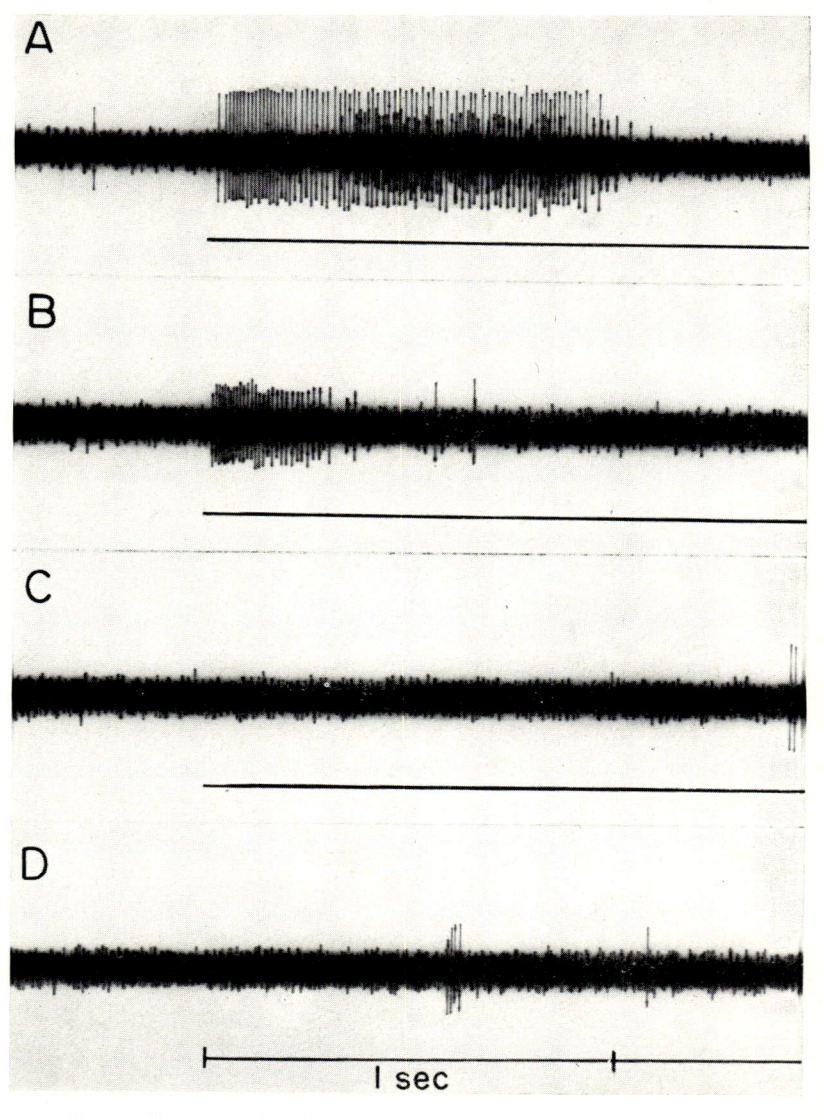

Fig. 1. The electrophysiological response of rat glossopharyngeal fibres to 0·3 M-sucrose (A), 0·01 N-HCl (B), 0·3 M-NaCl (C), and 0·001 M-quinine hydrochloride (D). Stimulation is indicated by the line beneath each record. One fibre (larger spike) responds to sucrose and the other (smaller spike) responds to sucrose and HCl; the former does not respond to acid and neither responds to NaCl or quinine. (Reproduced from Frank and Pfaffmann, 1969, copyright 1969 by the American Association for the Advancement of Science.)

per cent, multiple sensitivity was still observed in about 25 per cent of the fibres.

In Fig. 1, the response of one fibre to sucrose (large spikes) and of another fibre (smaller spikes) to sucrose and to acid is shown. The spike height difference is probably only apparent and is not due to an intrinsic difference in fibres but to the conditions at the recording electrode. We have found no systematic relation of spike height to sensitivity of individual units. Fibres which responded to two of the four basic stimuli were the most common, those which responded to only one next, those to three next and those responsive to all four basic qualities least common. Sixty per cent of the fibres responded to sodium chloride, 60 per cent to hydrochloric acid, 40 per cent to quinine and 40 per cent to sucrose. In attempting to account for the number of single, double, triple and quadruple sensitivities, we (Frank and Pfaffmann, 1969) examined the possibility that the four sensitivities were independent of each other and their occurrence in combinations was determined probabilistically. The obtained and theoretical number of fibres that respond to only one quality, that is, are narrowly tuned, and those

TABLE I

NUMBERS OF SINGLE FIBRES RESPONDING TO 1, 2, 3 OR 4 TASTES IN THE GLOSSOPHARYNGEAL NERVE

Number of qualities (N)	Number of responses Predicted ($p_n \cdot T$)	Observed
1	7.1	8
2	11.2	12
3	7.1	5
4	1.7	2

T: number of fibres in sample.

that are more broadly tuned, responding to two, three or four qualities, is given in Table I. If these four sensitivities are mutually independent with known probabilities p_x, and are distributed randomly among taste fibres, the probability p_n of obtaining responses to one, two, three or four stimuli ($n = 1, 2, 3, 4$) can be determined by the successive terms of the binomial expansion $(p_x + q_x)^N$. The probabilities of response to the four stimuli are estimated from the proportion of fibres in the sample which responded to each stimulus.

Similar calculations were performed by Ogawa, Sato and Yamashito (1968), who examined chorda tympani fibre responses of both rats and hamsters. They found a positive correlation between the responses to quinine and hydrochloric acid for the rat and a negative correlation between responses to sucrose and salt in the hamster. These correlations indicate a lack of independence among sensitivities and accordingly a departure from the probabilistic model of the four independent sensitivities. On the other hand, Sato applied the same analysis to data from two other organisms, the cat and the catfish, and found that the probability model applied (Sato, Yamashita and Ogawa, 1969). Thus, there is still some empirical discrepancy in the extent to which the combination of different sensitivities to the four basic taste stimuli occurs in a probabilistic manner. Sato and his colleagues, while documenting strongly the existence of multiple sensitivity, have argued that the departure from the probability model implies a complex kind of specificity of function. It can also be argued, as we have, that independence of the effects of taste stimuli may be one way of specifying the primary taste receptors physiologically (Frank and Pfaffmann, 1969). When responses to two stimuli such as sodium chloride and lithium chloride are highly correlated, the sensory unit fibre that responds to sodium chloride will also respond to lithium chloride. There is presumably a common receptor mechanism for these two.

The evidence discussed so far does not address itself to the question of how this random circuitry can mediate taste discrimination. In fact, the electrophysiological evidence deals only with the peripheral level, between the end organ and the connexions with the afferent fibres. From such data, we cannot decide whether multiple sensitivity occurs because there is much branching as a fibre enters the papilla on the tongue, as indicated by recent studies of neural innervation such as those of Beidler (1969), or because single receptor cells in the papillae have multiple sensitivities, as shown by Kimura and Beidler's (1961) microelectrode recordings. Either mechanism could account for multiple sensitivity.

Surprisingly, until recently most electrophysiological investigators have used whole tongue stimulation with a flow chamber or a flushing of the tongue with fluids from a medicine dropper. Rapuzzi and Casella (1965) were among the first to study electrophysiologically the multiple or branching interconnexions among the papillae on the frog's tongue and to map the neural connexions among papillae. They were able to record action potentials in the stalks of the individual papillae of the frog's tongue using a suction microelectrode. Electrical stimulation of the nerve fibres in a papilla was observed to elicit action potentials antidromically in several neighbouring papillae, presumably through collateral branches of the sensory fibres. They concluded that several receptors are, therefore, connected to a single afferent fibre with a mean of six receptors per fibre. Further work on stimulation of single papillae in mammals with the recording of electrophysiological responses is obviously important. I. J. Miller, working in Beidler's laboratory, has recently used single papilla stimulation in conjunction with single fibre recording and Dr Beidler will mention some of that work (see p. 55). I would like to give a preliminary account of recent work by Wang and Frank at the Rockefeller University who have used electrical and chemical punctate stimulation to map the "receptor field" of individual afferent nerve fibres in the rat chorda tympani.

After a single fibre's responses were isolated, the approximate position

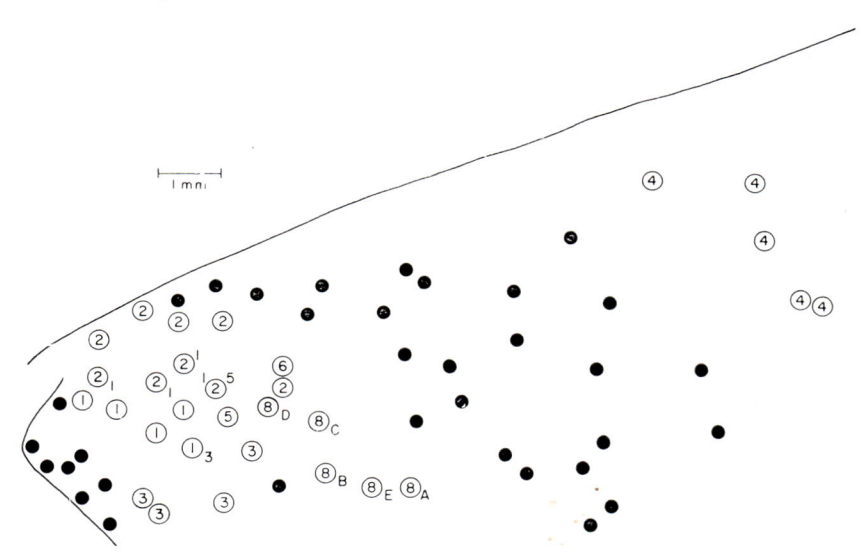

FIG. 2. Map of dorsal half surface of rat's tongue. Black dots represent papillae which were not innervated by fibres dissected in this sample; open circles are papillae innervated by such fibres. Numbers indicate the particular fibre innervating that papilla. When papillae are innervated by more than one fibre, the identifying numbers are placed outside the circle. Letters A-E of papillae innervated by fibre 8 correpond to panels A-E of Fig. 3 when this papilla was stimulated by chemicals. (From Wang and Frank, unpublished observations.)

of its receptive field was located by dropping small amounts of taste solutions on various parts of the tongue. Then individual papillae in the sensitive area were stimulated electrically. A fine gold wire (tip approx. 0.3 mm) was placed on a papilla and short anodal pulses were introduced (0.1 millisecond in duration at 1-second intervals). The intensity of the pulses was reduced until stimulation was effective on papillae but not between papillae. Electrical stimulation, in this way, typically produced one or two spikes with each pulse and latencies of about 20 milliseconds.

Fig. 2 gives a map of the tongue receptive field by this method of exploration. Each fungiform papilla of the rat's tongue usually contains one taste bud on the dorsal surface. The papillae can be seen with the aid of a binocular dissecting microscope. Judging from the electrical stimulation, most single afferent fibres are distributed to more than one papilla. Of 17 fibres,

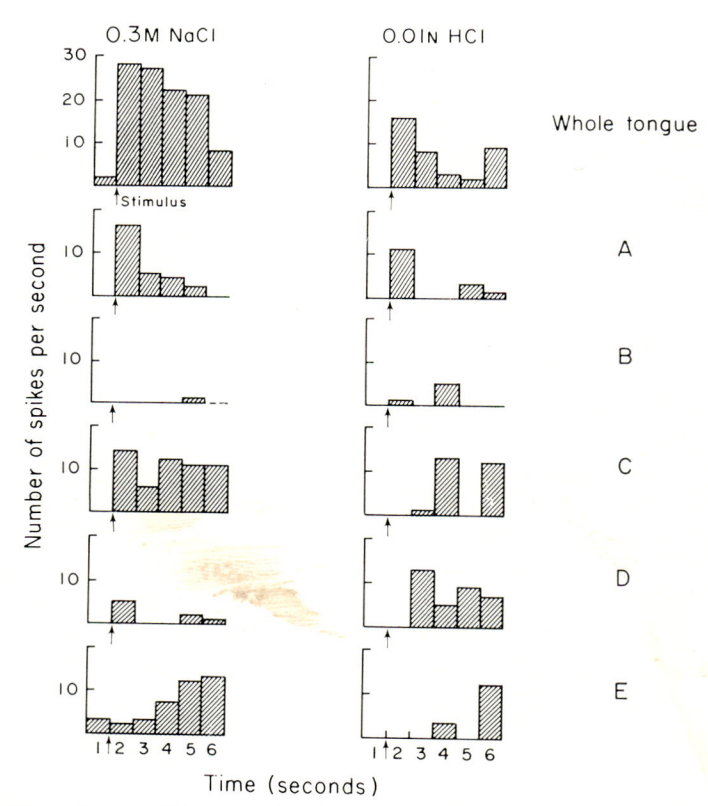

FIG. 3. Response histograms in spikes per second for stimulation of whole tongue and of single papillae (A-E) in the receptive field of fibre 8, Fig. 2. (From Wang and Frank, unpublished observations.)

two innervated only one papilla and the maximum number of papillae innervated by one fibre was nine. The average was 4·5 papillae per fibre. As seen in the figure the papillae and taste buds were more densely packed toward the tip and the larger numbers of papillae per fibre came from this region. Farther back there tended to be fewer papillae per fibre. The maximum distance spanned was 3·5 mm. The figure shows one papilla served by three fibres, others by two and still others by one. But, of course, there were many fibres innervating these papillae which were not sampled. Beidler has found as many as 50 fibres close to the base of the taste bud in his morphological studies. It remains to be determined how many fibres per papillae can be detected functionally.

The evidence on chemical selectivity in these studies is still most preliminary but it was possible for Wang and Frank to stimulate chemically several multi-ending preparations using a small ground-off stimulating capillary tube (diameter 0·5 mm) surrounded by a larger glass tube (1 mm diameter) to provide a vacuum seal. The following solutions were used: 0·01 M-hydrochloric acid, 0·3 M-sodium chloride, 1·0 M-sucrose, 0·01 M-quinine hydrochloride. Fig. 3 shows the response of a single NaCl-HCl sensitive fibre to a whole tongue stimulation by these two solutions. The fibre is more reactive to NaCl than to HCl, as shown by the frequency histograms for the first five seconds of discharge. Whole tongue stimulation in general gave the best response. Panels A-E compare the responses at five different papillae of the same fibre to HCl and NaCl. Single papillae never responded to stimuli which were not effective with areal stimulation of the single fibre's total receptive field.

Both NaCl and HCl were effective at the separate sensory endings but quantitative differences in responses were observed. In two cases, the receptive site was more reactive to HCl than to NaCl, suggesting that local receptor characteristics might influence the sensitivity. It is not possible to go beyond this preliminary description of chemical sensitivity patterns except to say that these observations tend to confirm the idea that multiple sensitivity does not result solely from the branching of nerve fibres among papillae. The multiple response character, which in this example was duplex, can be detected at the individual papilla loci.

So far, I have avoided a direct discussion of sensory coding in taste. The evidence discussed indicates that there is diversity in afferent neural patterning. Some units are specific in responding to only one of the four basic stimuli. Others are more broadly tuned to two, three or four of these stimuli.

It is now becoming clear that individual gustatory receptors are part of the receptive field of more than one afferent fibre. We have yet to learn how these interact and the nature of their excitatory and/or possible inhibitory relations. The gustatory sensory sheet is not static but contains components for integrative action. One would assume that this is the case for man as well. Although stimulation of punctate areas may arouse a clear and well-recognized taste quality in man, "the papilla with only one taste" may be the result of some integrative action of receptors, afferent nerve fibres and central nervous function. There is still a gap between animal and man and perhaps we should be wary of jumping the gap too readily. But other sensory systems display no physiological discontinuities, except perhaps colour vision. But even here the discontinuity does not occur between animal and man.

I should like now to turn to certain behavioural studies of the same species for which we have physiological data. The work of Erickson (1963) showed that the correlation in neural discharge patterns across fibres could predict behavioural similarities in taste "quality" in rats. Substances that excited the same neurons in a similar way (yielding a high positive correlation between the responses in the same fibres to two different stimuli) were confused in discrimination tests. In generalization, a behavioural test of similarity, they showed greater generalization with two stimuli that were correlated electrophysiologically.

But I should like to go beyond that and present evidence suggesting that the lack of a sharp specificity in the afferent neural discharge and a diversity of sensitivities aid in discriminating a variety of substances. I shall change species, to the squirrel monkey, and change stimuli to a series of sugars. Unlike in the rat, the chorda tympani of the squirrel monkey responds well to sugars and it is easy to record integrated responses of the total nerve to sugars of different concentrations and different types. When the magnitudes of response to concentration series of different sugars are compared, a clear seriation is seen. For equimolar concentrations, fructose produces the largest integrated response, sucrose the next, and dextrose next (maltose, lactose, etc. are all indistinguishable from dextrose).

Such a seriation in sensory effectiveness is well known in man's psychophysical reaction. Professor Zotterman, Dr Andersen and their colleagues have studied this in dog and man electrophysiologically. In the dog fructose gives a higher integrated chorda tympani response than sucrose, which in turn gives a higher response than glucose. However in man, sweetness is ordered in the same way as the chorda tympani response is ordered: sucrose, fructose, maltose and glucose (Andersen, Funakoshi and Zotterman, 1963; Borg *et al.*, 1967).

Since animals have strong taste preferences and aversions it is possible to test the efficiency of their sense of taste by a relatively simple procedure: the two-bottle preference test, which was first popularized by C. P. Richter. Since monkeys (like men, rats, insects and a wide variety of other species) like sugar, we applied this method to assay the relative efficacy of several sugars, especially fructose, sucrose and dextrose. In our initial tests, we followed a simple plan of 24-hour *ad libitum* testing with two bottles on the cage, one containing the sugar and the other the water on day 1 (and then reversing the positions of the sugar and water on day 2). The concentration of the sugar was increased every two days. The squirrel monkeys found the sugars so palatable that they began to ingest large amounts of fluid. Our intake measures became excessive and variable and we could see little difference among the sugars. We succeeded better in obtaining measures by the two-bottle method when we tested the animals for a shorter period

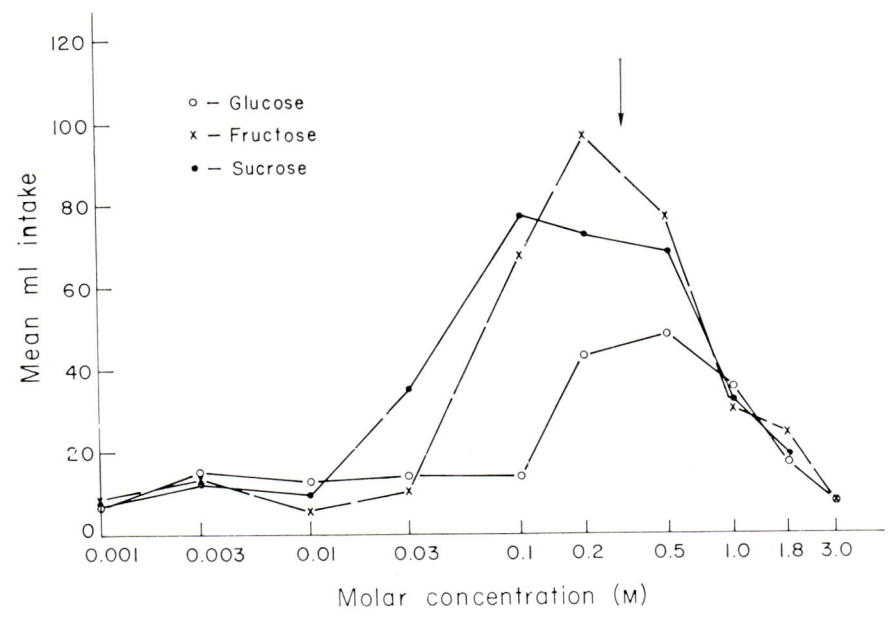

FIG. 4. The preferences of the squirrel monkey for different sugars compared to water. Two-bottle preferences were determined during daily one-hour test periods after 16 hours of food and water deprivation. Arrow indicates isotonic concentration.

each day. They were given no water over night, test fluids were available for one hour from 10.30 to 11.30 a.m., water was available from 3.00 to 6. p.m. and food from 2.00 to 5.00 p.m.; water was taken away at 6.00 p.m. until the test the next morning. Fig. 4 shows the results. Sucrose shows a lower preference threshold and somewhat greater intake up to a peak preference, at which point a slightly greater volume of fructose is taken before the usual turn-down or aversion is apparent. Glucose is clearly a less effective stimulus. The preference test suggests a reversal in preference of sucrose and fructose from the chorda tympani response. Other behavioural measures were taken to determine whether more precise methods would confirm this difference.

One procedure widely used for testing animal sensory functions is some adaptation of operant conditioning procedures. Instead of measuring simple intake, the animal is trained to press a bar for reinforcement or reward, in

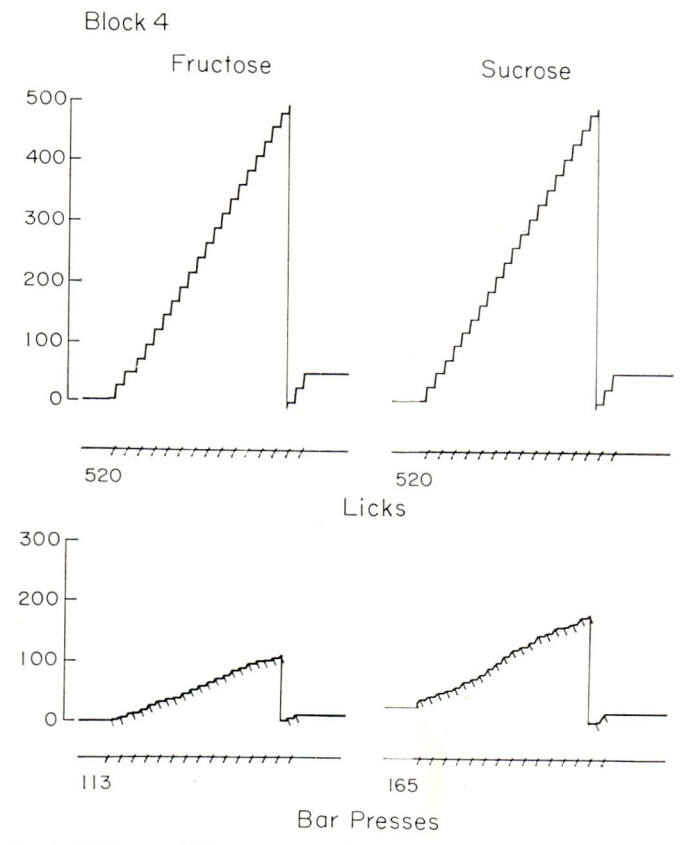

FIG. 5. Lick rate and bar presses to fructose and sucrose for one animal. (Reproduced from Pfaffmann, 1969, by permission of Academic Press.)

this case a drop of sugar water. To avoid the effect of satiation from drinking too much sugar water, we employed a schedule of reinforcement, a fixed-interval (30) schedule on which the animal presses the bar but receives reinforcement only every 30 seconds. Fig. 5 shows the record of a monkey responding to two sugars. Bar presses and licks were both monitored by electronic counters and converted to the cumulative graphs shown. The recorder advances upward on the record at each bar press or lick. The step-like form of the function represents the pause in responding when each reinforcement is given. The monkey works more, presses more often, for sucrose than fructose, even when his rate of licking to the solutions is equal. At the point when he's licking as fast as he can, the bar presses still indicate an advantage of sucrose over fructose. Fig. 6 shows a more complete bar

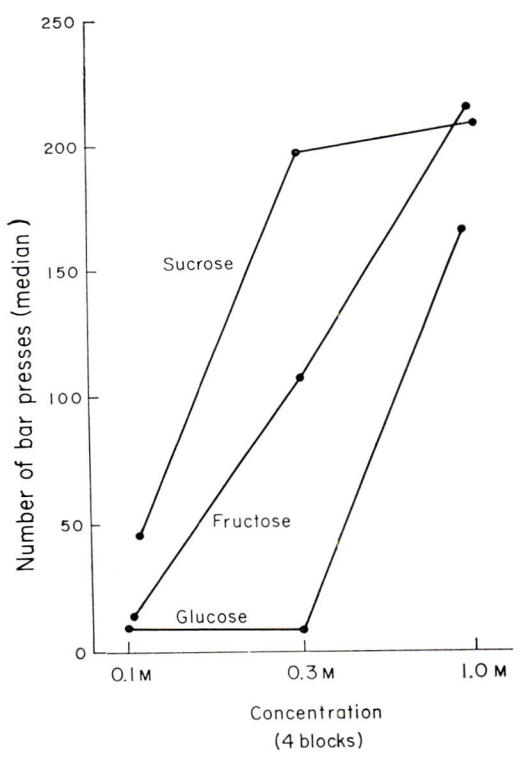

FIG. 6. Graph of bar press function for one squirrel monkey for three sugars. (Reproduced from Pfaffmann, 1969, by permission of Academic Press.)

press rate function for the three sugars, sucrose, fructose and glucose. At concentrations below the maximum, bar pressing is a monotonic function of concentration. Lick rate shows a similar relation. These measures have an advantage over simple preference intake because they avoid post-ingestive and metabolic factors which lead to the so-called aversion turn-down in simple preference tests. Still we find the same order: sucrose > fructose > glucose, in contrast with the electrophysiological results where fructose > sucrose > glucose.

We next used another version of the two-bottle preference test, modelled after the psychophysical "method of limits". Using the same deprivation and feeding schedule we now gave a standard sucrose solution (0·1 M) and presented an ascending series of fructose solutions followed by a descending series in order to determine the concentrations which were equally preferred. We then did the same for glucose. Fig. 7 shows the results. By taking the 50 per cent point (by interpolation) it is possible to determine the equally effective concentration. This figure presents the data for one animal and shows the same order as in the other tests.

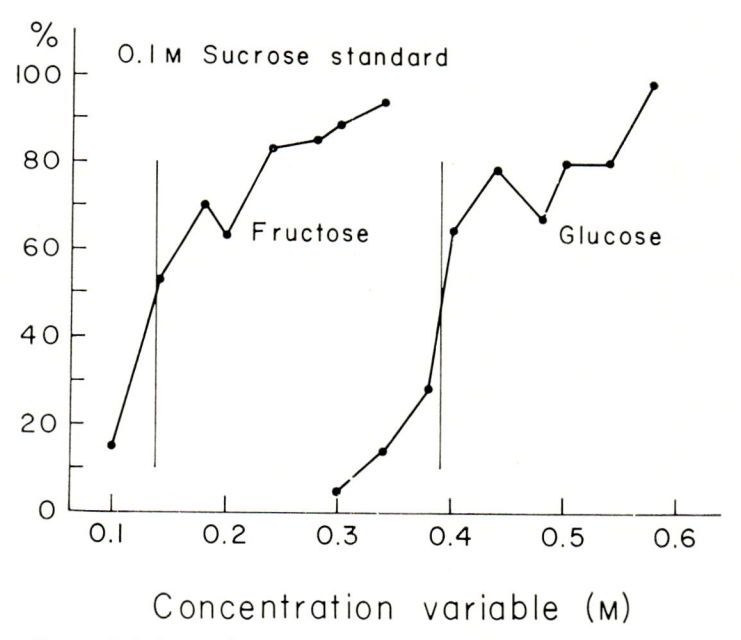

FIG. 7. Relative preference of fructose and of glucose concentrations compared with 0·1 M-sucrose. The 50 per cent point marks the "crossover" or point of equal effectiveness for each test sugar compared with the standard 0·1 M-sucrose.

At this time, we are convinced that all the behavioural measures are in agreement—namely, that sucrose is the more effective stimulus behaviourally than fructose in spite of the greater neural response to fructose. In confirmation of this is the further finding that no fructose concentration could be found that was ever preferred to 0·3 M-sucrose. All our tests depend on some measure of preference or reinforcement. There is thus an intrinsic motivational quality. It appears that the animal can distinguish the difference among the sugars not merely by their intensity but also by the particular kind of sweetness.

Indeed, if we return to a single unit study of the afferent input of the squirrel monkey we discover that although most single units show a higher frequency of responding to fructose than to sucrose, there are some units which respond more to sucrose than to fructose (Pfaffmann, 1969). In other words, not all sugar receptive units give the same responses. There are slight differences in reactivity within the class of sugar receptors so that a different pattern of relative activity to different sugars occurs which provides the theoretical possibility of identifying and recognizing the different kinds of sugar. The animal can recognize sucrose and prefers it over fructose of the same concentration. All our measures partake of some aspect of preference. Hence, some other learned behavioural procedure must be developed to allow the monkey to report sensory intensity (sweetness) as well as do Professor Zotterman's human subjects.

The main point is that the afferent "fuzziness" of the mammalian taste system, that is, its mixture of more specific and less specific receptors, may indeed be a virtue in providing a wider range of possibilities for discriminating fine nuances among closely related stimuli. Multiple sensitivity, combined with multiple branching of afferent fibres, all points to a diversified yet integrative sensory sheet. By contrast, invertebrate chemoreception, so elegantly analysed in insects by Dethier (1956) and his students, conforms to a model that is easier for us to understand. In the blowfly, stimulation of the chemoreceptive hairs by sapid solutions occurs by way of a small pore at the tip of the hair where contact is made with dendrites of the sensory cells located at the hair. One cell is specifically reactive to electrolytes, one to water, and one to some but not all sugars. In behavioural tests, however, this animal is unable to discriminate among different kinds of sugars. Electrophysiologically, all sugar receptors are alike and so all sugars are alike to the fly. As I have tried to show, this is not the case for the monkey. Man too can tell one sugar from the next, often with a clear preference for one over the other. This more complicated mammalian sensory coding may make it more difficult for the physiologist to decipher. For the taster, it provides the variety and detail of sensory experience.

SUMMARY

Electrophysiological studies of afferent taste units in the chorda tympani and glossopharyngeal nerves have revealed a multiple sensitivity to the four basic taste stimuli: sodium chloride, quinine, hydrochloric acid and sucrose.

Some afferent fibres respond to one, others to two, others to three and some to all four taste stimuli. We previously showed that these sensitivities to the four basic tastes in IXth nerve units were distributed in a manner that could be predicted from the binomial expansion by treating each basic taste as an independent process which combined probabilistically with each of the other basic taste processes. Other investigators have noted similar results, with exceptions.

Frank and Wang have extended the analysis of multiple sensitivity by determining the extent to which individual chorda tympani fibres branch to innervate fungiform papillae. Punctate electrical stimulation with brief anodal pulses proved most convenient for this purpose. Stimulation was effective on papillae and ineffective on the surface between papillae. Individual chorda tympani fibres responded to stimulation of from one to nine papillae, the mean number per fibre being 4·5.

These indications of multiple innervation and multiple sensitivity are discussed in relation to behavioural discrimination, especially of sugars. Electrophysiological data are presented on single taste units for sucrose and other sugars, and also the behavioural evidence on the ability to discriminate among different kinds of sugars. The hypothesis is developed that a system of highly specific receptors cannot mediate fine discrimination as effectively as one with overlapping sensitivities.

Acknowledgements

The research described in this report was supported by The Rockefeller University, a National Science Foundation grant GB-4198, a National Science Foundation predoctoral fellowship to M. Frank and a National Institutes of Health postdoctoral fellowship to M. Wang.

REFERENCES

ANDERSEN, H. T., FUNAKOSHI, M., and ZOTTERMAN, Y. (1963). In *Olfaction and Taste* (Proceedings of the First International Symposium, Stockholm, 1962), pp. 177–192, ed. Zotterman, Y. Oxford: Pergamon Press.

BEIDLER, L. M. (1969). In *Olfaction and Taste III* (Proceedings of the Third International Symposium, New York, 1968), ed. Pfaffmann, C. New York: Rockefeller University Press.

BORG, G., DIAMANT, H., OAKLEY, B., STROM, L., and ZOTTERMAN, Y. (1967). In *Olfaction and Taste II* (Proceedings of the Second International Symposium, Tokyo, 1965), pp. 253–264, ed. Hayashi, T. Oxford: Pergamon Press.

COHEN, M. J., HAGIWARA, S., and ZOTTERMAN, Y. (1955). *Acta physiol. scand.*, **33**, 316–332.

DETHIER, V. G. (1956). *Biol. Bull. mar. biol. Lab., Woods Hole*, **111**, 204–222.

ERICKSON, R. P. (1963). In *Olfaction and Taste* (Proceedings of the First International Symposium, Stockholm, 1962), pp. 205–213, ed. Zotterman, Y. Oxford: Pergamon Press.

FISHMAN, I. Y. (1957). *J. cell. comp. Physiol.*, **49**, 319–334.

Frank, M. (1968). Ph.D. Thesis, Brown University.
Frank, M., and Pfaffmann, C. (1969). *Science*, **164**, 1183–1185.
Kimura, K., and Beidler, L. M. (1961). *J. cell. comp. Physiol.*, **58**, 131–139.
Ogawa, H., Sato, M., and Yamashita, S. (1968). *J. Physiol., Lond.*, **199**, 223–240
Pfaffmann, C. (1941). *J. cell. comp. Physiol.*, **17**, 243–258.
Pfaffmann, C. (1955). *J. Neurophysiol.*, **18**, 429–440.
Pfaffmann, C. (1969). In *Reinforcement and Behavior*, pp. 215–240, ed. Tapp, J. New York: Academic Press.
Rapuzzi, G., and Casella, C. (1965). *J. Neurophysiol.*, **28**, 154–165.
Sato, M., Yamashita, S., and Ogawa, H. (1969). In *Olfaction and Taste III* (Proceedings of the Third International Symposium, New York, 1968), ed. Pfaffmann, C. New York: Rockefeller University Press.

DISCUSSION

Andersen: What was the temperature of the solutions that you applied to the tongue in your studies of single units in the chorda tympani in the rat? It is important that the temperatures of the solutions used are the same, and also which temperature range they are in.

Pfaffmann: The temperatures of all solutions were the same, room temperature.

Wright: Did you consider the effect of heat of dilution, which could be important in a concentrated solution? The heat of dilution depends upon the identity of the chemical compound; the heat of dilution of 1 M or 0·5 M-sucrose would not be the same as the heat of dilution of 1 M or 0·5 M-glucose or fructose.

Pfaffmann: Quite true, but how does this affect the observations when one is using the same concentrations of solute under constant conditions, yet the different receptors give different responses?

Wright: It has just been suggested that temperature is a factor in the response, and although you apply the solution to the tongue at a known temperature, the further dilution of the solution by saliva could be producing a local and possibly important change in the temperature of the actual receptor surface.

Pfaffmann: This would then be one way of accounting for a difference in effectiveness of different substances at the same receptor.

Andersen: Working in Professor Zotterman's laboratory several years ago (Andersen, Funakoshi and Zotterman, 1962) we observed, in a few cases, that maltose produced a very large burst of spikes in certain fibres. I now suspect that our solution of maltose had become hydrolysed to produce this great response. You had a case in which sucrose produced a similar large effect compared to that of fructose. Are you sure that you didn't have hydrolysis of sucrose in your experiments?

Pfaffmann: All I can say is that the conditions under which stimulation by sucrose or fructose occurred were identical, so the only thing we were varying was the particular fibre being recorded.

Andersen: Yes, but the only disaccharide that you tested was sucrose; all the others were monosaccharides which cannot undergo hydrolysis to yield a higher concentration of sugar.

Pfaffmann: We didn't test for hydrolysis directly.

Zotterman: I am delighted with Professor Pfaffmann's statistical treatment of multiple sensitivities. This makes sense, not only because it fits in with our single-fibre work in monkey, pig and dog, where we find between 8 and 12 per cent of what I call one-quality fibres, and two-quality or three-quality fibres at higher frequencies. The question is now, what is the perceptual response? If we could stimulate one of these two-quality fibres which responded to say salt and acid in man, would you expect to experience a modified salt taste, as you would expect if you applied salt and acid substances to the tongue, or would you experience what I think is most probable, namely pure salt or pure acid taste? I raise this because of our experience in skin, where H. Hensel and I described bimodality fibres which responded to cooling and also to touch, or pressure, as well. When the skin is stimulated by purely mechanical stimuli you always get the perceptual experience of touch or pressure although the fibre is capable of responding to cooling as well. For taste, this will be very difficult to establish until we can do experiments on single fibres in man. The idea, which I think is most probable, that you perceive only one kind of quality is based on George von Békésy's results when he stimulated a single taste bud electrically in humans with just suprathreshold stimulation. The young ladies reported for instance "heavenly sweet": a sweetness they had never experienced before! When you stimulate with sucrose solution generally over the tongue you stimulate several kinds of fibre, both one-quality fibres and other fibres which respond to one or two other qualities as well. Now sucrose is generally considered to taste "pure sweet" although it may also stimulate at least a few other fibres which respond to bitter substances.

Pfaffmann: Papillae which give a single quality might be innervated by a single fibre which responds only to one of the classes. We haven't any direct evidence; we can only suggest that the fibre with a single sensitivity produces a pure quality and that the multiple responders may not elicit pure qualities; perhaps they give a modified quality of sweetness. In the case of multiple responders to four classes, a generalized taste quality might be produced which is not readily described in words. My main point is that useful sensory information is provided by all responding fibres.

Døving: Professor Pfaffmann, you used a model to indicate that the four sensory modalities are independent of each other. But you can use a chi-square test to show that any two of the stimuli are independent. Dr P. Mac Leod and I have previously applied these analyses to results from the olfactory system (Døving, 1965; Leveteau and Mac Leod, 1969).

Pfaffmann: We could try this other way of doing it. Ogawa, Sato and Yamashita (1968) did a test of the probability of occurrence of pairs of tastes and in the rat found only that the HCl–quinine pair were not independent. In the hamster, on the other hand, NaCl–quinine and NaCl–sucrose were not independent and he argued for some degree of relative specificity.

Wright: There is a possible pitfall here. In electrophysiological work the concentration of the stimulus is enormous compared with the concentration used in studying behavioural responses. If you have a large number of end organs specialized for sensing sweet, the stimulation of any one or any small number of them will register a sensation of sweet in the central nervous system, so that at low concentration only a few of them are firing. If you put an electrode into one particular end organ the probability is that it won't fire, so you would increase the concentration until it began to fire. But at that stage it is not the only end organ firing, and some of these apparent "dual" sensitivities may really be spill-over from a neighbouring neuron of different specificity which is firing and inducing a change in one which would not normally fire.

Pfaffmann: The very point I wish to make is that these sugar concentrations are not enormous compared with the behavioural range and capacities of the animal. They are also well within the range used in psychophysical experiments in man on judgements of strength of sweetness. I agree that we must watch the intensity of stimuli in the physiological preparation.

Beidler: Dr Pfaffmann, you said that different single taste fibres gave different relative responses to different sugars; so the situation with sugars is like that with salt?

Pfaffmann: The majority of our fibres are more reactive to fructose than to sucrose but there are some in which this order is reversed.

Beidler: This poses some difficult problems for those interested in the relationship of molecular structure and taste. A model receptor site that accounts for the relative activities of a series of sugars will have to be modified to account for a different sequence of relative activities observed for other "sugar" nerve fibres.

Pfaffmann: I should like to say something more about the sugar response and the blocking action of potassium gymnemate. Fig. 1 from Dr Linda Bartoshuk's work (personal communication) shows the responses of five

48   DISCUSSION

single units in the hamster, a species giving a good response to sugar in the chorda tympani. The discharges in impulses per second for a 10-second period of each of 5 fibres, A to E respectively, are shown from left to right on each line before and after blocking by gymnemate. Stimuli were presented in order from left to right, with a water rinse between each stimulus, but only the last response to water is shown following the last stimulus, hydrochloric acid.

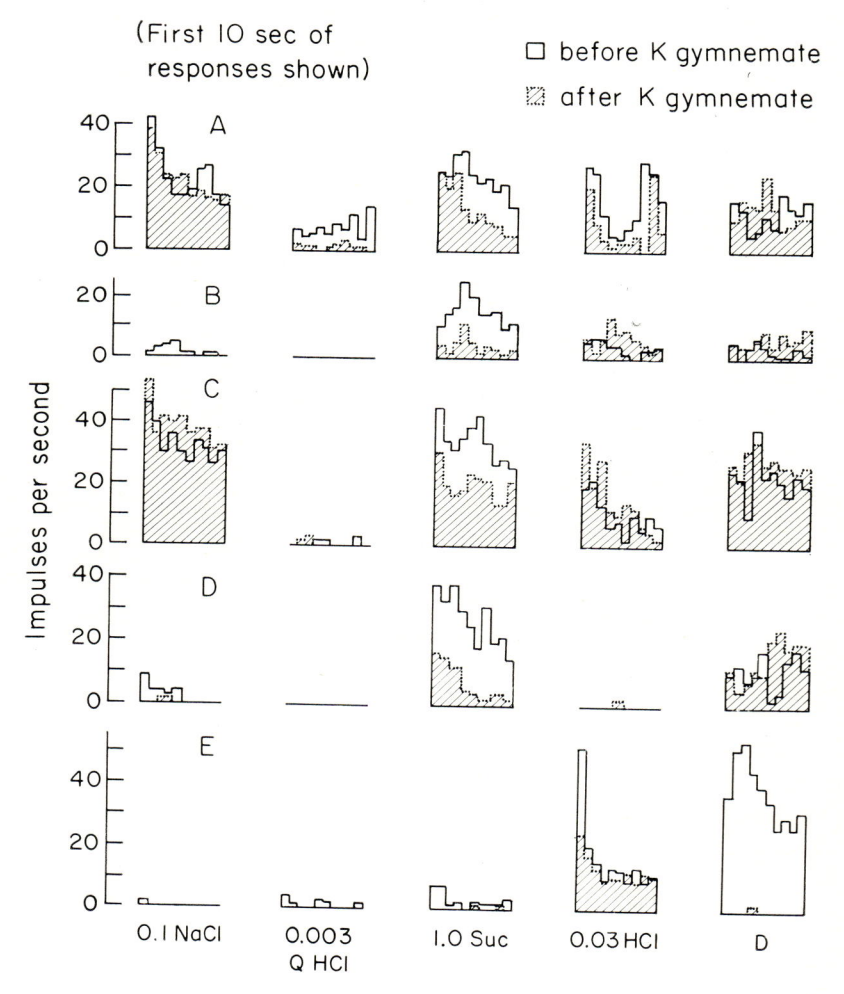

FIG. 1 (Pfaffmann). Responses of single units in chorda tympani nerve of hamster to taste stimuli, showing blocking action of potassium gymnemate. See text for details.

Fibre A is a non-specific general responder and although there is some reduction of the response to quinine and HCl, the response to sugar is markedly reduced. Fibres B, C and D show decrements almost exclusively to the sugar stimulus. Fibre E is of interest; it responds primarily to HCl and also gives a large "after acid" water response. Gymnemate blocks this water response. In man, water after acid has a sweet taste.

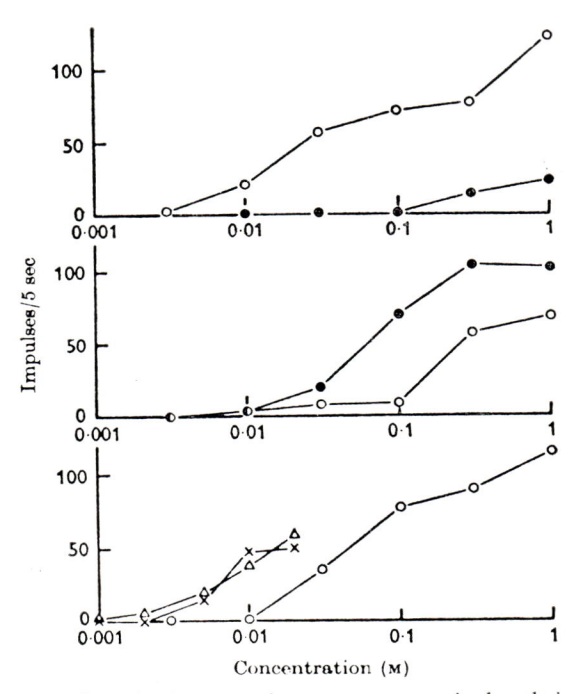

FIG. 2 (Pfaffmann). Concentration–response magnitude relationships in three typical chorda tympani fibres of rats. Top, unit predominantly sensitive to NaCl; middle, unit sensitive to sucrose more than to NaCl; bottom, unit sensitive to NaCl, quinine and HCl. ○, responses to NaCl; ●, responses to sucrose; ▲, responses to quinine hydrochloride; ×, responses to HCl. Ordinate indicates number of impulses elicited in the first 5 seconds after stimulation. (From Ogawa, Sato and Yamashita, 1968, by permission of Professor Sato and *The Journal of Physiology*.)

These results show that the gymnemate blocking action does not interfere with transmission in taste nerve fibres, but acts only on the presumed specific sugar receptive sites at the periphery. This evidence suggests that the afferent nerve fibres with their multiple sensitivity must be regarded largely as parallel transmission channels. They may be multimodal even when the peripheral sites are specific. The peripheral quality is reconstructed or reconstituted, perhaps by some central decoding process. I agree with

Dr Andersen that there is no necessary conflict between electrophysiological evidence and psychophysical studies, but there is a difficulty in interpretation if only single-quality specific input afferent fibres are postulated.

Adey: What is known about the effects of anaesthesia on the pattern of discharges? I understand that most of your experiments have been done under anaesthesia, but isn't it likely that you won't be able to consider coding problems until you consider the unanaesthetized animal?

Pfaffmann: That is a very good point. The use of unanaesthetized preparations is our next priority, to see whether there is anything equivalent to the sharpening in the central nervous system which occurs in the other sensory systems as a result of inhibitory mechanisms.

Wright: On my earlier point of the possibility of signals being picked up from neurons other than those initially stimulated when the stimuli are of high enough intensity, it is clear that if both lots of information are going into the central nervous system, it will recognize a difference. But surely it is hazardous to say that a "sweet" receptor is sensitive to both "acid" and "sweet", when in fact it may be responding to "spill-over" from a nearby acid receptor. The same error can be made when interpreting experiments on odour, when relatively high concentrations are used.

Pfaffmann: One figure from Ogawa, Sato and Yamashita's 1968 paper is relevant. They were recording the frequencies of responses to HCl, quinine hydrochloride and NaCl in one fibre which responded to all three stimuli (cf. Fig. 2, from Ogawa, Sato and Yamashita, 1968). The concentration function curves show that these multiple sensitivities occurred at 0·003 to 0·03 M concentrations. I agree with your point, which should be considered; however, many of these multiple sensitivities do occur at *low* concentrations.

REFERENCES

ANDERSEN, H. T., FUNAKOSHI, M., and ZOTTERMAN, Y. (1962). *Acta physiol. scand.*, **56**, 362–375.
DØVING, K. B. (1965). *Revue Lar. Otol. Rhinol.*, **86**, 845–854.
LEVETEAU, J., and MAC LEOD, P. (1969). *J. Physiol., Paris*, **61**, 5–16.
OGAWA, H., SATO, M., and YAMASHITA, S. (1968). *J. Physiol., Lond.*, **199**, 223–240.

PHYSIOLOGICAL PROPERTIES OF MAMMALIAN TASTE RECEPTORS

Lloyd M. Beidler

Department of Biological Science, The Florida State University, Tallahassee, Florida

I SHALL take this occasion to summarize current concepts of the physiological properties of mammalian taste receptors. Professor Murray has summarized present knowledge of the fine structure of the taste receptors (pp. 3-25). It should be stressed that these receptors are not merely transducers, but are biological cells with many of the characteristics of all other cells. Forty to sixty cells are collected together to form a taste bud which is associated with a papilla. The cells of the taste bud of the fungiform papilla are well protected by sheets of other cells which are laid down much like shingles on a roof (see Fig. 1). The cells' microvilli project into the taste pore which connects with the outside surface of the tongue (see Fig. 2). This organization in many mammals contrasts with that of frogs where the surfaces of most of the cells of the taste bud are in direct contact with the external environment (see Fig. 3).

LIFESPAN OF CELLS OF TASTE BUD

It has been shown that the cells of the rat fungiform taste buds have an average lifespan of about 250 ± 50 hours (Beidler *et al.*, 1960; Beidler, 1963; Beidler and Smallman, 1965). A similar lifespan was estimated from the data of de Lorenzo (1963) to be about 300 hours in the rabbit foliate taste bud. Robbins (1967) showed that the cells of the fungiform taste buds of frogs are also being renewed. The most recent and most rigorous analysis of this rapid turnover was undertaken by Conger and Wells (1969) using mouse vallate taste buds. They measured the number of cells labelled with tritiated thymidine in the taste bud as a function of time after injection of the isotope. The average lifespan of these cells was found to be about $10\frac{1}{2}$ days, in good agreement with the value previously determined in the rat. Note in Fig. 4, adapted from their paper, that there is no indication of two separate populations of cells turning over at two different rates. If one of the populations is as low as 10 per cent of the total number of cells of the taste bud, the turnover of this group would not be easily determined by the methods of analysis. The value obtained for the average lifespan using tritiated thymidine is independent of the number of cells in the taste bud.

X-rays at the proper dosage will kill the germinal cells, and can therefore also be used to study the properties of the taste bud. Conger and Wells

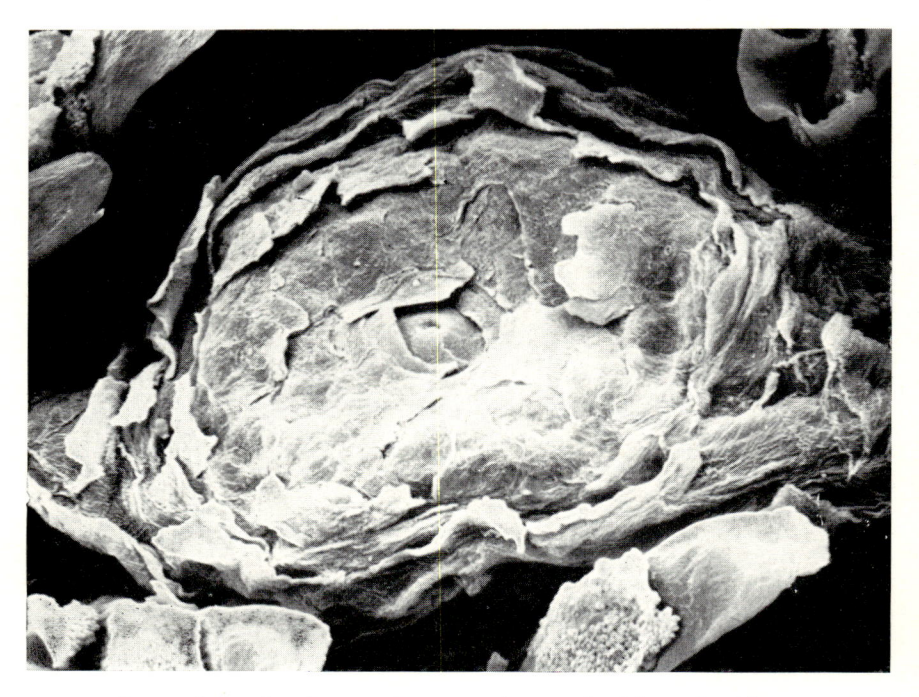

FIG. 1. A scanning electron microscope photograph of the rat fungiform papilla. Notice the sheet-like cells covering the top of the papilla with the taste bud penetrating the surface. ×745

(1969) showed that 50 per cent of the cells of the mouse taste bud disappeared in $5\frac{1}{2}$ days after X-ray irradiation with adequate dosages. This again is a good indication that the vast majority of the cells of the taste bud indeed have an average lifespan of $10\frac{1}{2}$ days.

Another recent method of analysing the lifespan of cells is to inject a pregnant rat continuously with tritiated thymidine so that all the cells of the taste buds of the newborn rats will be labelled. The number of unlabelled cells is then counted over various periods of time after birth, in order to determine the average lifespan. Such experiments have been performed by Mr J. F. Metcalf in our laboratory using the rat fungiform taste bud, and the average lifespan of about 10 days obtained is in good agreement with values determined earlier. No labelled cells were found in 63 tissue sections from taste buds taken 44 days after birth. Thus there is no evidence of long-living cells within the taste bud.

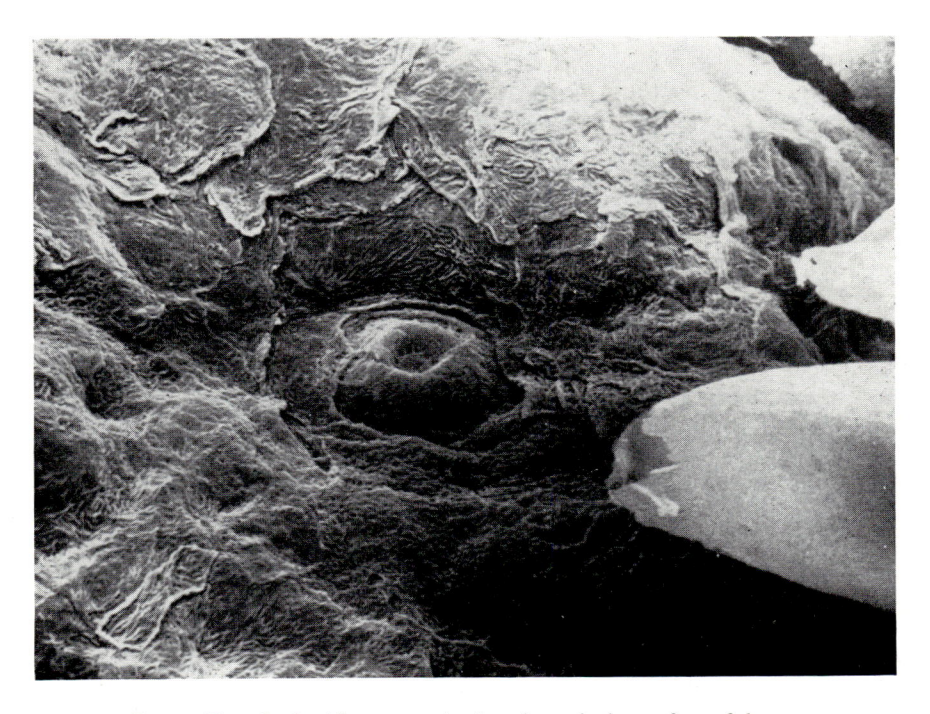

FIG. 2. Taste bud with pore projecting through the surface of the rat fungiform papilla. × 1677

We have shown that entire taste cells are replaced. It should be emphasized that parts of receptors may also be replaced. For example, protein is continually manufactured in the inner segment of the frog, mouse and rat retinal rods and passes upward through the outer rod segment and eventually leaves the cell (Young, 1967). This protein was recently identified as rhodopsin (Bargoot, Williams and Beidler, 1969). Information on similar replacements of components in taste and olfactory cells is still scanty.

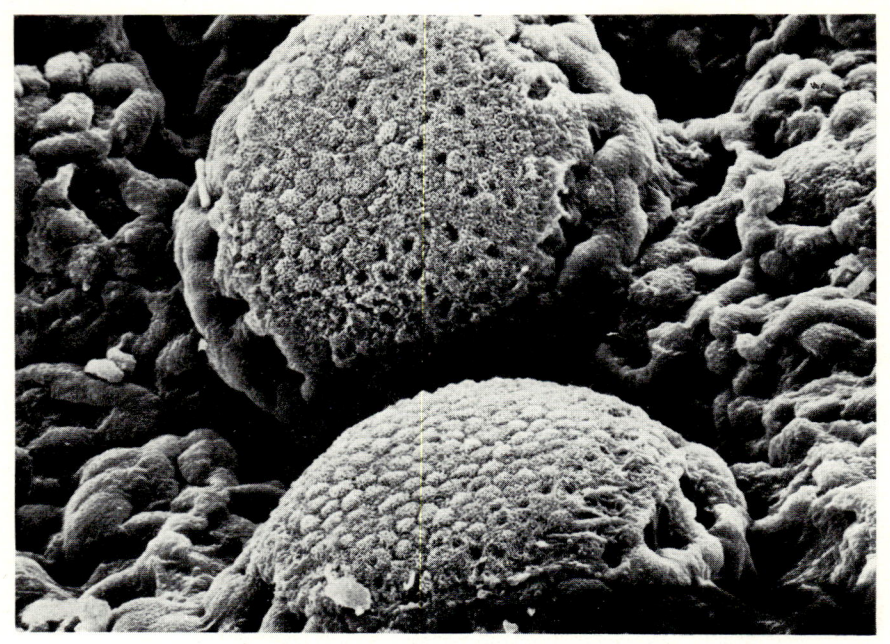

FIG. 3. Two frog fungiform papillae, showing the numerous cells exposed to the external environment. (By courtesy of P. Graziadei.) × 700

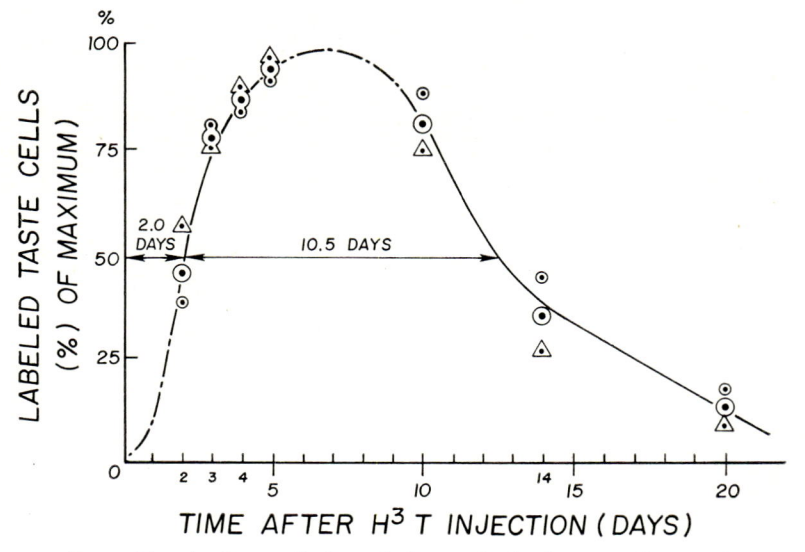

FIG. 4. Transit of taste cells through the taste buds of mouse vallate papillae, determined by labelling with tritiated thymidine. Ordinate is for first-generation descendants. (Adapted from Conger and Wells, 1969.)

TASTE BUD INNERVATION

Mammalian taste buds are usually associated with papillae. They are highly innervated. Our study with the rat fungiform papilla shows that about 200 unmyelinated nerve fibres enter the base of the papilla, and 20 or so myelinated fibres (Beidler, 1969). After lingual nerve degeneration, about one-quarter of the original number of unmyelinated fibres still enter the taste bud but the other three-quarters are now absent and no longer enter the papilla. Thus, the vast majority, if not all, of the nerves that enter the taste bud originate from the chorda tympani nerve and most of the remainder of the papilla is innervated by the lingual nerve. The 50 or so nerves entering the taste bud branch profusely to increase to about 200 or more branches.

The innervation of the cells of the taste bud was studied with an electron microscope on near-serial sections. It was determined that there are 50 or 60 cells per taste bud and that all of these cells, with the possible exception of one or two, are in anatomical contact with nerve fibres. Unfortunately the taste cells of the rat fungiform papilla do not show typical and apparent synaptic contacts and therefore it was not determined how many of these cell–nerve contacts are functional.

TASTE BUD INTERACTIONS

The total number of rat nerve fibres entering the base of the fungiform papillae and innervating the taste buds is far greater than the total number of nerve fibres in the chorda tympani nerve bundles which may have taste functions (Beidler, 1965, 1969). Therefore, it is very likely that these nerves branch and that one taste fibre innervates several taste buds. Rapuzzi and Casella (1965), using electrical stimulation of single frog papillae and recording from neighbouring papillae, showed such branching in the frog. Miller chemically stimulated single rat fungiform papillae while recording from a single taste nerve fibre (Miller, 1968). He proved that such a taste fibre may indeed innervate several taste buds (see Table I of Miller, 1968). In addition, he showed that stimulation of neighbouring taste buds could have either an enhancing or a depressing effect on the response of a given isolated taste bud (Beidler, 1969; Miller, 1968).

LOCALIZATION OF TASTE RECEPTOR SITES

Where does the stimulus interact with the taste cell? Latency to chemical stimulation of the tongue can be as low as 15–20 milliseconds. Responses of single taste fibres to moderate concentrations of substances such as sodium chloride start with a maximum frequency and then decrease over a period of about two seconds to a steady state, which is then maintained as long as the stimulus is applied to the surface of the tongue (Beidler, 1953). Note that there is no build-up in the response, as would be expected if the transport of the stimulus to the receptor site was slow; that is, if the receptor response reflected a build-up in concentration over a period of several seconds or more at the level of the receptor site.

In early days it was thought that the taste hairs were the sites of the interaction between stimulus and receptor. Today it is commonly thought that the receptor sites are contained on the microvilli of the taste cell. There is no direct evidence for this. However, Professor Murray now tells us that the sensory cells are those cells with the least number of well-developed microvilli (see pp. 3–25). In fact, the cells showing a supporting nature have the best developed microvilli which extend well into the taste pore. If this is correct, one must conclude that the microvilli are not as important to the taste mechanism as was once thought. This is similar to stimulation of receptors by odours. Mammalian olfactory receptors contain cilia. The vomeronasal receptors are anatomically very similar to the olfactory receptors and respond quite well electrophysiologically to many odours (Tucker, 1963). They may contain no cilia, however, but do contain microvilli (see Fig. 5). Olfactory receptors of birds, on the other hand, contain microvilli as well as cilia (Brown and Beidler, 1966).

Is only the apex of the taste cell sensitive to chemical stimulation? It should not be forgotten that there are methods of stimulating the taste receptors other than applying a stimulus to the surface of the tongue. Intravenous taste is often used as a measure of circulation time. Mr R. M. Bradley, in my laboratory, is currently studying intravenous taste using a heart–lung machine and artificial blood to perfuse the rat's tongue. The responses to substances such as sodium saccharin are clear indeed. There is some indication that the responses can be increased over and above that already induced by applying the chemical stimulus to the surface of the tongue. This then raises the question, where are the receptor sites for intravenous stimulation? If the surface of the tongue is contained in a flow chamber one can eliminate the salivary glands as a possible source of stimulus transport. In such experiments it is difficult also to eliminate slow diffusion through the tongue epithelium. However, the response is quite fast by way of intravenous stimulation and Professor Murray has already indicated that there are tight junctions at the apical surfaces of the taste bud, which are barriers to transport at this region. Mistretta, who studied the penetration of

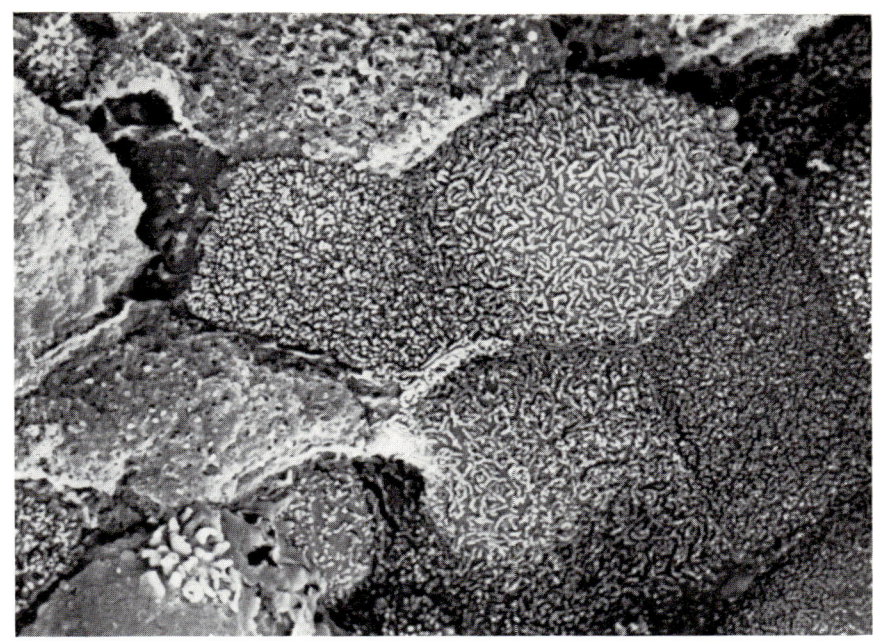

Fig. 5a. Scanning electron microscope view of four vomeronasal receptor cells of the box turtle, showing microvilli but no cilia. (By courtesy of D. Tucker.) × 3333

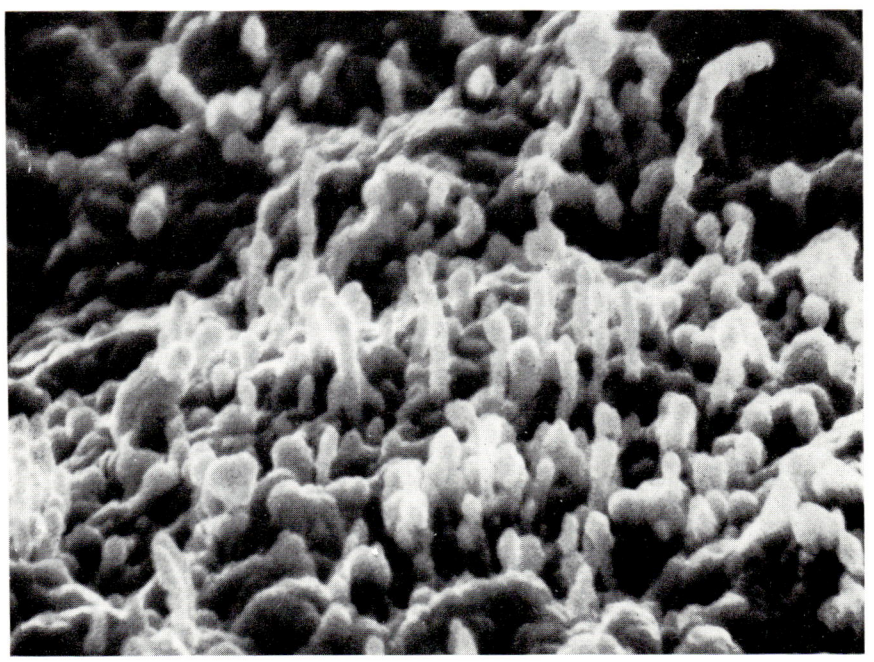

Fig. 5b. Scanning electron microscope photograph of surface of vomeronasal receptor of gopher tortoise, showing numerous microvilli. (By courtesy of P. Graziadei.) × 14 400

a large number of isotopically labelled substances through the tongue surface, has shown that the surface of the tongue of the rat presents a barrier to chemical penetration that is not dissimilar to that, for example, of the belly skin of the rabbit (Mistretta, 1968). Penetration is quite difficult for most chemicals. We are thus forced to consider receptor sites on taste cells other than at the apical end. Perhaps the chemical sensitivity of the taste receptor membrane is rather similar over most of its surface.

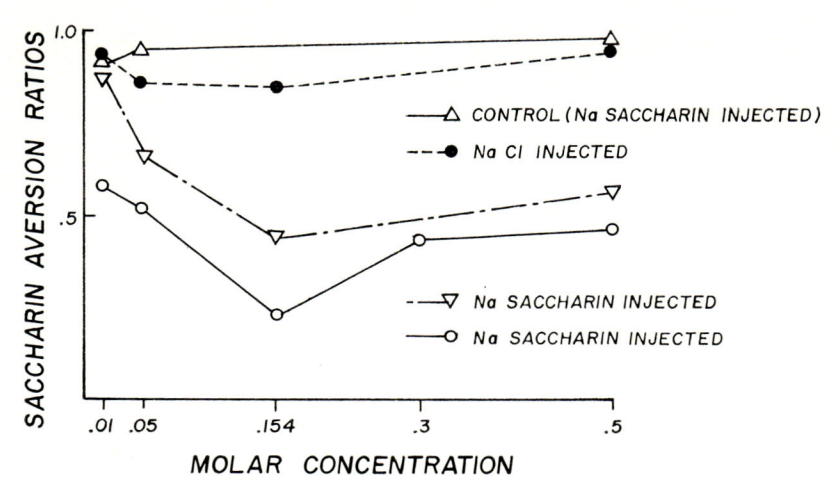

FIG. 6. The aversion to drinking sodium saccharin solution by rats is shown in this graph. The conditions are explained in the text. Triangles indicate rats intravenously injected with sodium saccharin but not exposed to X-rays (controls). Solid circles indicate rats injected intravenously with sodium chloride and conditioned with X-rays. Inverted triangles, rats injected intravenously with sodium saccharin and conditioned with 100 R of X-rays. Open circles, rats injected intravenously with sodium saccharin, as for the third group, but 200 R of X-rays were given. Abscissa, molar concentration of sodium chloride or sodium saccharin injected into the rats. The ordinate is the ratio of the intake of sodium saccharin solution to the total volume of sodium saccharin solution and water ingested.

One might ask the question, can the rat really react to the sweet quality of the sucrose or saccharin when so injected? In other words, is the taste quality the same when the stimulus is applied to the surface of the rat tongue as when it is applied intravenously? In order to obtain more information, Mr Bradley exposed the rat to one application of 100-200 R X-irradiation a few minutes after injecting various concentrations of sodium chloride or saccharin into the vein of the rat's tail. In this manner the rat is conditioned to avoid subsequent orally administered solutions of the same taste quality. The alteration of normal preferences is shown in Fig. 6. In these experiments the rats were given a choice of water and the normally preferred 0·125 per cent sodium saccharin solution in a two-bottle situation.

The aversion to saccharin was expressed as the ratio of the volume of saccharin solution ingested to the total volume of saccharin and water consumed. Fig. 6 shows the results. Eight rats were used to obtain each point shown on the curves, or a total of 128 rats for this series of experiments. Note that the control rats (injected with saccharin but not exposed to X-rays) and the rats injected with sodium chloride and conditioned with X-rays strongly preferred drinking the sweet saccharin (0·125 per cent saccharin) to water. However, the rats that were injected with sodium saccharin and conditioned with X-rays tended to avoid drinking the 0·125 per cent saccharin solution that is normally preferred. This again is evidence that the rat can taste intravenously injected saccharin as being sweet, as does man. Incidentally, Professor Pfaffmann has told me that when he was injected with saccharin he found the sweet sensation to be rather cyclical, indicative of the fact that the pulse of saccharin reappeared at the tongue each time the blood was recycled. Since the pulse is less sharp and the concentration decreases with each recycling, the sweetness declines.

REACTION TIME

I have mentioned that one may stimulate the receptor by applying the stimulus either directly to the surface of the tongue or intravenously. There is also a third method of stimulating taste, and that is electrically. This has been well studied, particularly in the laboratory of Bujas (Bujas and Chweitzer, 1936; Bujas, 1949). I want to bring to your attention the result of a series of experiments conducted by Nejad (1961) and repeated by Warren (1965). If the stimulating current per unit area is progressively increased and the electrophysiological response recorded, one finds evidence of two separate curves, which indicates two different methods of stimulation (see Fig. 7). The receptor cell is stimulated at low current densities and the innervating axon at a higher current density. The difference of about 8 milliseconds in the two minimum latencies can be attributed to the time between the presentation of the stimulus and the excitation of the nerve by the receptor cell. Note that the minimum latency, 14 milliseconds, shown by the first curve, is of the same order of magnitude as that obtained by applying high concentrations of sodium chloride to the tongue surface. Electrical stimulation of taste receptors differs in two ways from stimulation of most other nervous tissue. First, stimulation best occurs when the surface of the tongue is made anodal rather than cathodal. Second, the response is dependent upon the type of ionic solution in which the electrode is placed. The latter is due to the fact that the current is carried by cations, the nature of which determines the intensity of stimulation at the receptor surface.

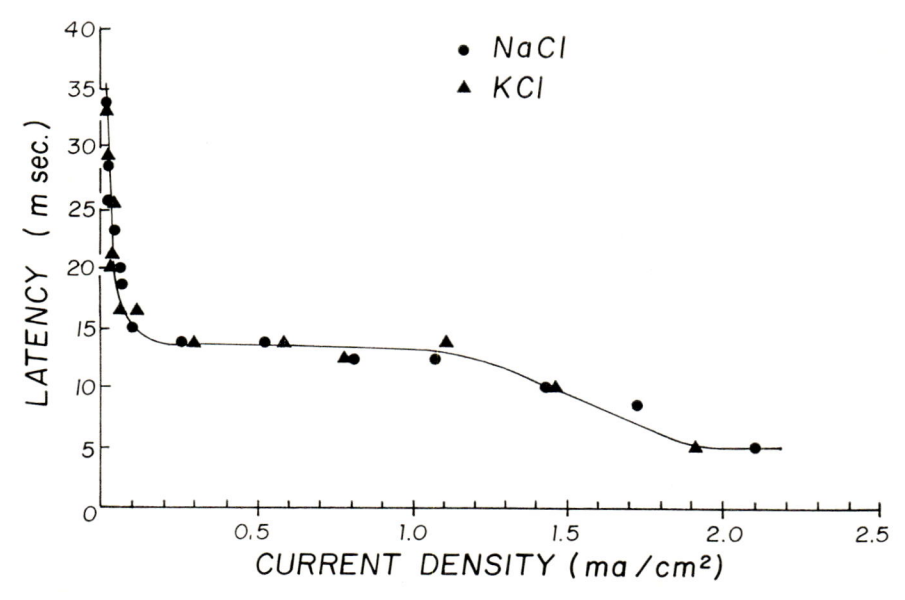

FIG. 7. The latency of neural activity recorded from a small chorda tympani nerve bundle in the rat was measured as the current density of stimulation was increased. The agar electrode was in either sodium chloride or potassium chloride solution.

MECHANISM OF TASTE RECEPTOR STIMULATION

How is the taste cell stimulated by a chemical change of its environment? A series of experiments led us to conclude that both electrolytes and non-electrolytes are weakly adsorbed to the surface of the taste cell (Beidler, 1954, 1961). Further analysis led us to believe that a conformational change also occurs in the receptor site molecule to which the stimulus substance is adsorbed. It is commonly thought that such conformational changes would be sufficient to change the permeability of the membrane, and thus initiate an ionic exchange between the inside and the outside of the cell. This would lead to the usual electrical depolarization of the cell. However, it should be noted that almost all chemical stimuli tested showed an increase in response as the stimulus concentration is increased. Certainly this is true for both potassium and sodium chloride. Thus, if the change in electrical potential is due to a change in the permeability to potassium or sodium ions, one would not expect the taste response to increase as the concentration of these substances is increased from zero to more than 1·0 M. From this observation one must conclude that a conformational change in the receptor molecule must produce an effect that is transmitted to areas of the cell distant from the receptor sites by means other than electrical depolarization associated with changes in ionic fluxes. The process by which such effects can be propagated is not yet known. Changes in the resting potential of the cells of the taste bud have been recorded in response to chemical stimulation of the taste cell (Kimura and Beidler, 1961; Tateda and Beidler, 1964; Sato, 1969). Therefore, membrane permeability must be changing at some portion of the cell membrane, presumably near the synaptic junction, where the cell is stimulated.

Taste proteins

It was assumed that binding of electrolytes and non-electrolytes to the cell surface involves interactions between the stimuli and proteins, phospholipids or polysaccharides of the cell membrane surface. This theory led others to search for specific molecules that might be associated with taste. Dastoli and Price (1966) isolated various protein fractions obtained from the bovine tongue surface, and found a particular "sweet" protein which complexed with sugars and saccharin. In some ways this protein has properties expected of a receptor protein that would interact with sugars. The strengths of binding parallel the relative sweetness of the sugars and saccharin and are of the same order of magnitude as the binding strength found previously by Beidler (1961) using electrophysiological methods with rats. It is interesting to note that, in order to measure the stimulus–protein interaction, Dastoli and Price used alterations in ultraviolet spectra and refractive indexes which are measures of conformational changes in the taste proteins. Specific "bitter" proteins have also been found (Dastoli, 1968).

Sweet and bitter tastes

If the taste stimulus is adsorbed to the molecules of the taste cell membrane, and if the binding forces are indeed weak, then one might look at a series of molecules that possess the same taste quality and try to determine what structural characteristics they have in common which might be involved in weak binding to the receptor sites. Shallenberger (1963) studied the structural properties of nine sugars and concluded that the intensity of their sweetness is inversely related to the ability of the hydroxyl groups of a sugar to hydrogen-bond to one another. This intramolecular hydrogen bonding was thought to restrict the ability of the sugar molecules to bind to receptor sites by way of the sugar hydroxyl groups. Shallenberger and Acree (1967) later examined a large variety of molecules that produce sweetness. Structurally unrelated molecules such as chloroform, glycol, alanine, fructose and saccharin were among those studied. They concluded that a molecular feature common to all these sweet substances was an AH–B system where A and B are electronegative atoms and the H is a hydrogen atom covalently bonded to the A atom. Atoms A and B are often oxygen or nitrogen atoms. The average distance between the AH and B atom is about 0·3 nm (3 Å). They postulated that the receptor site contains a similar AH–B system and binds the taste molecule by two hydrogen molecules or:

$$\text{receptor site} \begin{bmatrix} -A-H------B- \\ -B------H-A- \end{bmatrix} 0·3 \text{ nm, unit of sweet stimulus}$$

They later included a spatial barrier at the receptor site to explain differences in sweetness of D and L amino acids (Shallenberger, Acree and Lee, 1969).

Similar concepts have been proposed for some bitter substances. Kubota and Kubo (1969) studied the chemical structures of certain bitter substances found in plants, mostly diterpenes or their derivatives. They concluded that all the substances they studied had an AH–B system with a separation of about 0·15 nm (1·5 Å), which allowed intramolecular hydrogen bonding. They postulated that the receptor site contained a similar AH–B system which formed two hydrogen bonds with the bitter molecules after severing the intramolecular hydrogen bonds of the taste molecule. Certain basic amino acids, carboxylic acids and tyrosine may be part of the postulated receptor site system. As did Shallenberger for sweet taste, Kubota and Kubo mentioned that more than the AH–B system may be the necessary condition for bitter taste, the additional requirement possibly being a special size or shape of the molecule. Perhaps the greatest tool for the study of molecular characteristics important in taste will come from genetic studies, such as those undertaken by Kalmus (Kalmus, 1958; Harris and Kalmus, 1949).

Salty and sour tastes

Early in the quantitative studies of taste it was realized that both cations and anions played a role in taste and that the former stimulated whereas the latter had a more regulatory role (Beidler, 1954, 1963). This fact was already obvious from studies of acids which suggested that the hydrogen ion was responsible for sourness but the anions regulated their effectiveness.

Monovalent cations with the same anion (NaCl, KCl, NH_4Cl, etc.) are quite different in their stimulating efficiency. Since all these cations have the same charge, the only property that differs is their size. It was shown that the relative stimulating efficiency of the cations varies with the animal species and also with the particular taste nerve fibre stimulated within a given species (Beidler, Fishman and Hardiman, 1955). Therefore it appears that the size of a given hydrated ion also changes, depending upon its environment or the nature of the receptor site. The binding of ions to various substances has been well described by Ling (1962) and by Eisenman (1961). These descriptions appear adequate to form a theoretical basis for our understanding of receptor site binding of salts and the variation with species where the character of the receptor sites differs, particularly in their electric field intensity. Thus, quantitative differences in the number of the various carboxylic groups contained in membrane proteins or differences in the relative number of phosphate or sulphate groups available for cation binding may occur from one species to another.

Acids are not equally sour at equimolar concentrations or at equal pH. A possible reason for this is that the receptive membrane becomes more and more positively charged as the positive hydrogen ions are bound, so that further H^+ binding becomes difficult because of electrostatic repulsion. However, if the anions are also bound to the membrane, then the relative sourness of various acids at equal pH is regulated by the ease with which the anions can be bound. Thus, the theoretical basis for an understanding of sourness is not too different from the theoretical basis of acid binding to wool protein, for example (Beidler, 1967). The taste equation shown in the next paragraph can be rewritten and the logarithms taken of both sides of the equation to obtain:

$$\log \frac{R}{R_s - R} = -\mathrm{pH} - \mathrm{p}K_i$$

where

$$\frac{R}{R_s - R}$$

is a measure of the fractional number of receptor sites filled and K_i is the association constant of the binding of H^+ to the receptor site. If the additional factor of electrostatic repulsion is considered, then a relation is obtained that is similar to that found by Tanford (Tanford and Wagner, 1954; Tanford, 1961) for H⁺ binding to protein,

$$\log \frac{r}{n-r} = \text{pH} - \text{p}K_{int} + 0.868\, Zw$$

where r is the number of dissociated sites at a given pH, K_{int} is the intrinsic dissociation constant, Z is the net charge of the protein at the given pH and w is an electrostatic factor. The additional factor $0.868\, Zw$ describes the effect of the positive surface charge of the protein on H⁺ binding. The charge Z is dependent upon the numbers of both hydrogen ions and anions bound.

MATHEMATICAL DESCRIPTION OF TASTE RESPONSES

Mathematical descriptions of biological phenomena are useful not only to describe relationships simply but also in predicting new relationships and thus in helping to design new experiments. A simple quantitative description of the relation between concentration and magnitude of response was developed by Beidler (1954). It was first developed to describe responses to salts. The mass action law was applied to the binding of the salt to the receptor site and a typical hyperbolic function was obtained,

$$\frac{C}{R} = \frac{C}{R_s} + \frac{1}{KR_s}$$

where C is the stimulus concentration that produces a magnitude of response R, by binding to the receptor site with a binding constant of K; R_s is the maximum response obtainable with a high concentration of the stimulus. This equation satisfactorily describes the response to many compounds, both electrolytes and non-electrolytes, particularly those of simple tastes. Substances of mixed tastes are described by a modification of this equation obtained by considering the binding of the stimulus molecules to two different receptor sites (Beidler, 1961).

The above equation was derived using the most simple physicochemical assumptions. Thus, the equation is very simple and generally useful. However, it is usually true that the more precisely one wants to describe a phenomenon, the more complicated the mathematical relationship. Furthermore, as the number of different stimuli to be described is increased, the number of additional requirements also increases. For example, Tateda (1967), in his study of response of rats to amino acids and sugars, found it necessary to consider that a given receptor site binds two or more stimulus molecules rather than one.

From the above taste equation it is clear that no single point on the curve, including the threshold, can adequately describe the stimulus–response relationship. For example, a comparison of the taste effectiveness of two different chemical stimuli based upon their taste thresholds can be very misleading. Threshold depends inversely upon two independent factors: K, a measure of the binding strength and R_s, a measure of both the number of receptor sites available for a particular stimulus and the effectiveness of producing a response when the stimulus molecule is bound to the receptor site. It should be noted that the development of the adsorption theory of taste together with its mathematical description paralleled similar developments in the description of pharmacological activity (Ariëns, 1964).

TASTE MODIFIERS

I cannot conclude this paper without considering taste modifiers. They have importance not only for a better understanding of the mechanism of taste stimulation but also for their potential economic impact. Two particular taste modifiers, gymnemic acid and miracle fruit, have been intensively studied recently although both have been known for many years.

Miracle fruit is obtained from the plant *Synsepalum dulcificum*, found in Ghana and Nigeria. It modifies taste so that sour substances are tasted as sweet. The active principle has been isolated and characterized as a basic glycoprotein with a molecular weight of about 44000 (Kurihara and Beidler, 1968; Brouwer et al., 1968). It has been postulated that the protein can bind near a sweet receptor site and that a decrease in pH causes a conformational change in the receptor membrane that allows the sugar component of this molecule to bind to the receptor site in a manner similar to that of other simple sugars (Kurihara and Beidler, 1969).

Gymnemic acid is contained in the leaves of the plant *Gymnema sylvestre* found in India. Stocklin, Weiss and Reichstein (1967; Stocklin, 1967) isolated the active ingredient that destroys sweetness and proposed the structure as the D-glucuronide of hexahydroxy-triterpene which is esterified with acids. These acids appear to be acetic acid, isovaleric acid and tiglic acid (Kurihara, 1969). Removal of all the ester groups destroys the sweetness of molecules of such diverse structures as cyclamate, D-amino acids, beryllium chloride and lead acetate. The effectiveness of this taste modifier appears not to be as great in the rat as it is in man. Further studies on the mechanism of action should help to elucidate the mechanisms of taste stimulation as well as the neural coding of taste.

SUMMARY

A large majority of the cells of mammalian taste buds have a lifespan of about 10 days. There is no evidence yet for another population of cells within the same taste bud having a different turnover time. The cells of the taste bud are innervated by about 50 nerves entering the base of the taste bud and branching profusely to about 200 or more branches. All except one or two of the cells are in anatomical contact with the nerve fibres. Some of these fibres branch before entering the rat fungiform papilla and innervate other fungiform papillae. Miller has shown that both facilitation and depression can occur by the simultaneous stimulation of several papillae innervated by the same single taste nerve fibre, depending upon the character of the stimulus.

Bradley has shown that a rat can respond behaviourally to the sweetness of saccharin perfused intravenously into the tongue. Neural activity in the chorda tympani also increases with such stimulation.

Electrical stimulation indicates a latency of about 8 milliseconds between stimulus presentation and the excitation of nerves by the receptor cell. The adsorption theory of taste correlates well with the characteristics of recently isolated taste proteins and also with modern theories of molecular events postulated by Shallenberger. A number of taste modifiers have been described which differentially modify one of the taste qualities without appreciably affecting others.

REFERENCES

ARIËNS, E. J. (ed.) (1964). *Molecular Pharmacology: the Mode of Action of Biologically Active Compounds*, vol. 1. New York: Academic Press.
BARGOOT, F. G., WILLIAMS, T. P., and BEIDLER, L. M. (1969). *Vision Res.*, **9**, 385–391.
BEIDLER, L. M. (1953). *J. Neurophysiol.*, **16**, 595–607.
BEIDLER, L. M. (1954). *J. gen. Physiol.*, **38**, 133–139.
BEIDLER, L. M. (1961). *Prog. Biophys. biophys. Chem.*, **12**, 107–151.
BEIDLER, L. M. (1963). In *Olfaction and Taste* (Proceedings of the First International Symposium, Stockholm, 1962), pp. 133–148, ed. Zotterman, Y. Oxford: Pergamon Press.
BEIDLER, L. M. (1965). *Cold Spring Harb. Symp. quant. Biol.*, **30**, 191–200.
BEIDLER, L. M. (1967). In *Olfaction and Taste II* (Proceedings of the Second Internationa Symposium, Tokyo, 1965), pp. 509–534, ed. Hayashi, T. Oxford: Pergamon Press.
BEIDLER, L. M. (1969). In *Olfaction and Taste III* (Proceedings of the Third International Symposium, New York, 1968), ed. Pfaffmann, C. New York: Rockefeller University Press.
BEIDLER, L. M., FISHMAN, I. Y., and HARDIMAN, C. (1955). *Am. J. Physiol.*, **181**, 235–239.
BEIDLER, L. M., NEJAD, M., SMALLMAN, R., and TATEDA, H. (1960). *Fedn Proc. Fedn Am. Socs exp. Biol.*, **19**, abstr. 302, 749.

BEIDLER, L. M., and SMALLMAN, R. (1965). *J. Cell Biol.*, **27**, 263-272.
BROUWER, T. N.,WEIL, H. VAN DER, FRANCKE, A., and HEKNING, G. J. (1968). *Nature, Lond.*, **220**, 373-374.
BROWN, H. E., and BEIDLER, L. M. (1966). *Fedn Proc. Fedn Am. Socs exp. Biol.*, **25**, abstr. 786, 329.
BUJAS, Z. (1949). *Année psychol.*, **50**, 159-168.
BUJAS, Z., and CHWEITZER, A. (1936). *Année psychol.*, **36**, 137-145.
CONGER, A. D., and WELLS, M. A. (1969). *Radiat. Res.*, **37**, 31-49.
DASTOLI, F. R. (1968). *Nature, Lond.*, **218**, 884-885.
DASTOLI, F. R., and PRICE, S. (1966). *Science*, **154**, 905-907.
DE LORENZO, A. J. D. (1963). In *Olfaction and Taste* (Proceedings of the First International Symposium, Stockholm, 1962), pp. 5-18, ed. Zotterman, Y. Oxford: Pergamon Press.
EISENMAN, G. (1961). In *Symposium on Membrane Transport and Metabolism*, pp. 163-179, ed. Kleinzeller A., and Kotyk, A. Prague: Publishing House of Czechoslovak Academy of Sciences.
HARRIS, H., and KALMUS, H. (1949). *Ann. Eugen.*, **15**, 32-45.
KALMUS, H. (1958). *Ann. hum. Genet., Lond.*, **22**, 222-230.
KIMURA, K., and BEIDLER, L. M. (1961). *J. cell. comp. Physiol.*, **58**, 131-140.
KUBOTA, T., and KUBO, I. (1969). *Nature, Lond.*, **223**, 97-99.
KURIHARA, K., and BEIDLER, L. M. (1968). *Science*, **161**, 1241-1243,
KURIHARA, K., and BEIDLER, L. M. (1969). *Nature, Lond.*, **222**, 1176-1179.
KURIHARA, Y. (1969). *Life Sciences*, **8**, 537-543.
LING, G. N. (1962). *A Physical Theory of the Living State.* Waltham, Mass.: Blaisdell.
MILLER, I. J., JR. (1968). *Peripheral Input from Rat Fungiform Papillae to Single Chorda Tympani Fibers.* Doctoral Dissertation, Florida State University.
MISTRETTA, C. M. (1968). *The Permeability of Rat Tongue Epithelium.* Master's Thesis, Florida State Universtiy.
NEJAD, M. S. (1961). *Factors Involved in the Mechanism of Stimulation of Gustatory Receptors and Bare Nerve Endings of the Tongue of the Rat.* Doctoral Dissertation, Florida State University.
RAPUZZI, G., and CASELLA, C. (1965). *J. Neurophysiol.*, **28**, 154-165.
ROBBINS, N. (1967). *Expl Neurol.*, **17**, 364-380.
SATO, T. (1969). *Experientia*, **25**, 709-710.
SHALLENBERGER, R. S. (1963). *J. Fd Sci.*, **28**, 584-589.
SHALLENBERGER, R. S., and ACREE, T. E. (1967). *Nature, Lond.*, **216**, 480-482.
SHALLENBERGER, R. S., ACREE, T. E., and LEE, C. Y. (1969). *Nature, Lond.*, **221**, 555-556.
STOCKLIN, W. (1967). *Helv. chim. Acta*, **50**, 491-503.
STOCKLIN, W., WEISS, E., and REICHSTEIN, T. (1967). *Helv. chim. Acta*, **50**, 474-490.
TANFORD, C. (1961). *Physical Chemistry of Macromolecules.* New York: Wiley.
TANFORD, C., and WAGNER, M. K. (1954). *J. Am. chem. Soc.*, **76**, 3331-3336.
TATEDA, H. (1967). In *Olfaction and Taste II* (Proceedings of the Second International Symposium, Tokyo, 1965), pp. 383-397, ed. Hayashi, T. Oxford: Pergamon Press.
TATEDA, H., and BEIDLER, L. M. (1964). *J. gen. Physiol.*, **47**, 479-486.
TUCKER, D. (1963). *J. gen. Physiol.*, **46**, 31-49.
WARREN, J. F., JR. (1965). *A Study of the Responses of Taste Receptors of Rat Tongue to Electrical Stimulation.* Master's Thesis, Florida State University.
YOUNG, R. (1967). *J. Cell Biol.*, **33**, 61-72.

DISCUSSION

Lowenstein: Is it possible that you stimulated the dendritic endings rather than the gustatory cells in your perfusion experiments?

Beidler: Our first idea was that we were stimulating the nerves and not the taste cells, but if that is true, the various nerves must be highly specific and able to discriminate sodium saccharin from a number of other taste substances. In other words, you are asking nerve endings to do everything we believe taste cells do.

Reese: I wonder whether you can eliminate the possibility that the substances injected intravenously were being secreted in saliva bathing the taste buds?

Beidler: Mr R. Bradley uses a flow chamber over the tongue in order to eliminate an effect of saliva from all the large salivary glands.

Hellekant: You say that Mr Bradley can perfuse the tongue for several hours. Did he ever experience any increase of vascular resistance?

Beidler: Changes in resistance can be observed, particularly in response to adrenaline or acetylcholine.

Hellekant: Is the natural circulation going on all the time? And do you cannulate the tongue?

Beidler: The tongue and head region are circulated by an artificial medium.

Zotterman: Dr Hellekant has shown that alcohol on the tongue has to be at rather high concentration (18 per cent) in order to stimulate the gustatory cells, to be tasted. It's about the strength of port wine! But it penetrates the mucus and goes down in the skin and stimulates all kinds of nerve endings, which are also stimulated by noxious stimuli, and by warmth and cold and by pressure. So the perception of alcohol from the tongue is a very complex one.

Beidler: C. Mistretta showed that alcohol penetrates the tongue epithelium very fast (see Beidler, 1969). I showed many years ago that a high concentration of alcohol is needed to stimulate the tongue and a lower concentration to stimulate lingual nerve endings.

I might mention that Mr Bradley put a sweet stimulus on the tongue to produce a maximum response, and at the same time perfused sodium saccharin intravenously to see if the effects would be additive or not, and they were. That started us wondering about other receptor sites that could not be reached by surface application. Did you give alcohol intravenously, Dr Hellekant?

Hellekant: No, it was applied to the surface of the tongue, but I found that it probably penetrated the tongue epithelium, because non-gustatory fibres were also responding to alcohol (Hellekant, 1965).

Murray: Possibly the alcohol is getting through the epithelium and stimulating the taste buds from below.

Beidler: No; the alcohol penetrates the epithelium and stimulates lingual nerve fibres.

Duncan: Just between friends, Professor Beidler, where exactly do you think that intravenously perfused saccharin is acting?

Beidler: Between friends, on the taste cell!

Duncan: Whereabouts on the taste cell?

Beidler: We normally think of specific receptor sites on given taste cells as discriminating sweet substances from others. If we retain this idea for intravenous stimulation, I suspect that the base of a taste cell may not be too dissimilar from the apex.

Zotterman: The question is whether the nerve endings could be stimulated directly by the substance you put on the tongue and not indirectly via the epithelial taste cells.

Beidler: I see no reason why not, but if so it must stimulate one specific nerve ending that codes sweetness and not others.

Zotterman: We know that nerve endings have specific properties in other parts of the nervous system. The non-medullated C fibres are mostly highly specific and there are no receptor cells there. We still don't know whether the nerve ending of a gustatory nerve fibre has the particular property of being stimulated by sapid substances directly.

Beidler: I doubt if a taste substance applied directly to the taste bud would penetrate through to the nerve fibres in the short time allowed.

Zotterman: I agree, for dilute concentrations, but we don't know for higher concentrations.

Ottoson: How do you think that the surrounding inhibition is produced, Professor Beidler?

Beidler: Potassium benzoate, applied to the surface of the tongue, will inhibit the very small amount of spontaneous activity present in the rat taste fibres. This effect is probably due to the fact that the inhibitory anion, benzoate, is adsorbed to the surface of receptor. This receptor is normally biased at a certain level, and the response goes down in response to a strong inhibitory anion coupled with a weakly stimulating cation. How an inhibitory action at one taste bud affects the activity of a neighbouring bud innervated by the same fibre is not clear. An electrotonic effect over several millimetres is doubtful.

Murray: It might be that the poorly differentiated endings on type II cells are synaptic and inhibitory.

Andres: There are similarities here to Merkel's touch receptor. The touch cell of the Merkel's corpuscle is very like taste cells in its innervation; the multiple innervation of several corpuscles from one fibre is the same, as Iggo and Muir (1969) have shown. If there is excitation in one corpuscle there will be inhibition in another supplied by the same fibre. The corpuscles may be 5 mm apart.

Zotterman: The antidromic impulses might be releasing inhibitors so that the cells don't respond. If there are vesicles on both sides of the cell membrane there could be an antidromic stimulus coming up to one receptor cell and causing hyperpolarization of that cell. And Professor Murray has found vesicles on both sides of the synaptic membrane of the gustatory cells in the taste buds.

Murray: This is true in the rabbit foliate taste bud, but we are less certain in the case of fungiform buds.

Beidler: Don't forget that with most taste stimuli you get enhancement, not inhibition.

Zotterman: You can have that too! It would depend upon the transmitter substance.

Lowenstein: The antidromic idea is certainly not to be excluded from a role in lateral inhibition.

REFERENCES

BEIDLER, L. M. (1969). In *Olfaction and Taste III* (Proceedings of the Third International Symposium, New York, 1968), ed. Pfaffmann, C. New York: Rockefeller University Press.
HELLEKANT, G. (1965). *Acta physiol. scand.*, **65**, 243–250.
IGGO, A., and MUIR, A. R. (1969). *J. Physiol., Lond.*, **200**, 763–796.

PROBLEMS OF TASTE SPECIFICITY

Harald T. Andersen

Nutrition Institute, University of Oslo, Blindern, Oslo

The concept of specificity in taste mechanisms is associated with our recognition of four primary gustatory qualities, namely bitter, salt, sour and sweet. This fundamental principle of taste sensation—established by Fick in 1864—has never been seriously challenged. Indeed, von Békésy, who investigated the possibility of finding for instance six or eight fundamental taste qualities in man rather than just four, concluded that only four exist (von Békésy, 1964b, 1966).

Among electrophysiologists it is widely held that a continuous spectrum of gustation may be based on these four taste elements. This idea has been supported particularly by numerous demonstrations of multiple sensitivity in gustatory afferents.

It is easy already at this point to anticipate a controversy between the psychophysical and the electrophysiological approach to specificity problems of taste. The former approach invokes a high degree of chemical selectivity at taste receptors, as has been emphasized by von Békésy (1966), whereas the latter approach is more concerned with peripheral coding and central decoding of compound sensory messages as a basis for gustatory perception.

One difficulty is shared equally by psychophysicists and electrophysiologists: a poor understanding of the transducer mechanisms underlying chemoreceptor stimulation, even to the extent of being ignorant of the significant properties of the adequate stimuli. With respect to the design of experiments and interpretation of results the electrophysiologists are additionally at a disadvantage in being left with subjective evaluations by the human brain to explain relationships between specific taste sensations and electrical activity in gustatory afferent nerves. Psychophysical methods at least allow direct communication between investigator and subject. Obviously the two approaches complement each other. Thus, any serious disparity between psychophysical and electrophysiological interpretations of fundamental taste phenomena may be attributed to our lack of understanding.

I believe it would be worthwhile to discuss first current concepts of gustatory specificity differently entertained by electrophysiologists and psychophysicists. Then I would like to turn to a discussion of some significant properties of chemical stimuli, particularly those that elicit taste sensations of sweet and bitter.

TASTE SPECIFICITY PROBLEMS IN ELECTROPHYSIOLOGY AND PSYCHOPHYSICS

Nervous activity in primary taste neurons of higher vertebrates is recorded from single functional fibres in the chorda tympani or the glossopharyngeal nerve upon stimulation of the tongue surface with solutions tasting bitter, salty, sour and sweet to humans. Frequently the chemical stimuli used are applied in one concentration of the solute only.

This procedure is limited in two respects. First, there is no way of controlling whether the population of nerve fibres tested for specificity during an experiment is representative of the total number of fibres present. This will be amended only by chance and by doing a large number of experiments. Second, if the stimulus strength is not varied for each of the four taste qualities studied, the possibility of any one fibre transmitting more than one modality remains unsettled.

Pfaffmann (1941, 1955) reported multiple sensitivity in single gustatory nerve fibres of the rat and suggested that specific sensations of taste are elicited by an "across fibre pattern" input into the gustatory afferents. This hypothesis has been further developed by Erickson (1963) who made an analysis of a signal system with three fibre types responding within a hypothetical stimulus continuum. His theoretical analysis supported the conclusion that "there are many fiber types representing gustatory quality as for pitch discrimination, rather than a few fiber types as in color vision," and moreover, that "the neural message for gustatory quality is a pattern made up of the amount of neural activity across many neural elements."

Attractive as this hypothesis may appear the evidence presented in its support is not conclusive. Moreover, the idea is based entirely on receptor activation by gustatory stimuli whereas the concept of inhibition has not been considered. The "across fibre pattern" hypothesis needs to be modified, therefore, in order to allow for the peripheral inhibition which has been shown to take place (Andersen, Funakoshi and Zotterman, 1963): a solution of salt applied to the surface of the tongue of the dog was shown to depress or completely to block the response to a sugar solution that was to follow. When the sodium chloride solution was preceded by one of sugar, however, no inhibition of the neural response to the second stimulus was detected. Admittedly, evidence was not presented to ascertain whether the inhibition is a surface phenomenon at receptor sites or whether it is nervous in origin. In any event, a satisfactory explanation of peripheral taste mechanisms must take the possibility of inhibition or interaction into account.

Peripheral interaction, summation as well as inhibition, has been demonstrated in man by von Békésy (1964b). He also presented evidence to show that four specific types of taste papillae conforming to the four basic qualities exist on the tongue and even reported that they could be separated into four different morphological categories. This, of course, would not prevent terminal branches from one nerve fibre from making contact with several different types of receptors so that multiple sensitivity might be recorded from a single gustatory afferent. However, von Békésy (1964b) observed that the four basic taste qualities may not be the same in man, rats and cats, and he argued that the stimuli used are monogustatory in man but perhaps not in animals. These objections cannot be readily accepted. First, monogustatory sensations are usually dependent on concentration in humans (Rengvist, 1919). Second, a monogustatory sensation may be due not entirely to peripheral mechanisms but equally well to central nervous discrimination. The important conclusion to draw from these differences of opinion is that compound signals in single gustatory afferents are not incompatible with the concept of specific receptors—not even specific taste buds for that matter—since they may receive sensory innervation from collateral branches of the same nerve fibre. In fact, Rapuzzi and Casella (1965) have amply demonstrated that several taste papillae are usually connected to a single gustatory afferent.

Recently, electrophysiological evidence has been presented to show that fibres transmitting activity elicited by taste stimuli may also carry information arising from thermal variations at the tongue surface (Ogawa, Sato and Yamashita, 1968). Their material includes observations on 48 single units from rats and 28 fibres from the hamster, all of which displayed multiple sensitivity transmitting two or more gustatory modalities and, in addition, responses to warm or cold or both. These findings certainly add to the complexity of specificity problems of the peripheral gustatory apparatus. Also, they support the "spectrum theory". Peripheral specificity is necessary only for the transducer process at the site of stimulation. Signals from different receptors may well be transmitted along the same cables and sorted out again at higher levels of the central nervous system.

The complicating factor that gustatory afferents relay information from thermoreceptors as well did not emerge quite unexpectedly. A few years ago von Békésy (1964a) investigated the interaction between solutions tasting bitter, salty, sour and sweet, and added the sensations of warm and cold to the gustatory stimuli. He was able to demonstrate interaction between members of one group of sensations including warm, bitter and sweet; and similarly an interaction between a second group of stimuli including salty, sour and cold. No interaction took place between sensations belonging to different groups. These results, presented as the "duplexity theory" of taste, permit an analogy between taste on one hand and vision, hearing and equilibrium on the other. So far neither psychophysicists nor electrophysiologists seem to understand how to use this information that gives promise of a much-needed simplification of taste specificity problems.

In my opinion the electrophysiological and psychophysical data available—taken separately or combined—do not provide sufficient evidence at the present time to solve the problems of specific taste sensations. Possibly we need to know a great deal more about the primary transducer mechanism at the receptors before we can proceed any further.

INTERACTION BETWEEN RECEPTORS AND GUSTATORY STIMULI

The concept of four gustatory primaries implies discrimination among taste sensations at high levels of the central nervous system. The focal points of reference of this analytical system are the tastes of substances commonly used to produce a certain gustatory sensation, namely sucrose for sweet, sodium chloride for salty, quinine for bitter and, usually, hydrochloric acid for sour. However, it is well known that substances tasting bitter, salty, sour or sweet respectively do not necessarily evoke the sensation elicited by the corresponding reference substance. This is reflected in certain curious statements which have found their way into the literature. For instance, it is commonly said that sucrose is the only substance eliciting a pure sweet taste, or likewise, that common table salt, sodium chloride, is the only one that tastes salty. Actually, this is only to state that sucrose tastes like sucrose, which no other sugar really does, or that sodium chloride when tasted elicits the sensation of sodium chloride. This, then, goes to show how much our evaluation of a taste stimulus depends on its being referred to the focal point of the corresponding gustatory primary.

Since we do not know which molecular properties determine stimulation characteristics we are unable to decide whether or not the stimuli within each group of basic taste qualities form a continuum. Similarly, if spectral bands exist within each group of gustatory primaries we may ask whether they are discontinuous among themselves or whether they form a truly continuous spectrum for the sense of taste.

During recent years data have been obtained which taken together support the idea of a stimulation continuum including at least the sensations of sweet and bitter.

Three different observations suggest an intimate relationship between these two gustatory primaries. They belong to the same group of interacting substances, as has been shown by von Békésy (1964a). The sensations of bitter and sweet are both affected by extracts of the plant *Gymnema sylvestre*, whereas sour and salty tastes are not (Warren and Pfaffmann, 1959).

TABLE I

RELATIVE SWEETNESS OR STIMULATORY ABILITY OF SUGARS

Results obtained by

Sugar	Becker and Herzog (1907)	Biester, Wood and Wahlin* (1925)	Fabian and Blum (1943)	Walton (1926)	Andersen, Funakoshi and Zotterman (1963)
D-fructose	II	173 (I)	I	103–150 (I)	120 (I)
D-glucose	III	74 (III)	III	50–60 (III)	70 (III)
D-galactose	VI	32 (IV)	—	—	59 (V)
D-mannose	—	—	—	—	77
L-sorbose	—	—	—	—	86
Lactose	V	16 (VI)	V	27–28 (V)	53 (VI)
Maltose	III	32 (VI)	IV	60 (III)	67 (IV)
Sucrose	I	100 (II)	II	100 (II)	100 (II)

*Equal weights of sugars used in test solutions.
(From Andersen, Funakoshi and Zotterman, 1963)

And in a series of chemical homologues increasing molecular weight is frequently associated with a change in taste from sweet to bitter. This latter fact will be especially important in the discussion to follow.

Several years ago it was demonstrated that the stimulating ability of different monosaccharides and disaccharides varied according to their solubility in water (Andersen, Funakoshi and Zotterman, 1963). The higher the solubility in water of any particular sugar, the greater its power to elicit electrical activity in the chorda tympani nerve of the dog. These results closely parallelled those obtained by psychophysical methods in human subjects asked to rate the relative sweetness of the same sugars (Table I).

Our conclusion did not imply any simple relationship between water solubility on the one hand and stimulatory strength on the other. We suggested, however, that the molecular characteristics responsible for water solubility would play a decisive role in chemoreceptor stimulation also.

Admittedly, in limiting our work to certain monosaccharides and disaccharides which are stereochemically similar our conclusion could not be extended to sweet-tasting substances in general, first, because compounds evoking a sensation of sweet taste may be entirely dissimilar in the skeleton of the molecule, and second, because a single and very simple substitution into a complex molecule may change the taste sensation remarkably (Hamor, 1961). However, taste receptor stimulation may be not so much a function of the configuration of the molecular skeleton but rather may be dependent on certain groups protruding from the core and their reactivity in forming hydrogen bonds.

Thus, Shallenberger and co-workers (Shallenberger and Acree, 1967; Shallenberger, Acree and Lee, 1969) have suggested that the sweetness of a molecule depends upon an AH–B grouping where AH is a proton donor and B a proton acceptor. The critical distance between the AH and the B groups was shown to be approximately 0·3 nm (3 Å) which is too great a distance for the formation of intramolecular hydrogen bonds. Recently, the Japanese workers Kubota and Kubo (1969) have shown in a series of diterpenes that bitter taste depends upon AH–B systems in such close proximity, 0·15 nm (1·5 Å), that intramolecular formation of hydrogen bonds may take place. The idea, then, is that the AH and B groups find complementary sites on the chemoreceptor surface where certain chemical properties of proteins such as free basic amino groups, free carboxylic acid groups and phenolic hydroxyl groups may be available to form hydrogen bonds with stimulating molecules.

This information may prove very helpful in the search for explanations of several electrophysiological and psychophysical observations. For instance, a clue is provided to our understanding of changes in taste sensations from sweet to bitter that accompany increasing molecular weight within a series of chemical homologues. Thus, in small and relatively rigid molecules the AH–B systems are situated too far apart to allow intramolecular formation of hydrogen bonds, leaving the AH–B systems free readily to establish such bonding with taste receptors. Larger molecular size obviously increases the possibility of twisting and bending the molecular skeleton. Consequently, the possibility of proton donors (AH) and proton acceptors (B) moving closer together increases also. When intramolecular hydrogen bonds are formed, then, the immediate reactivity of the molecule towards complementary sites on the receptor surface decreases, apparently with the result that the taste sensation produced changes from sweet to bitter.

The hydrogen-bonding hypothesis may be applied also to the observation that a single substitution with a simple group into a complex molecule frequently changes its taste or renders it tasteless. Thus, the substituted group may not fit into an AH–B system or it may block an existing one. This seems to be the case if one scrutinizes the compounds studied by Hamor (1961).

The concept of proton donor and proton acceptor groups being crucial to sensations of sweet and bitter may be extended even further. It has been known for half a century (Rengvist, 1919) that certain inorganic salts such as NaCl, KCl and $CaCl_2$ may taste sweet or bitter at threshold concentrations. If one takes into account the water shells surrounding the ions in such dilute solutions, it is not inconceivable that hydrogen bonding in the water aggregates surrounding the ions contributes more to the taste elicited than the hydrated ion itself. Thus the presence of threshold concentrations of ions may influence hydrogen and hydroxyl groups to form proton donor and proton acceptor systems to evoke sweet or bitter sensations, whereas the gustatory characteristics of the hydrated ions themselves are ineffective when the solution is very dilute.

It may be interesting also to consider the water taste that certain vertebrates display in relation to these new findings. Liljestrand and Zotterman (1954) suggested that the water fibres that they had recorded from might actually have been fibres responding to sweet substances because the electrical activity in the water fibres increased very much when sugar was added to the water. They rejected the idea, however, since the response to sugar nearly disappeared when Ringer's solution was used as solvent. Taking into account the demonstration that salt stimuli are inhibitory to sweet it may well be that they were right in their original assumption of a close relationship between sweet fibres and water fibres. If threshold conditions are considered we may argue that animals showing the specific water taste may be equipped with extremely low threshold fibres for sweet-tasting substances. Consequently, these sweet fibres may be activated by the formation of hydrogen bonds with AH–B systems found in pure water. I cannot recall having seen a "water fibre" in the dog that did not also respond to sugar, whereas "sweet fibres" do not always distinguish themselves as being sensitive to pure water.

Finally, it should be pointed out that the hypothesis that bitter and sweet tastes are dependent on proton donor–proton acceptor systems separated to make intramolecular hydrogen bonding possible or impossible, respectively, does not seem to be generally valid. Ethylene glycol (1,2-ethanediol) and glycerol both taste sweet although these molecules have rather strong intramolecular hydrogen bonding. Likewise, the aldohexose D-mannose evokes a definite sweet taste but also a strong and unpleasant bitter taste. Therefore, this monosaccharide should have at least two conformations between which to alternate if the hydrogen bonding hypothesis is to apply as presently stated. When I first accepted the invitation to this symposium I had strong hopes that I would be able to tell you everything about the double taste sensation elicited by D-mannose. Unfortunately, nuclear magnetic resonance studies give D-mannose only one stable conformation, and its ability to produce two different taste sensations is as puzzling as ever.

SUMMARY

Although we do not understand what are the significant properties of chemical stimuli, taste specificity mechanisms are associated with four primary gustatory qualities—bitter, salty, sour and sweet. In order to explain the multitude of taste sensations recognized by the human brain it is widely held that a continuous gustatory spectrum exists between these four primaries.

The idea that just four taste qualities should be regarded as fundamentals has been much debated but attempts to find additional gustatory primaries, for instance by stimulating taste papillae electrically, have confirmed that only four exist.

Most research on taste specificity problems has focused on peripheral mechanisms. Consequently there has been a tendency to interpret the perception of different taste sensations solely in terms of events taking place in the periphery. Also, reactions leading to activation of peripheral nervous elements have dominated our thinking about gustatory functions whereas inhibition has hardly been considered.

A better understanding of the sense of taste depends upon knowledge about the nature of the chemical stimuli and their interaction with peripheral structures, the data-processing operations in the central nervous system and the possible influence exerted by central nervous mechanisms upon peripheral receptors.

REFERENCES

ANDERSEN, H. T. FUNAKOSHI, M., and ZOTTERMAN, Y. (1963). In *Olfaction and Taste* (Proceedings of the First International Symposium, Stockholm, 1962), pp. 177–192, ed. Zotterman, Y. Oxford: Pergamon Press.
BECKER, C. T., and HERZOG, R. O. (1907). *Hoppe-Seyler's Z. physiol. Chem.*, **52**, 496–505.
BÉKÉSY, G. VON (1964*a*). *Science*, **145**, 834–835.

Békésy, G. von (1964b). *J. appl. Physiol.*, **19**, 1105–1113.
Békésy, G. von (1966). *J. appl. Physiol.*, **21**, 1–9.
Biester, A., Wood, M. W., and Wahlin, C. S. (1925). *Am. J. Physiol.*, **73**, 387–396.
Erickson, R. P. (1963). In *Olfaction and Taste* (Proceedings of the First International Symposium, Stockholm, 1962), pp. 205–214, ed. Zotterman, Y. Oxford: Pergamon Press.
Fabian, F. W., and Blum, H. B. (1943). *Fd Res.*, **8**, 179–193.
Fick, A. (1864). *Lehrbuch der Anatomie und Physiologie der Sinnesorgane*, heft I, p. 85. Lahr: Schauenburg.
Hamor, G. (1961). *Science*, **134**, 1416–1417.
Kubota, T., and Kubo, I. (1969). *Nature, Lond.*, **223**, 97–99.
Liljestrand, G., and Zotterman, Y. (1954). *Acta physiol. scand.*, **32**, 291–303.
Ogawa, H., Sato, M., and Yamashita, S. (1968). *J. Physiol., Lond.*, **199**, 223–240.
Pfaffmann, C. (1941). *J. cell. comp. Physiol.*, **17**, 243–258.
Pfaffmann, C. (1955). *J. Neurophysiol.*, **18**, 429–440.
Rapuzzi, G., and Casella, C. (1965). *J. Neurophysiol.*, **28**, 154–165.
Rengvist, Y. (1919). *Skand. Arch. Physiol.*, **38**, 97–108.
Shallenberger, R. S., and Acree, T. E. (1967). *Nature, Lond.*, **216**, 480–482.
Shallenberger, R. S., Acree, T. E., and Lee, C. Y. (1969). *Nature, Lond.*, **221**, 555–556.
Walton, C. F., Jr. (1926). In *International Critical Tables of Numerical Data, Physics, Chemistry and Technology*, ed. Washburn, E. W. New York: McGraw-Hill.
Warren, R. M., and Pfaffmann, C. (1959). *J. appl. Physiol.*, **14**, 40–42.

DISCUSSION

Martin: I am puzzled by the continual reference to sweet, sour, bitter and salt as being the primary taste sensations. It seems to be perfectly obvious from personal observation that glutamic acid, various other things present in meat juice, fish, cheese and orange can all be appreciated in the mouth without any assistance from the nose. I recently completely lost my sense of smell but was nevertheless able to distinguish a great many things in my mouth. The mere fact that, as you say, sucrose tastes like sucrose and sodium chloride like salt, indicates that there are various gradations in these things other than the four "simple" tastes.

Andersen: That is exactly what I tried to say. But one may very well imagine a continuous taste spectrum based on the four "primaries", containing salty, sour, sweet and bitter as "spectral bands" with the substances commonly used to produce a certain taste quality as points of reference. Such a spectrum doesn't need to be one-dimensional. It may, for instance, be four-dimensional. Also, any sweet-tasting substance has both a detection and a recognition threshold to human experimental subjects. With increasing concentration of the substance you reach the point where they can tell you that "this is something other than just the solvent", but they can't tell you what it is, and as concentration is increased further they can recognize any sugar, or any other sweet-tasting substance for that matter, and can name it. Among these various sweet tastes there exists a reference point, that of the most used sweetening agent, to which the others are compared. All of these produce a sensation of sweet; nevertheless they taste different.

Zotterman: You mentioned our work on salt diminishing the response to sugar. You can abolish the response to water in the cat and monkey by 0·05 per cent sodium chloride but that concentration of sodium chloride doesn't affect the response to sweet if you mix that amount of salt in a sugar solution.

Andersen: I don't think that this runs contrary to my argument. In fact, your observation would equally well support it. As long as we are talking about the possibility of specific water fibres actually being low-threshold sweet fibres, a salt solution may very well abolish the response in these particular fibres without inhibiting the whole population of fibres responding to sweeteners.

Zotterman: Another point. In the frog there is no response to sweet substances or sugars, but there is a response to tap water but very little response to distilled water. G. Rapuzzi showed that tap water contains a low percentage of calcium chloride; and if you remove the calcium or if you increase it, you abolish the response. Dr J. Konishi found that ferri-ammonium cyanide in concentrations of about 0·001 M evoked vigorous responses in carp fishes but evoked no response at higher concentrations.

There are of course receptors with very different responses than to just four qualities, as we know; the carp fish responds to a lipoprotein which is present in a millionth amount in our saliva (which is why fisherman spit on their bait!) and there are specific fibres which don't respond to anything other than this substance, which is found in all palatable food for this species of fish.

Finally, it is frequently stated that *Gymnema sylvestre* extract abolishes the sweet and the bitter taste. We have tested it on ourselves and on patients electrophysiologically, recording from the chorda tympani in the middle ear, and we have never found any effect on the bitter taste; it completely abolishes the response to sugars and to saccharin and other sweeteners.

Andersen: On this point of the water-specific fibre possibly being a low-threshold sweet fibre, I wouldn't have thought of the idea if it hadn't been specifically stated in your paper with Liljestrand in 1954; you originated the idea, Professor Zotterman!

On your other point, when a fish like a carp responds to this lipoprotein, for instance, do we know what the fish tasted from it? Isn't it possible that it could have been sweet, sour, salty or bitter? I am not particularly interested in defending the idea of four primary tastes as such, but as long as we are unable to communicate with the subject of the experiment we are forced to use our own brains to interpret what an animal may have been tasting.

Pfaffmann: I can support Professor Zotterman's statement that water extracts of gymnemic acid have no effect on the bitter taste. The effect on bitter is certainly reported for the *alcohol* extracts. Richard Warren (personal communication) has been doing further work with water-soluble extracts and it doesn't have much effect on the bitter taste, but primarily affects the sweet. I think there may be a confounding effect of the bitterness of the crude extract itself which produces bitter adaptation, whereas the true blocking action is on the sweet mechanism.

Zotterman: We have tested both direct extracts from the plant and the crystalline preparation which I was given by Dr Beidler.

Hellekant: I am very glad to hear that it is the alcohol in the extract of *Gymnema* that diminishes the bitter taste, because I found that the electrical response recorded to quinine in the cat was almost completely abolished when alcohol was given beforehand (Hellekant, 1967).

Beidler: To return to the subject of four taste primaries, our information on the neural coding is rather deficient. The concept of four taste qualities is very useful for the experimenter because it simplifies his experiments, but when you consider flavour in the real world it is an inadequate hypothesis. What we need is an experimental method for looking at a number of single units simultaneously over long periods of time. We need better techniques, in other words.

Pfaffmann: The moment one speaks of a quality one is making an assumption of a correlation of the stimulus with the sensation; if one simply stated the stimulus used, that would be an improvement. The other point is that referring to four basic tastes is an effort to simplify our conception of the way the taste system works, in analogy with say the primary colours of vision.

Wright: If there is no response, however, how do you know it was a stimulus?

Pfaffmann: I mean that instead of using sodium chloride but calling it a "salty stimulus", or using sucrose and saying that we gave a "sweet" stimulus, we should specify the stimulus by the name of the substance, not its presumed quality. Whether it elicits a response, and which quality, is an empirical question.

Wright: This then corresponds to the question of terminology that I brought up in the Istanbul Congress, where half the people felt that an odour was the sensation and the other half thought it was the substance!

They were therefore talking completely at cross purposes, and so I suggested that we use "olfactory" as the adjective relating to the sensation and "osmic" as the adjective relating to the stimulus (Wright, 1968).

Lowenstein: Professor Pfaffmann has just said that we want to simplify matters and we therefore think in terms of four fundamental taste qualities. Historically speaking, how did it come about that we talk about these four qualities?

Pfaffmann: At first there were held to be a great many primary tastes and gradually they were reduced by a series of observations. These included punctate stimulation of the human tongue, showing that thresholds for bitter are lower on the back of the tongue and the intensity of sensations there is stronger; acid is best on the edges of the tongue, sugar on the tip and salt along the tip and sides. The hypothesis was developed that there are separate receptors that occur with different densities. In single-papilla stimulation, if people are instructed to report in terms of four basic qualities, they report just these four. The action of gymnemic acid and other blocking agents seems to dissociate tastes into four. On the other hand if you look at subjective descriptions and use a wide array of stimuli the evidence for only four primary qualities is less strong.

REFERENCES

HELLEKANT, G. (1967). In *Olfaction and Taste II* (Proceedings of the Second International Symposium, Tokyo, 1965), pp. 465–479, ed. Hayashi, T. Oxford: Pergamon Press.
LILJESTRAND, G., and ZOTTERMAN, Y. (1954). *Acta physiol. scand.*, **32**, 291–303.
WRIGHT, R. H. (1968). In *Theories of Odor and Odor Measurement* (Proc. NATO Summer School, Istanbul, 1966), pp. 459–478, ed. Tanyolaç, N. Istanbul: Robert College.

THE INFLUENCE OF THE CIRCULATION ON TASTE RECEPTORS AS SHOWN BY THE SUMMATED CHORDA TYMPANI NERVE RESPONSE IN THE RAT

GÖRAN HELLEKANT

Department of Physiology, Kungl. Veterinärhögskolan, Stockholm

NUMEROUS studies in the field of gustation have been based on electrical recordings of the nerves which mediate information from the gustatory receptors. However, few studies have been concerned with the physiological parameters which may influence the gustatory receptors and their response. The effect of the temperature of the stimuli has been investigated by some authors, for example Sato (1963); Yamashita, Yamada and Sato (1964); Yamashita (1964); Sato (1967). The change in the response of the cut chorda tympani nerve due to lysis of the nerve and to arrest of its neurotropic influence on the taste receptors has been touched upon by Bartoshuk (1965). The effect on the responses of the time-intervals between stimulations has been studied and described (Hellekant, 1968, 1969), but there remain several factors to be studied which may be of importance in this respect.

This preliminary report of experiments that will be carried further and more thoroughly discussed and described later, will focus attention on another parameter which influences the response of the gustatory receptors and therefore must be taken into consideration in future experiments, as well as in the evaluation of earlier ones. It will also suggest a way in which the organism may modify its gustatory sensitivity.

METHODS

About 60 adult male rats (Sprague-Dawley) weighing 270–330 g were used. The techniques for dissecting and stimulating the chorda tympani nerve were essentially the same as in earlier studies (Hellekant, 1968, 1969). The systemic blood pressure was recorded in the right femoral artery, which also during the perfusion experiments supplied blood for a loop to the tongue. A pump forced the blood through the loop. A glandular branch from the external maxillary artery was connected to the exit of this loop.

Thus the tongue could be supplied with blood in the normal way until perfusion started. Then the carotid artery was clamped between the heart and the branching of the lingual artery and the pump started. All other arterial branches were permanently occluded. The perfusion pressure was recorded in the external maxillary artery. The arterial oxygen tension was recorded by means of a Clark electrode (Beckman 315780 Microelectrode) projecting out of the common iliac artery into the lumen of the abdominal artery. The blood flow was measured with an electromagnetic flow meter (Statham M-4000) in the ipsilateral common carotid artery, since the flow through the lingual artery is too small to measure with the instruments available today. The femoral vein was always cannulated. About 400 international units of heparin were injected before the perfusion experiments to prevent coagulation. Muscular relaxant, succinylcholine, was sometimes administered during artificial respiration. Haemaccel (Hoechst) was used as a plasma substitute throughout these experiments. Blood was sometimes collected from one animal and injected into another.

RESULTS AND DISCUSSION

Fig. 1 shows the result of an experiment in which the blood flow through the common carotid arteries was cut off. In this animal the contralateral carotid artery had been permanently occluded during the surgical pro-

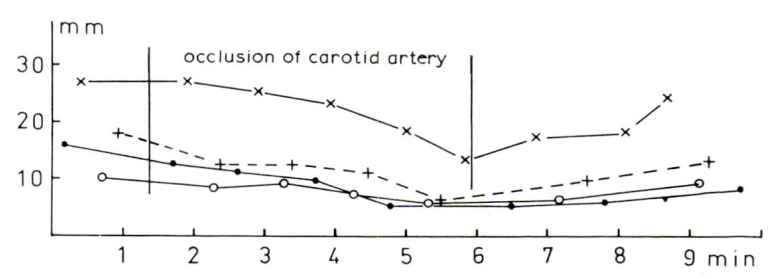

FIG. 1. The effect on the chorda tympani nerve response of occlusion of the ipsilateral common carotid artery of a rat. The contralateral carotid artery had earlier been permanently occluded. ×, 0·5 M-NaCl; +, 0·05 M-acetic acid; ●, 0·05 M-quinine hydrochloride; ○, 0·5 M-sucrose.

cedure, while the ipsilateral one was occluded during the time-period indicated by the vertical bars in Fig. 1. The figure shows that the summated chorda tympani response to all four stimuli used decreased during this temporary occlusion and that less than 50 per cent of the original response remained after about four minutes. It can be inferred that the remaining part would also have disappeared if the occlusion had been extended, as

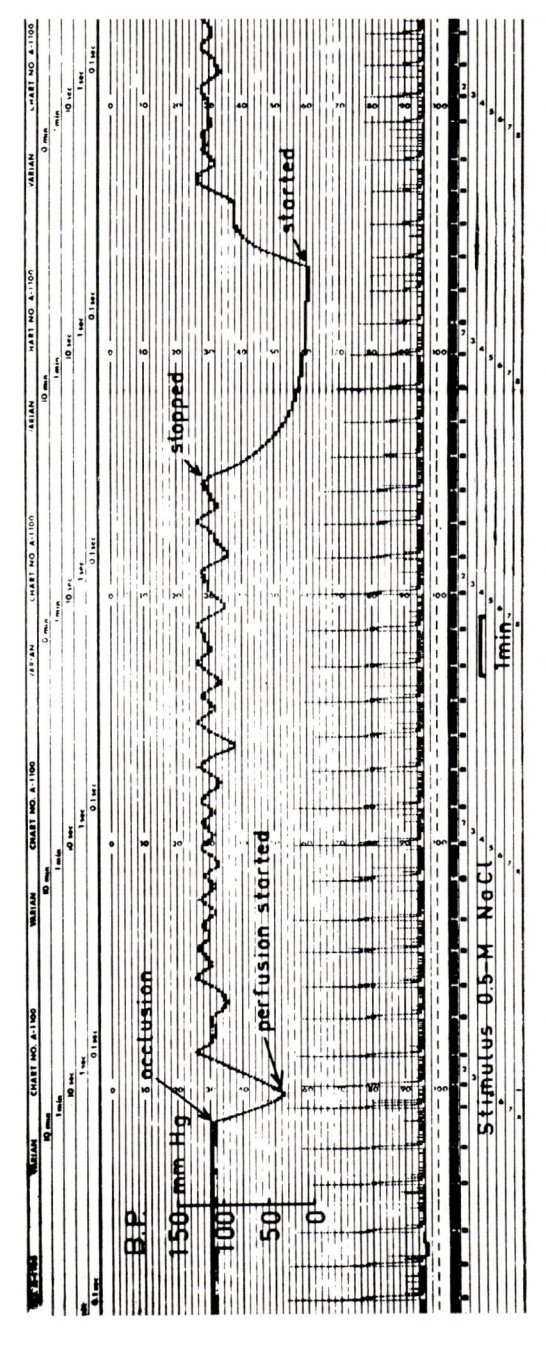

FIG. 2. The upper trace shows the blood pressure in the lingual artery before and during perfusion of the tongue with blood from the femoral artery of the same animal. The lower trace shows the summated chorda tympani nerve response to 0·5 M-NaCl on the tongue.

86   GÖRAN HELLEKANT

this was shown in other experiments. Occlusions repeated within relatively short time-periods, of about five minutes, shortened the time between the occlusion and the onset of the decrease of response from 3–4 minutes the first time to 1–2 minutes the third time.

It is possible that permanent occlusion of the contralateral carotid artery before the recording of Fig. 1 was made diminished the total blood supply to the tongue and may have affected the result. This would not change the general observation, because similar results were obtained in animals with both carotid arteries open prior to occlusion. Effects of occlusion of the contralateral artery alone were, however, not noticeable as long as the other side was left open and the general circulatory status remained relatively

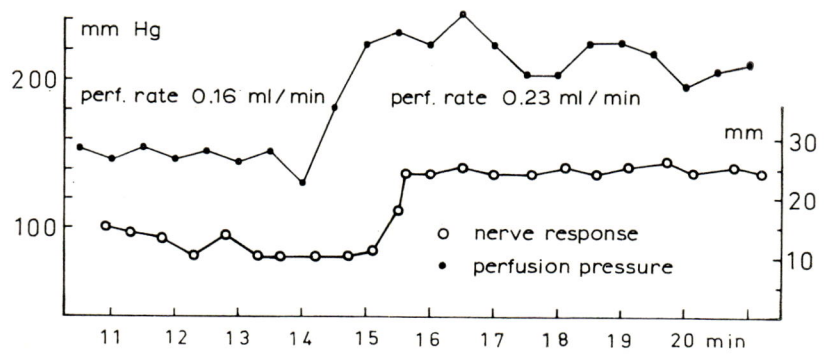

FIG. 3. The effect of increasing the perfusion rate to the tongue on the blood pressure of the lingual artery and on the summated nerve response to stimulation with 0·3 M-NaCl.

unaffected. Further, most animals were able to maintain their gustatory sensitivity for a considerable time even with their ipsilateral carotid arteries occluded, as long as the contralateral artery was left open.

It is not likely that this can be attributed to any direct effect on the chorda tympani nerve fibres of cessation of the circulation, because the excitability and conductivity of the nerve fibres remain unchanged for much longer times than those described here (Wright, 1947). Fig. 1 indicated then that the gustatory receptors cells are relatively sensitive to circulatory changes, but also raised the next question of whether ischaemia or anoxia or both are the primary cause.

Attempts to ventilate rats with gas mixtures of different oxygen content gave no conclusive results, because it proved to be impossible to change the oxygen tension without affecting the cardiovascular system. The method of perfusing the rat's head described by Thompson, Robertson and Bauer

(1968) did not offer a solution, since it involves the much larger vascular bed of the brain and does not avoid the effect of vascular reflexes elicited by changes of the blood supply to the brain.

The technique briefly outlined under Methods was therefore developed and recordings like that of Fig. 2 could then be obtained. In Fig. 2 the pump kept the blood pressure of the lingual artery at approximately the same level after occlusion as before. The blood flow produced was apparently large enough to sustain the sensitivity of the receptors, as the response would otherwise have declined within a few minutes. Fig. 2 demonstrates also that the sensitivity of the receptors depended upon the perfusion flow, since the response diminished when the pump was turned off.

Fig. 3 demonstrates that the neural response to the same stimulus could be increased by increasing the rate of perfusion. In this case about a 40 per cent increase of the perfusion flow rate more than doubled the magnitude of the summated response to 0·3 M-sodium chloride solution.

The idea that blood flow is a more important regulator of gustatory sensitivity than oxygen tension was further supported by experiments like that shown in Fig. 4. Fig. 4 is composed of four simultaneous recordings, the pulsatile blood flow in the ipsilateral common carotid artery, the arterial oxygen tension measured in the abdominal aorta, the systemic blood pressure recorded in the femoral artery and the summated neural response to automatically repeated rinses of 0·3 M-NaCl over the tongue. The first arrow in Fig. 4 indicates when the rat was switched from spontaneous respiration of air to artificial ventilation with pure oxygen. The figure shows that this caused a diminution of the cardiac output and a fall of blood pressure. The fall of blood pressure and diminution of cardiac output were much less when the animal was artificially ventilated with for example 7 per cent carbon dioxide in air. This excludes the possibility that this effect was caused by increased intrapulmonary pressure resulting from the artificial respiration. It may be concluded that the decrease of flow and blood pressure was responsible for the decrease of neural response, since the arterial oxygen tension rose. Artificial respiration was then stopped at the third arrow. This must have caused an increase of metabolites, especially carbon dioxide. They stimulated vascular receptors, which reflexly accelerated the heart and contracted the vessels. This was recorded as an increase in cardiac output and a rise in the systemic blood pressure. The recording indicates that the oxygen tension rose for a short time after ventilation with pure oxygen had ceased. This was probably caused by the faster flow of oxygen-rich blood around the oxygen electrode, due to the

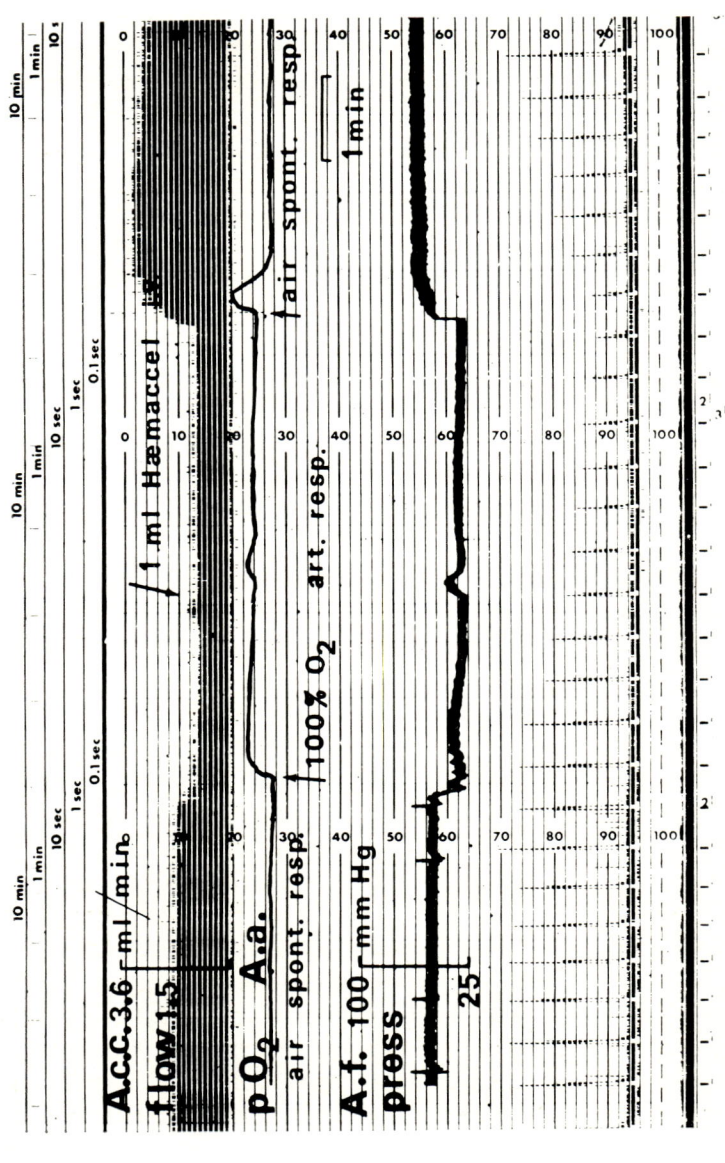

Fig. 4. The effect of artificial ventilation with pure oxygen on the blood flow of the common carotid artery, the arterial oxygen tension, the arterial blood pressure and the summated chorda tympani nerve response to stimulation with 0·3 M-NaCl.

increased cardiac output, before less saturated blood had reached the electrode.

The second arrow indicates the effect of 1 ml Haemaccel injected intravenously during artificial respiration with 100 per cent oxygen. Its effect on blood pressure and blood flow was slight at that time. The absence of an effect indicates that the injected volume was mainly pooled in the dilatated

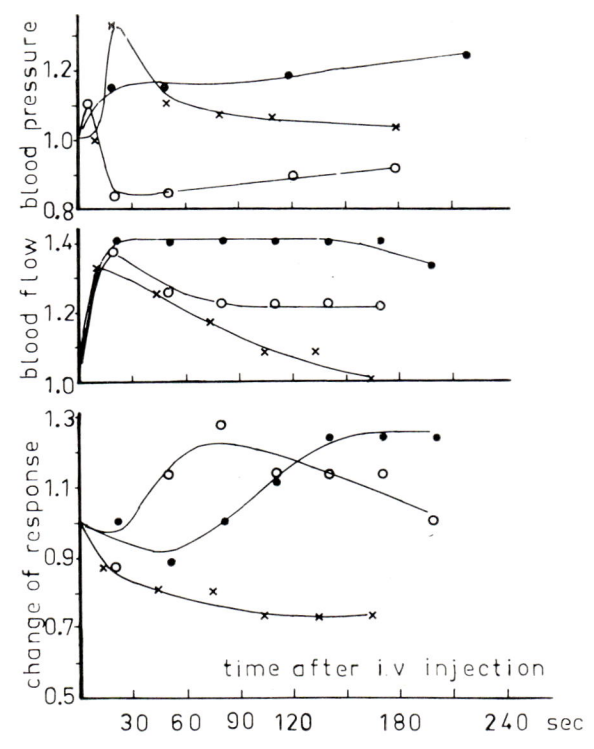

FIG. 5. The effect of the intravenous injection of 1 ml Haemaccel (●), 0·5 μg noradrenaline (×) and 0·2 μg isoproterenol (○) on blood pressure, blood flow and responses of the chorda tympani nerve to 0·3 M-NaCl.

vessels. It exerted its effect first when the vessels contracted and it then gave the strong increase in cardiac output recorded in Fig. 4. Similar experiments, but without any injection of Haemaccel, showed no increase in cardiac output after artificial respiration above the level observed before.

The results plotted in Fig. 5 support further the assumption that the blood flow around the taste buds is important to their gustatory sensitivity. Fig. 5 shows the effect of intravenous injections of noradrenaline, isoproterenol and a plasma substitute, Haemaccel, on blood pressure, blood flow and gustatory response in one rat.

It can be seen that injection with noradrenaline gave the well-known large increase of blood pressure, mainly due to an increase in peripheral vasoconstriction, partly to an increase in cardiac output. The net result was a diminished circulation to the taste receptors, which caused the observed decrease of response.

Isoproterenol acts almost exclusively on the beta-receptors (Goodman and Gilman, 1968), thus lowering the peripheral resistance and therefore the blood pressure, despite an enhancement of the cardiac output. In this experiment it may be concluded that the net effect on the taste receptors was a temporarily increased blood flow and consequently an increase of their sensitivity, but of rather short duration, since the effect of isoproterenol disappears fast.

The data of Fig. 5 show that the most efficient method tested for restoring gustatory sensitivity was to increase the blood volume. This can be accomplished by infusing blood, a blood plasma substitute like Haemaccel or, with less lasting effect, salt solutions like Ringer's solution. Their effects can be explained along the following lines. The increase of blood volume caused an increase of cardiac output and hence a rise of blood pressure which triggered a vascular dilatation and improved the circulation to the taste receptors. In fact, infusion with blood or Haemaccel proved to be the most efficient way of restoring gustatory sensitivity in all failing preparations.

In 1883 Heidenhain stimulated the peripheral part of the cut chorda tympani nerve electrically while he counted the number of blood drops from the sublingual vein (Heidenhain, 1883). He noted that stimulation substantially increased the blood flow from the tongue. In 1952 Erici and Uvnäs repeated his experiment with similar results (Erici and Uvnäs, 1952). From the stimuli used and the observations obtained, they concluded that this effect was caused by efferent nerve fibre stimulation and not by any antidromic effects.

I have repeated their experiments during perfusion of the tongue through the lingual artery. The result of such an experiment is shown in Fig. 6, which indicates, in accordance with the results of the authors mentioned above, that the vascular resistance of the tongue decreased when electrical stimulation was applied to the nerve. In the preparation of Fig. 6 the blood flow to the tongue had to be raised about 40 per cent after stimulation to obtain the same perfusion pressure as before. The effect disappeared within two minutes. Other experiments showed that the effect of electrical

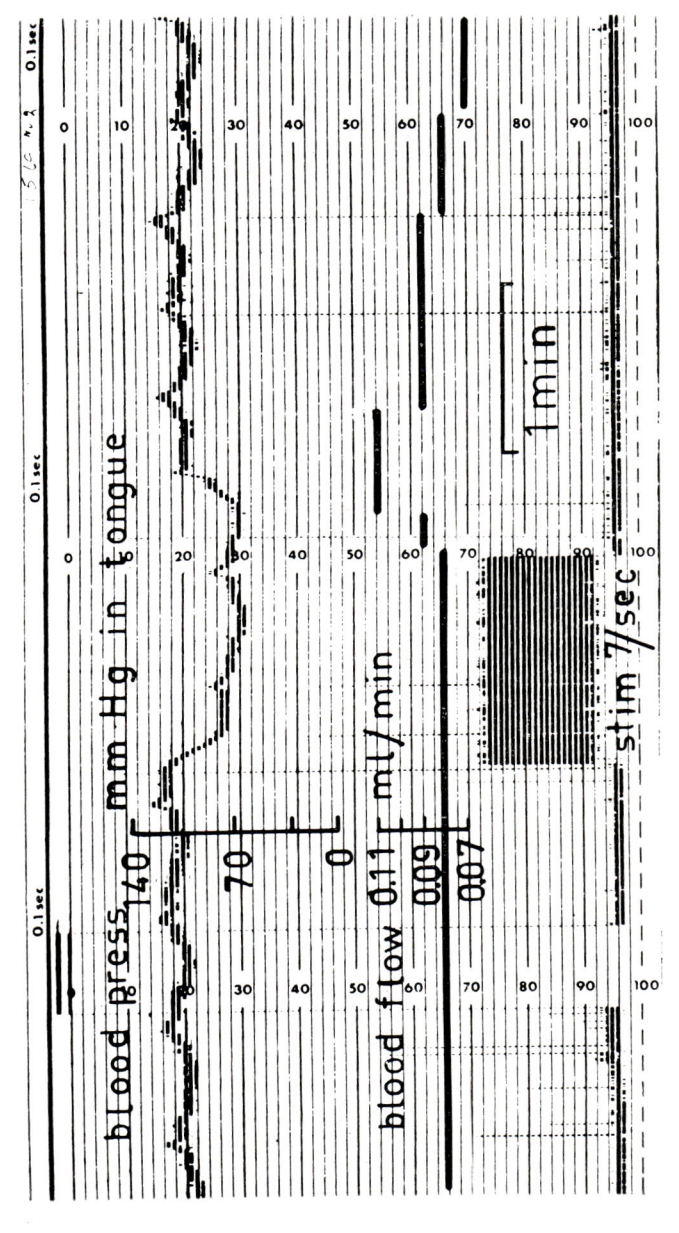

FIG. 6. The effect of electrical stimulation of the peripheral end of the cut chorda tympani nerve on the perfusion pressure and perfusion flow to the tongue.

stimulation declined when it was repeated. This has been noticed by others and can also be seen in the recordings published by Erici and Uvnäs (1952).

It was also possible to revitalize the taste receptors in some animals by stimulating the peripheral ends of their cut chorda tympani nerves. An example of this is shown in Fig. 7. The progressive diminution of the taste response of this animal was probably caused by a reflexly induced decrease of the blood flow to the taste cells, due to the falling systemic blood pressure.

The lingual artery of the rat does not supply the major sublingual or submaxillary glands (Greene, 1968). Injections of methylene blue into the lingual artery in this study stained neither these glands nor the minor sublingual. It may be concluded that the increase of blood flow through

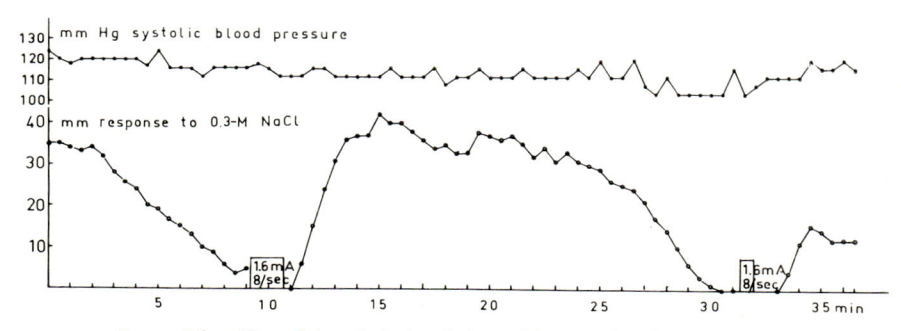

FIG. 7. The effect of electrical stimulation of the peripheral end of the cut chorda tympani nerve on the responses of the same nerve to repeated rinses with 0·3 M-NaCl over the tongue.

the lingual artery gave an increase of flow through the tongue only. It is known that glandular tissue can liberate vasoactive substances (Hilton and Lewis, 1955). Any effect on the vessels by substances from the sublingual or submaxillary glands during the perfusion is rather improbable for the reasons mentioned. There are, however, a number of small glands in the surface of the tongue which may liberate such substances during the stimulation of the chorda tympani nerve. It therefore cannot be excluded that they or vasoactive substances liberated from them might have been responsible for some of the observed increase of blood flow through the tongue, though it seems unlikely.

The existence of a rich vascular network underlying and extending into the fungiform papillae is well documented. For example Nemetschek-Gansler and Ferner (1964) stated that "there is a rich blood supply below the subgemmal nerve plexus" and Fish, Malone and Richter (1944) wrote "the fungiform papilla has a unique capillary network". It is also very

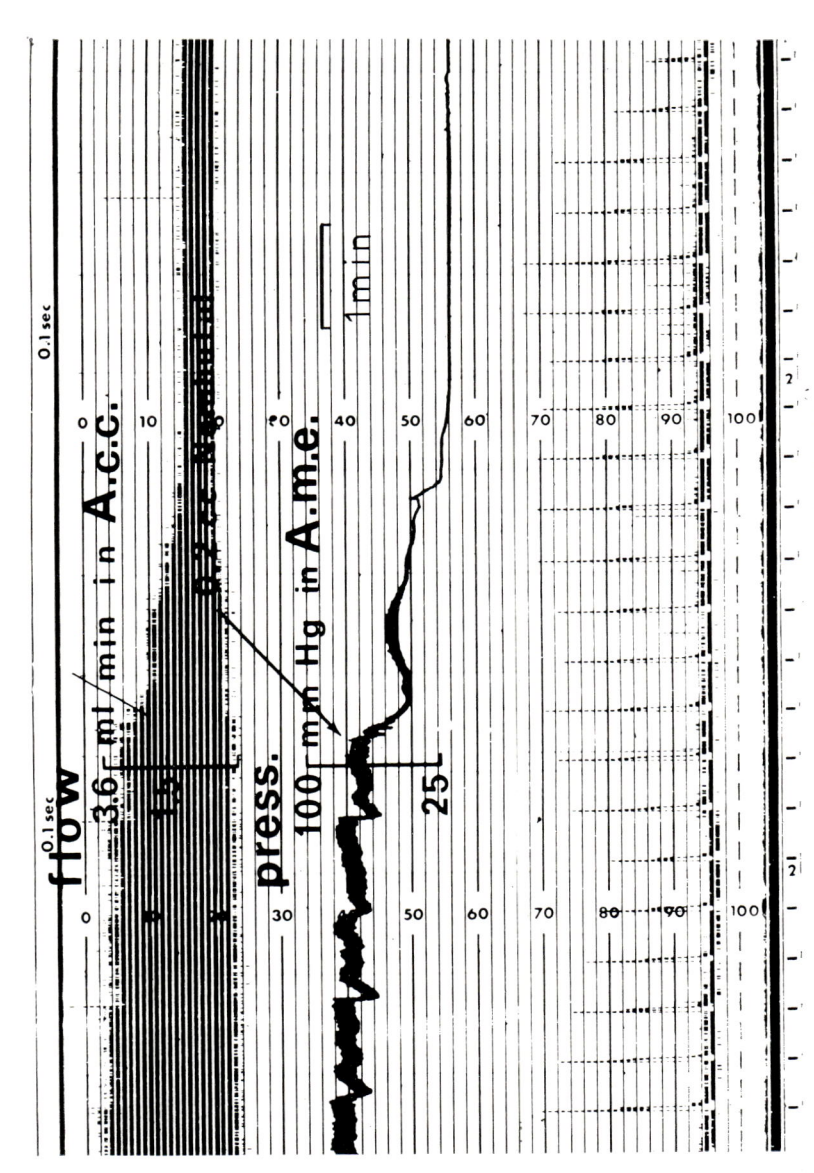

FIG. 8. The effect of intravenous injection of 0·2 ml 6 per cent Nembutal on blood flow, arterial blood pressure and the chorda tympani nerve response.

likely that the vessels of the tongue have rather high contractility. Baradi and Bourne (1959) remarked that "the blood vessels of the tongue showed deeply stained masses around the arterioles in the adventitia that presumably represent parasympathetic innervation". It may therefore be suggested from morphological evidence that the vertebrate possesses an ability to influence the sensitivity of its gustatory receptors by varying their blood supply.

It has been suggested to me by Professor Beidler that filling of the blood sinuses underlying the taste buds might have caused them to protrude, which could be responsible for the observed increase in response (see also p. 96). Such a possibility cannot be excluded, but it can account for only a part of the observations described here. Fig. 8 supports this statement.

A lethal dose of Nembutal was injected intravenously at the arrow of Fig. 8. This gave a rapid fall of cardiac output and of systemic blood pressure. Erectile tissue would have collapsed concomitantly with the drop of blood pressure and the gustatory response would have followed it. On the other hand, some of the effects observed in this study have a faster time-course than that of Fig. 8. It therefore seems possible that some of the effects I have described may have been caused by a combination of two factors, though the underlying basic mechanism is the same.

The effect of the first factor is most clearly demonstrated in Figs. 1 and 8. The diminution of the gustatory response in these figures was apparently caused by a deficit of constituents necessary for metabolism, probably oxygen. These changes of gustatory sensitivity had a latency of a few minutes.

The second factor was less evident but may be the factor suggested by Professor Beidler or it may be a local circulatory change close to the taste cells. The latencies of these effects were less than half a minute.

SUMMARY

In summary this study demonstrates that:
(1) Bilateral occlusion of the common carotid arteries in the rat extinguished the chorda tympani nerve response to gustatory stimulation within 4 minutes.
(2) The observed effects had nothing to do with changes in the fibres of the chorda tympani nerve.
(3) The primary reason for these effects was probably local anoxia, due to circulatory changes in the vicinity of the taste cells.

(4) The magnitude of the chorda tympani response to the same gustatory stimulus was closely related to the systemic blood pressure below a certain level.
(5) Intravenous injections of blood substitute or blood were the most efficient method of restoring the chorda tympani nerve response, if the systemic blood pressure was lowered.
(6) Pharmacological substances which raised the arterial blood pressure by increasing peripheral resistance, decreased the chorda nerve response.
(7) Pharmacological substances which decreased peripheral vascular resistance increased the chorda nerve response.
(8) It was possible to increase or decrease the chorda tympani nerve response to gustatory stimulation by increasing or decreasing the rate of perfusion of the tongue.
(9) Electrical stimulation of the peripheral part of the lingual or the chorda tympani nerves decreased the vascular resistance of the tongue during perfusion.
(10) Electrical stimulation of the peripheral chorda tympani nerve under normal circulation could revitalize its gustatory receptors.
(11) It is likely that the central nervous system, acting through the vascular system, can regulate the sensitivity of the taste receptors.

Acknowledgements

The author is indebted to Dr Arne Åström, Stockholm, Sweden, and Dr Bruce Oakley, Ann Arbor, Michigan, U.S.A. This study was supported by grants from Svenska Maltdrycksforskningsinstitutet, Swedish Medical Research Council grant No. B70-14X-2467-03, and Magnus Bergvalls Stiftelse.

REFERENCES

BARADI, A. F., and BOURNE, G. H. (1959). *J. Histochem. Cytochem.*, **7**, 2-7.
BARTOSHUK, L. M. B. (1965). Ph.D. Thesis, Brown University (unpublished).
ERICI, I., and UVNÄS, B. (1952). *Acta physiol. scand.*, **25**, 10-14.
FISH, H. S., MALONE, P. D., and RICHTER, C. P. (1944). *Anat. Rec.*, **89**, 429-440.
GOODMAN, L. S., and GILMAN, A. (1968). *The Pharmacological Basis of Therapeutics*, 3rd edn., p. 497. London: Macmillan.
GREENE, E. C. (1968). *The Anatomy of the Rat.* New York and London: Hafner.
HEIDENHAIN, R. (1883). *Arch. Anat. Physiol., Physiol. Abt.*, suppl., 133-177.
HELLEKANT, G. (1968). *Acta physiol. scand.*, **74**, 1-9.
HELLEKANT, G. (1969). *Acta physiol. scand.*, **75**, 39-48.
HILTON, S. M., and LEWIS, G. P. (1955). *J. Physiol., Lond.*, **128**, 235-248.
NEMETSCHEK-GANSLER, H., and FERNER, H. (1964). *Z. Zellforsch. mikrosk. Anat.*, **63**, 155-178.
SATO, M. (1963). In *Olfaction and Taste* (Proceedings of the First International Symposium, Stockholm, 1962), pp. 149-164, ed. Zotterman, Y. Oxford: Pergamon Press.

SATO, M. (1967). In *Contributions to Sensory Physiology*, vol. 2, pp. 223–251, ed. Neff, W. D. New York: Academic Press.
THOMPSON, A. H., ROBERTSON, R. C., and BAUER, T. A. (1968). *J. appl. Physiol.*, **24,** 407–411.
WRIGHT, E. B. (1947). *Am. J. Physiol.*, **148,** 174–184.
YAMASHITA, S. (1964). *Jap. J. Physiol.*, **14,** 488–504.
YAMASHITA, S., YAMADA, K., and SATO, M. (1964). *Jap. J. Physiol.*, **14,** 505–514.

DISCUSSION

Lowenstein: Have you any answer to why blood *flow* around the receptors affects their sensitivity?

Hellekant: The problem is not yet solved, because it could still be the oxygen tension. I want to emphasize that the organism usually meets metabolic demands in an area by increasing the blood supply to the area, not by increasing the oxygen tension of the blood. So the last link may very well be in the oxygen tension, but the first link is the change of blood flow.

Lowenstein: This makes me feel happier. In experiments where the responses from the utricular nerve of the vestibular organ are being recorded in an intact frog with simultaneous recording of the electrocardiogram, one sees that at the first faltering of the electrocardiogram the spontaneous discharge from the utriculus peters out and one understands how important oxygen tension is, for some receptors.

Beidler: Do you think it possible that a change in the relation between a taste bud and its environment could be produced merely by changing the blood supply beneath the bud? If the papilla were erectile, an increase of blood supply might elevate the bud and open the pore.

Hellekant: Your suggestion may account for a mechanism responsible for fine adjustments of the sensitivity of the taste cell. However, I don't think that the large changes I have demonstrated here can be explained by such a mechanism as yours alone. I think that changes of the blood vessels supplying the taste cells might have caused a deficiency of oxygen or another factor, which, by being essential to cell metabolism, also affects the sensitivity of the cell, and that this effect has a rather short latency because of the high metabolic rate.

Beidler: Have you looked at the papillae from the surface during your experiments?

Hellekant: No.

MacLeod: Is it possible that you are observing only the effect of clearance of the stimulus? The increased blood flow would increase the gradient of concentration across the receptor cells.

Hellekant: We have shown (Hellekant, 1968) that the first exposure to a gustatory stimulus depresses the following response if the time-interval between them is short enough. In this study, experiments not described here do not indicate any change in this relation when the blood supply is cut off. Both responses diminish to the same extent. This indicates to me that the blood flow has little to do with the clearance of the stimulus, provided that the stimulus does not penetrate the cell.

Zotterman: It is most interesting too that whether you rinse once, or twice at double speed, the recovery time is just the same.

REFERENCE

HELLEKANT, G. (1968). *Acta physiol. scand.*, **74**, 1–9.

NEURAL AND PERCEPTUAL RESPONSES TO TASTE STIMULI

G. Borg, H. Diamant and Y. Zotterman

Department of Otorhinolaryngology, Umeå University and Department of Physiology, Kungl. Veterinärhögskolan, Stockholm

By a freak of nature the gustatory nerve fibres from the anterior part of the tongue run through the middle ear in the chorda tympani nerve. During middle ear surgery this nerve is often exposed, which makes it possible to record the electrical responses of the nerve to the application of different sapid solutions to the tongue. In 1958 our first successful experiments were performed in the Ear Clinic of Karolinska Sjukhuset, Stockholm, during operations undertaken in an effort to mobilize the stapes (Diamant and Zotterman, 1959). In these experiments we recorded the summated electrical response of the nerve to touch and to various sapid solutions applied to the tongue. As will be seen from Fig. 1 there was a good response to 0.5 M-sodium chloride, 15 per cent sucrose, 0.04 per cent saccharine, 0.02 M-quinine and 0.2 M-acetic acid. The application of water to the human tongue was followed by a reduction in the spontaneous activity of the nerve in exactly the same fashion as we had previously found in the rat, which does not possess any taste fibres that respond to water as is the case with the rabbit, the cat and the rhesus monkey. In further experiments we were able to obtain valuable information about the relation between the strength of the gustatory stimulus and the summated electrical response from the nerve (Diamant et al., 1963, 1965).

Recording from human taste nerves offers, however, the further opportunity of comparing the relation between the stimulus strength and the electrical response with the corresponding relation between molar concentration and the perceptual (subjective) intensity as estimated by the patient in psychophysical experiments. The choice of the electrophysiological recording technique is particularly important. Is the rectified, summated and filtered transformation of the activity of the whole chorda tympani nerve related to the intensities and qualities of the taste sensations we experience? We do not know whether the central nervous system "sees" the responses from the chorda tympani in the same way as our electrical apparatus. What we record might be just an epiphenomenon that parallels in time the true and unrecorded information-bearing responses of the peripheral taste nerve. However, if there is a good correspondence between the summated chorda tympani discharge and the psychophysical responses, we can infer not only that we have chosen a meaningful recording technique to "tap into the peripheral input line" but also that the central mechanisms do not seriously alter the input. If such correspondence is lacking we must conclude either that an inappropriate recording technique was used, and/or the sensory input is modified in the central nervous system.

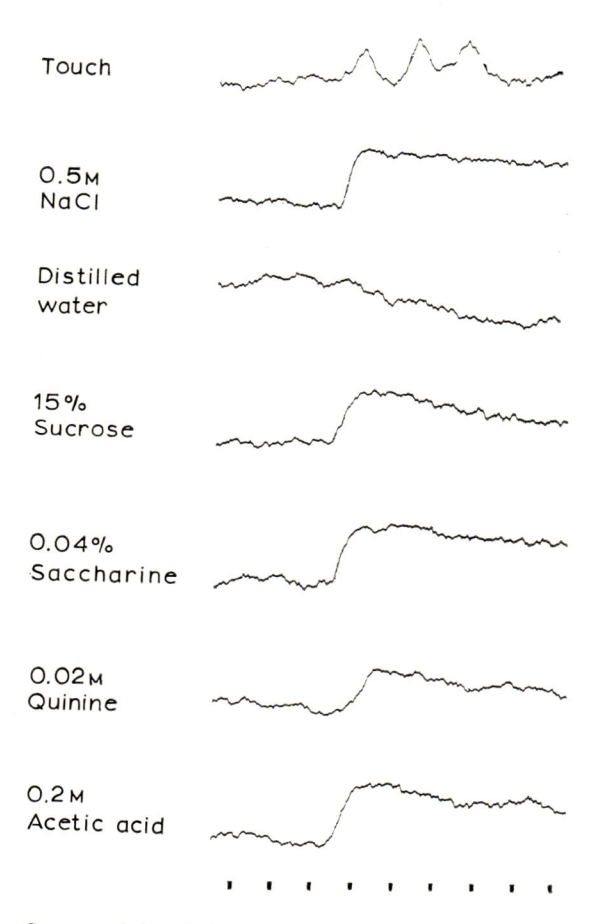

FIG. 1. Summated electrical response of the whole chorda tympani nerve of man to touch and to various sapid solutions applied to the tongue. (From Diamant and Zotterman, 1959, by permission of *Nature*.)

PSYCHOPHYSICAL EXPERIMENTS

Two days before the operation psychophysical taste experiments with citric acid, sodium chloride and sucrose were carried out on the patients. The method of magnitude estimation was used. This method, which was introduced by Stevens (1957), requires that the subject can handle numbers and make quantitative estimations in terms of ratios. In a trial experiment all patients had to make quantitative estimations of surfaces of different sizes

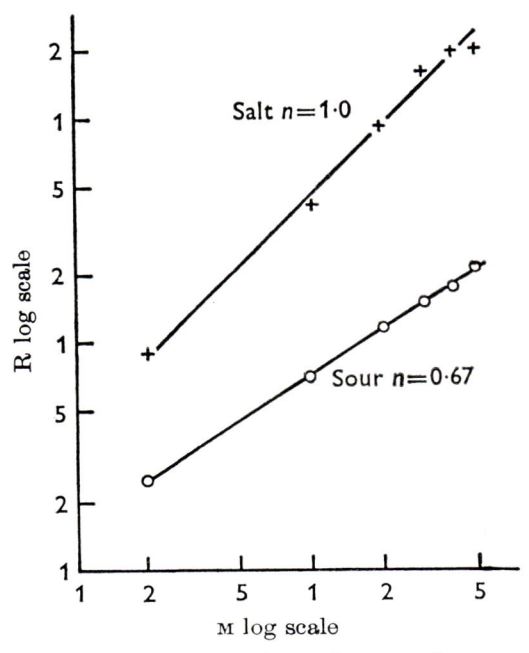

FIG. 2. Result of a psychophysical experiment on fourteen students. Perceptual intensity is plotted against molarity of citric acid on a log–log scale. (From Borg et al., 1967a.)

so that we could screen out those who obviously could not make magnitude estimations. The same stimuli and the same random order of presentation were used in the electrophysiological experiments as in the psychophysical experiments. The stimuli were presented in pairs, the standard with one comparison stimulus.

The results of psychophysical experiments with a group of fourteen young students are seen in Fig. 2. A straight line may be very nicely fitted to the psychophysical response to citric acid when plotted against molarity in log–log coordinates. A simple power function $R = cM^n$ with $n = 0.67$

describes the relation (R, reaction; M, molarity). Although the fit of a straight line to the values for salt is not so good, the relation may also be described with a power function of the form $R = a + cM^n$, with a rather high a value and $n = 1\cdot 0$. The result for salt is in accordance with previous results by Ekman, who introduced the additive constant a and found a power function of this form applicable to salt and sucrose (Ekman, 1961; Ekman and Åkesson, 1965) but with n slightly above $1\cdot 0$.

The results of the first successful experiments in which we obtained both

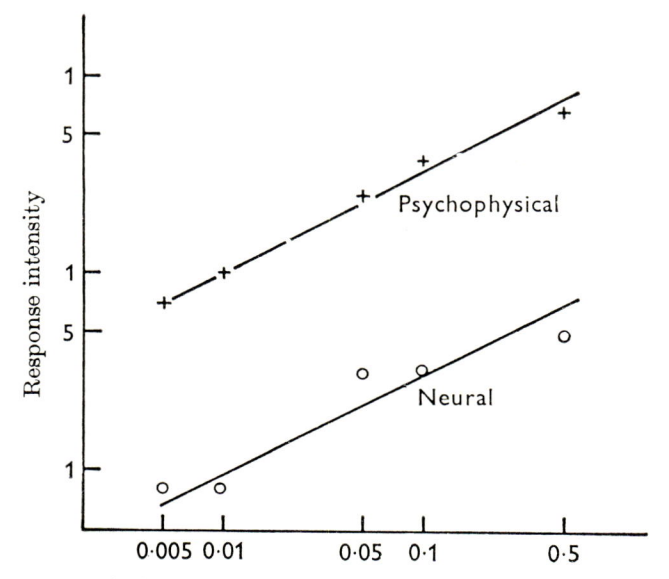

FIG. 3. Graphs from one patient (C.L.) showing subjective intensity and neural response plotted against molarity of citric acid on a log-log scale. (From Borg et al., 1967a.)

relative psychophysical and neural responses to citric acid from the same patient are presented in a log-log diagram (patient C. L., Fig. 3). Straight lines may be adjusted to the values, that is, a power function may describe the relations, although a Fechnerian log function may give a better fit to the variation in neural activity (Borg et al., 1967a). If as a first rough estimation we describe both the neural and the psychophysical responses with power functions of the simple form $R = cM^n$, we find an astonishingly good agreement. The exponent of the psychophysical function $n_R = 0\cdot 5$ is the same as that of the neurophysical function $n_N = 0\cdot 5$.

In November 1965 only one (V.R.) out of three patients was able to perform the psychophysical tests (Fig. 4). Unfortunately the chorda tympani response was very poor. We obtained, however, good responses to citric acid and salt solutions from another patient (H.N., Fig. 5) although she failed in the subjective tests. When plotted in a log-log diagram her neural responses gave a good fit to straight lines of about the same slopes as those of the subjective data from patient V.R. It is apparent that in all these cases the slope of the salt line is definitely steeper than that of the acid line.

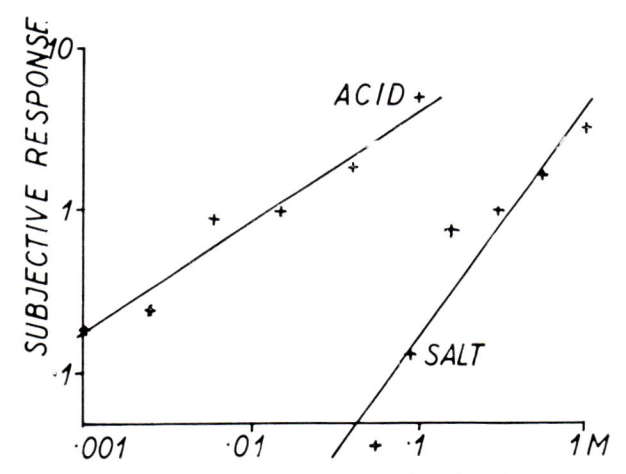

FIG. 4. Log-log diagram of relation between the subjective response and the molarity of salt and citric acid (patient V.R.). (From Borg et al., 1967a.)

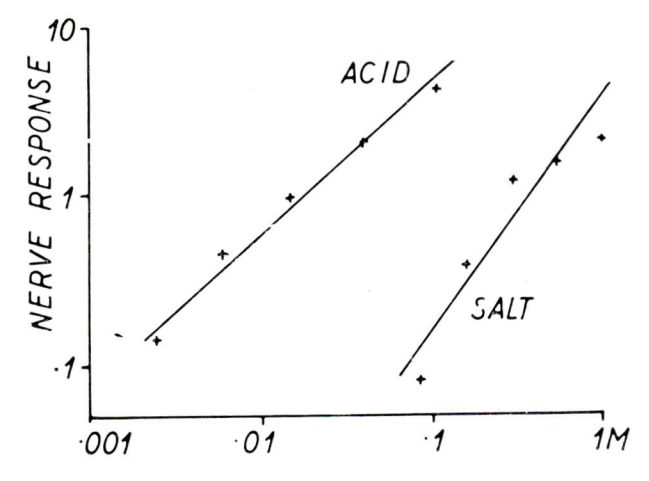

FIG. 5. Nerve responses to salt and citric acid plotted against molarity (patient H. N.). (From Borg et al., 1967a.)

The following year perceptual estimations were made and the electrical responses were recorded to sucrose and citric acid. In Fig. 6 both the relative psychophysical and electrical responses to citric acid are plotted in a log-log diagram (patient S.P.). The diagram in Fig. 7 gives the relations for sucrose for patient I.J. Each point is the median of three observations in each series of tests on the same individual. How well the psychophysical functions follow the neural functions will be seen in Fig. 8 where the diagram gives the functions of the mean values obtained for these two patients.

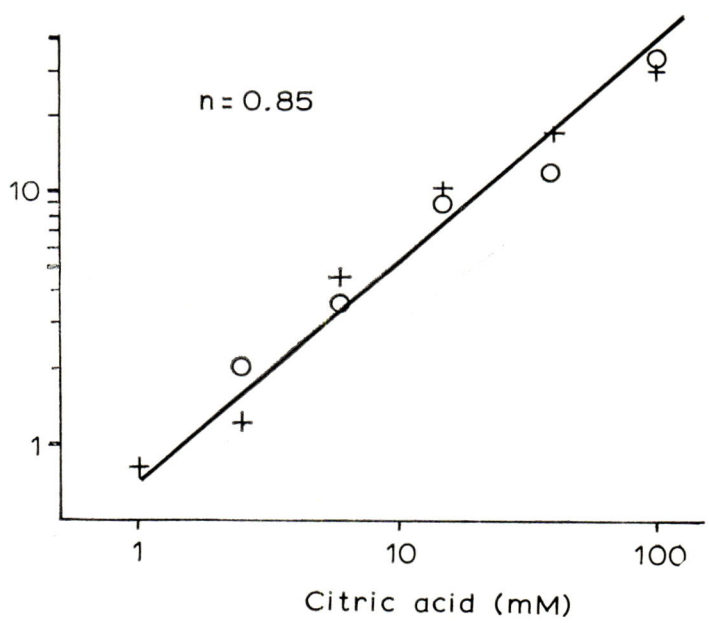

FIG. 6. Nerve responses to citric acid solutions of increasing molarity (open circles) and psychophysical estimations (crosses) plotted against the molarity on a log-log scale (patient S.P.). (From Borg et al., 1967a.)

Recently (15–16 September, 1969) we managed to obtain full perceptual and neural data in two patients (A.A. and G.Å.) on sodium chloride solutions of various strength. As will be seen from Figs. 9 and 10 there is a great difference in the general shape of the individual curves describing the perceptual and neural function. In spite of this there is a remarkable congruity between the perceptual and the neural function in each case.

Quite apart from the question of whether the function describing the relation between the strength of the sapid solution and the subjective estimation satisfies a Stevens power function or a Fechnerian log function, it

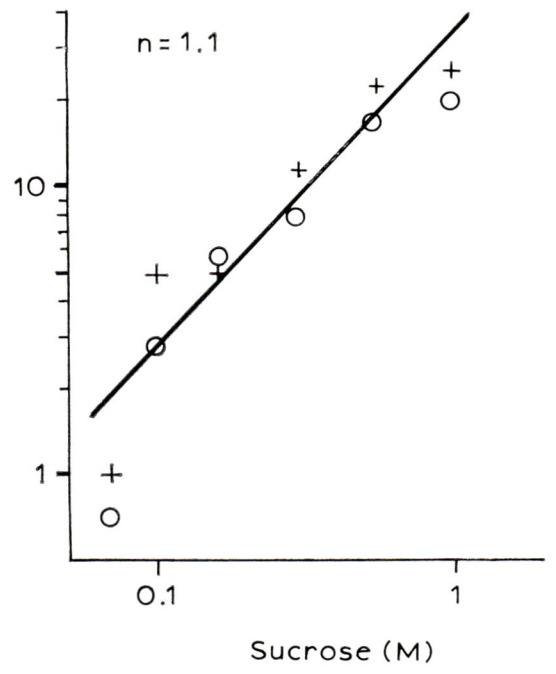

FIG. 7. Nerve responses (open circles) and subjective estimation plotted against molarity of sucrose solution (patient I.J.). (From Borg *et al.*, 1967a.)

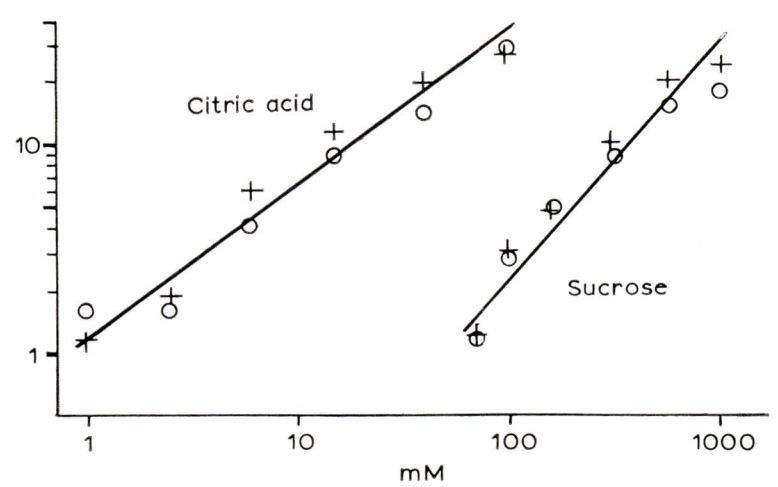

FIG. 8. Mean values of neural response (open circles) and of subjective response from two patients (I.J. and S.P.) plotted against molarity of citric acid and sucrose solution. (From Borg *et al.*, 1967b, by permission of *The Journal of Physiology*.)

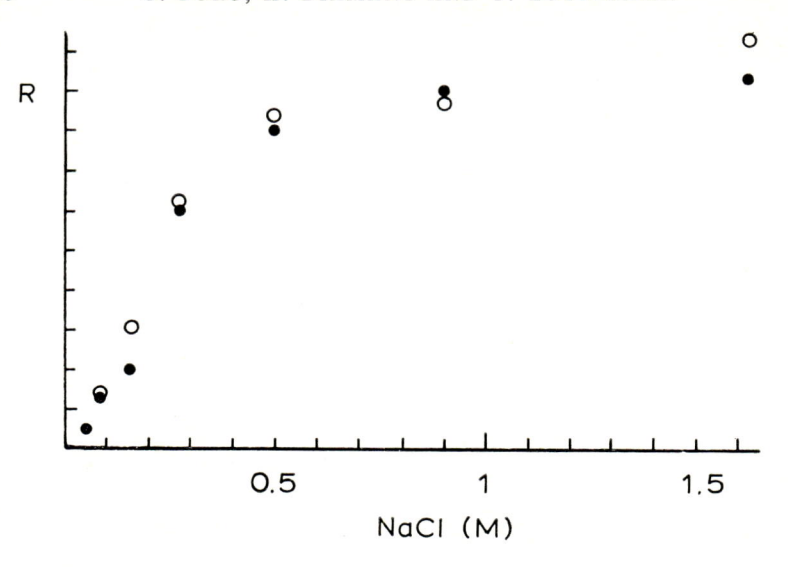

Fig. 9. Mean values of neural response (open circles) and perceptual estimations plotted against molarity of NaCl solution (patient A.A.).

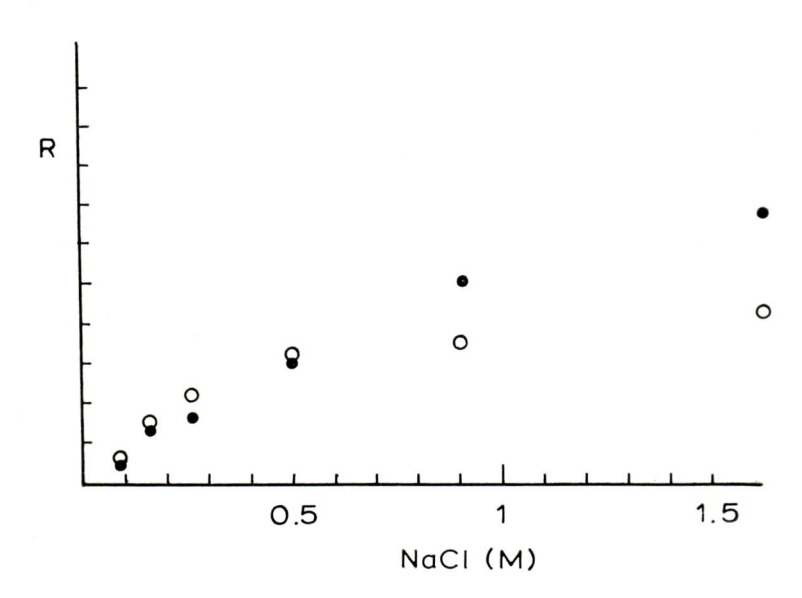

Fig. 10. Mean values of neural response (open circles) and perceptual estimations plotted against molarity of NaCl solution (patient G.Å.).

is apparent from the diagrams presented that there is a remarkably close correlation between the subjective and neural data. When describing the relation as a power function it is clearly seen that for each of the three sapid substances, acid, sucrose and salt, the exponent n of a simple power function, $R = cM^n$, will have a different value in such a way that the exponent for citric acid is always less than one while that for sucrose and NaCl is usually equal to or higher than one.

Previously (Diamant et al., 1965) we demonstrated that in spite of individual variations in the responses of the chorda tympani there was a good correspondence between the psychophysical and neural data on the sweetness of a series of biological sugars and synthetic sweeteners (Table I). The

TABLE I

COMPARISON OF PSYCHOPHYSICAL AND NEURAL
RESPONSES TO SWEET-TASTING SUBSTANCES

	Patient 3		Patient 4	
Stimulus	Psychophysical	Neural	Psychophysical	Neural
0·5 M-sucrose	100	100	100	100
0·5 M-fructose	100	100	80	80
0·5 M-maltose	—	40	75	60
0·5 M-galactose	40	45	45	40
0·5 M-lactose	45	45	30	30
0·5 M-glucose	25	45	35	40
0·004 M-Na saccharin	100	65	125	105
0·03 M-cyclamate	55	80	115	100

The values in each column are relative to 0·5 M-sucrose set at 100. The maximum height of the summator record was measured. (From Diamant et al., 1965.)

values in each column were rounded off to the nearest 5 per cent and are relative to the response to 0·5 M-sucrose, which has been set at 100. The subjective reports are the means of two determinations, while the neural values are based upon a single determination. The correspondence seems quite good for the sugars. The artificial sweeteners, saccharin and cyclamate, have quite different tastes from the sugars and this may have affected the judgements of their sweetness. The better agreement of patient 4 may depend upon the fact that this patient received the standard before each of the test substances whereas the other patient received the standard only at the beginning and end of the two runs.

Fig. 11 shows records from a fine strand of the peripheral part of the chorda tympani of the rhesus monkey (Gordon, Kitchell and Ström, 1959). In this preparation quinine elicits an electrical response which consists exclusively of small spikes, whereas sweet-tasting substances like sucrose, saccharin, glycerol and ethylene glycol produce large spikes. If the records are carefully scrutinized it will be found that in addition to the large spikes sucrose produces a few small spikes while saccharin and still more so glycerol and glycol elicit a large number of small spikes. It is of interest to compare these records, although they derive from the monkey, with the taste sensations which one experiences from these substances. Of these substances sucrose is

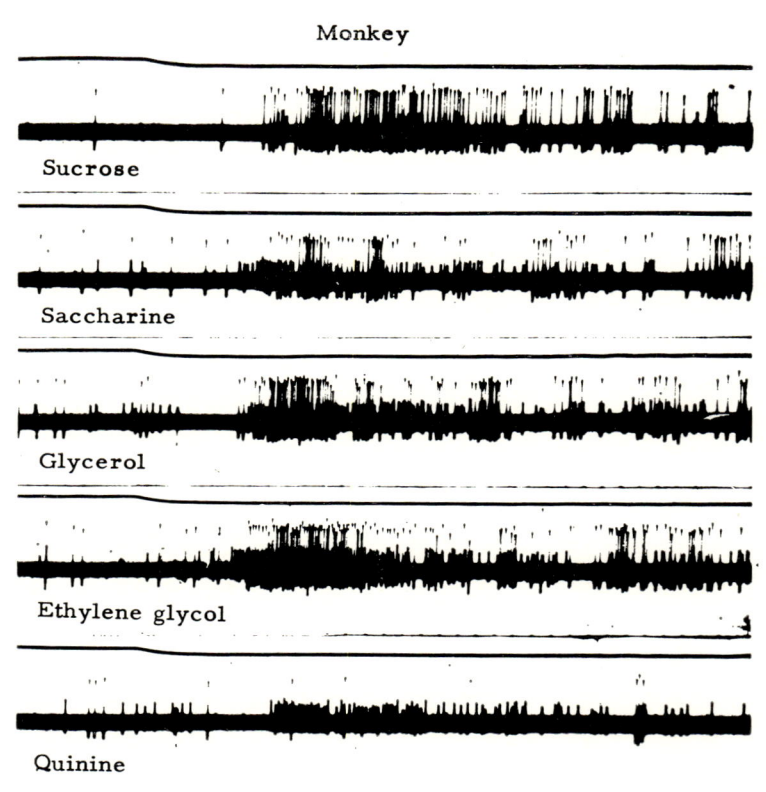

Fig. 11. Records from a small strand of the chorda tympani of a rhesus monkey which contained few active fibres. Note particularly the large spikes, and also those of intermediate size which project both above and below the baseline. All solutions made up in Ringer's solution. Time: 10 spikes per sec. (From Gordon, Kitchell and Ström, 1959.)

considered to give a pure sweet taste; saccharin, glycerol and glycol give in addition to the sweet taste a more or less bitter taste. If we assume the same kind of neural mechanisms in man as in the monkey it is thus conceivable that the different sensations experienced may depend upon the signalling of two specific kinds of nerve fibres, one carrying the information "sweet", the other, the smaller fibres, the information "bitter". On this assumption we may also explain a phenomenon described by G. von Békésy (personal communication). He uses a special technique of electrically stimulating single taste papillae in the human tongue; the quality of the gustatory sensation is reported by the patient. When the report is "sweet", the subject says that it is "heavenly sweet"—more pure sweet and pleasant than the taste of sugar. As von Békésy applies weak electrical pulses he may be stimulating the larger nerve fibres only, thus eliciting a sweet sensation of hitherto unknown quality which is more sweet than the taste of saccharose, which stimulates some "bitter fibres" as well. This assumption is strengthened by the observations reported by Andersson and co-workers (1950) that quinine on the dog's tongue stimulates only small gustatory nerve fibres and by Iriuchijima and Zotterman (1961) that "bitter fibres" in the dog conduct at velocities of only a few metres per second. If we assume that this is also the case in man it is conceivable that the summated electrical responses to saccharin containing a greater number of small spikes will attain relatively lower values than those of saccharose, containing more large spikes.

It is important to keep in mind that it is the largest gustatory nerve fibres which dominate the summated electrical response. So when we speak of a close relationship between the summated electrical response and the subjective response we mean the relation between the activity of large fibres and the perceptual estimation. Assuming that the activity of the tiny fibres has as strong an influence as the large fibres on the perceptual response, we should predict a relatively stronger subjective response to saccharin. The estimation of the subjective strength of saccharin is most likely built up by a summation of information carried in large "sweet fibres" and small "bitter fibres". Thus the failing correspondence between the neural and the perceptual responses to sweeteners as seen in Table I may be really due to the failure of our electronic recording system and not to any misjudgements by the subjects.

ADAPTATION

The summated chorda tympani response to a 0·2 M-sodium chloride solution is seen in Fig. 12. The decline in neural activity in response to a continuous flow of salt solution over the tongue for 3 minutes may be seen in the records A, B and C from three different human subjects. The initial peak response to the application of the salt solution (at 37°C) is indicated by an arrow, while the application of distilled water is indicated by a dot above the curve. The responses to water, here, were due primarily to the low temperature of the water (20°C) and also to some degree of mechanical stimulation of the tongue. Record A is the most satisfactory from the technical point of view, for the signal-to-noise ratio was good and the baseline quite stable. The response was 95 per cent adapted with 50 seconds. In addition to the three records shown here, the adaptation in two more patients was very similar to that shown in A and C. Our records of human neural adaptation to 0·2 M-NaCl contrast distinctly with record D which is taken from the rat chorda tympani. Here the response declines very slowly over a 3-minute period, and when the record is continued for some minutes more little or no further decline in amplitude is seen (Beidler, 1953; Zotterman, 1956). The patient whose neural response is shown in record C indicated in the psychophysical test that the salt taste disappeared after 90

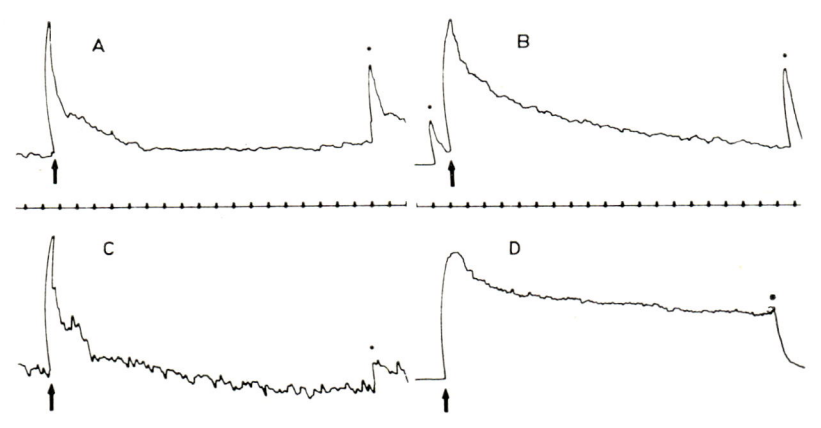

FIG. 12. The summated chorda tympani response to a continuous, 3-minute flow of 0·2 M-sodium chloride solution. A, B and C are human responses, from three different patients, and D is the response of a rat. Dots indicate onset of distilled water (20°C). Arrows, onset of 0·2 M-NaCl (37°C). Time-base in 10-second intervals. (From Diamant et al., 1965.)

seconds. A fourth patient indicated that he could no longer taste salt after 79 seconds, which corresponded to a 95 per cent reduction in the magnitude of his neural response.

Bujas (1953) studied psychophysical adaptation to sodium chloride in two subjects. Using 0·15 M-NaCl to stimulate a tongue area 1 cm in diameter, he found complete adaptation in 50 and 54 seconds. These values are of the same order of magnitude as those given by our patients (79, 90 and 122 seconds). Our records of neural adaptation to 0·2 M-NaCl solution suggest that adaptation is complete; that is, the activity decreases until it reaches the resting level of activity. In addition there is a reasonable correspondence between the neural and psychophysical records for the time necessary for complete adaptation. Thus we may conclude that the psychophysical observation in man of rapid and complete adaptation to the taste of salt can be accounted for by diminished activity in the chorda tympani nerve. There is no need to postulate the existence of central adaptation mechanisms.

The very close agreement between the subjective and neural function is not surprising considering that our subjective estimation, made by the neural analyser in our central nervous system, must work on the information it receives from the peripheral receptors. Katz (1950) found that there is a linear relation between the height of the receptor potential of the muscle spindle and the peak frequency of the nerve discharge, and Døving (1964) found such a relation in the frog between the peak amplitude of the electro-olfactogram, Ottoson's electro-olfactogram in the frog (1956), and the discharge of impulses from secondary neurons of the olfactory bulb. Such a linear relation between receptor potential and impulse frequency in the labellar sugar receptor of the blowfly was recently reported by Morita and Yamashita (1966). Thus the receptor potential evoked by the sapid solution is transformed into a volley of spikes propagated to the next neuron where this volley sets up a postsynaptic potential which is transmitted to the next relay station, in the way characteristic of impulse frequency modulation. Therefore it should not be surprising that the summated electrical response which we obtain from the chorda tympani varied linearly with the amplitude of the postsynaptic potential evoked in the cerebral cortex.

SUMMARY

During middle ear operations the chorda tympani nerve is often exposed, which makes it possible to lead off the summated electrical response of its gustatory fibres to the application of sapid solutions to the tongue. Thus data can be obtained for a quantitative study of the relation between the neural activity and the strength of the stimulus applied to the tongue which can be compared with the relation between the subjective estimation and the stimulus strength.

Psychophysical taste experiments were carried out according to the method of magnitude estimation introduced by Stevens. The same stimuli and the same random order of application were used in the electrophysiological experiments. Full comparative data, subjective responses as well as electrical responses from the nerve to the application of citric acid, sodium chloride and sucrose solutions on the tongue, were obtained in two pairs of patients.

Quite apart from the question whether the function describing the relation between the strength of the sapid solution and the summated electrical response satisfies a Stevens power function or a Fechnerian log function, it is apparent that there is a fundamental congruity between neural activity and perceptual intensity.

REFERENCES

ANDERSSON, B., LANDGREN, S., OLSSON, L., and ZOTTERMAN, Y. (1950). *Acta physiol. scand.*, **21**, 105.
BEIDLER, L. M. (1953). *J. Neurophysiol.*, **16**, 595.
BORG, G., DIAMANT, H., STRÖM, L., and ZOTTERMAN, Y. (1967a). In *Olfaction and Taste II* (Proceedings of the Second International Symposium, Tokyo, 1965), pp. 253–264, ed. Hayashi, T. Oxford: Pergamon Press.
BORG, G., DIAMANT, H., STRÖM, L., and ZOTTERMAN, Y. (1967b). *J. Physiol., Lond.*, **191**, 118P–119P.
BUJAS, Z. (1953). *Acta Inst. Psychol. Univ. Zagreb*, **17**, 1.
DIAMANT, H., FUNAKOSHI, M., STRÖM, L., and ZOTTERMAN, Y. (1963). In *Olfaction and Taste* (Proceedings of the First International Symposium, Stockholm, 1962), pp. 193–203, ed. Zotterman, Y. Oxford: Pergamon Press.
DIAMANT, H., OAKLEY, B., STRÖM, L., WELLS, C., and ZOTTERMAN, Y. (1965). *Acta physiol. scand.*, **64**, 67.
DIAMANT, H., and ZOTTERMAN, Y. (1959). *Nature, Lond.*, **183**, 191–192.
DØVING, K. (1964). *Acta physiol. scand.*, **60**, 150.
EKMAN, G. (1961). *Scand. J. Psychol.*, **2**, 185.
EKMAN, G., and ÅKESSON, C. (1965). *Scand. J. Psychol.*, **6**, 241–253.
GORDON, G., KITCHELL, R., and STRÖM, L. (1959). *Acta physiol. scand.*, **46**, 119.
IRIUCHIJIMA, J., and ZOTTERMAN, Y. (1961). *Acta physiol. scand.*, **51**, 283.
KATZ, B. (1950). *J. Physiol., Lond.*, **111**, 261.
MORITA, H., and YAMASHITA, S. (1966). *Mem. Fac. Sci. Kyushu Univ., Ser. E*, **14**, 83.
OTTOSON, D. (1956). *Acta physiol. scand.*, **35**, suppl. 122, 1–83.
STEVENS, S. S. (1957). *Psychol. Rev.*, **64**, 153.
ZOTTERMAN, Y. (1956). *Acta physiol. scand.*, **37**, 60–70.

DISCUSSION

Lowenstein: Have you any idea of the number of neural stations—that is, the number of synapses—between the gustatory receptor and the cortical end station?

Zotterman: We have the first relay in the medulla and the second in the thalamus, but it is still not certain where taste projects to the cortex.

Lowenstein: So the strict linearity is something which is inherent in such a communication system, provided, of course, that attention is focused on this modality and nothing else interferes, say via the reticular formation and so on?

Zotterman: There has been a lot of speculation about the gustatory connexion between the thalamus and the cortex. In a long series of experiments on more than 50 cats we didn't manage to find in the tongue area more than five cortical cells which responded to gustatory stimulation of the tongue (Cohen *et al.*, 1957).

Hellekant: I would just add that Funakoshi and Kawamura (1968) have recorded the summated cortical response to taste stimuli in man by means of an average computer, and they obtained results to tartaric acid and sodium chloride, but not to sucrose or quinine hydrochloride.

de Lorenzo: When the chorda tympani is cut, either deliberately or by mistake during surgery, patients frequently report a metallic taste on the anterior two-thirds of the tongue, months after the operation. Have you observed this?

Zotterman: This has been found in our cases too after division of the chorda tympani.

Lowenstein: Partial interference with the chorda tympani creates phantom taste sensations, but complete transection does not?

Zotterman: Our experience is that after one year these phantom sensations generally vanish when the chorda is not transected. After full transection only five out of 45 complained of phantom sensations. They have reported sensations in the tongue while resting on the operated ear or when exerting pressure on the ear (Wiberg, 1969).

Beidler: I looked at a number of Dr Samuel Rosen's patients several months after one of their chorda tympani nerves had been transected and many of them did not realize that they had lost the function of taste on one side of their tongue.

Zotterman: Very few realize it, only 17 per cent of cases according to Wiberg, and the complaints are not serious compared to the pleasure at recovering hearing.

REFERENCES

COHEN, M., LANDGREN, S., STRÖM, L., and ZOTTERMAN, Y. (1957). *Acta physiol. scand.*, **40**, suppl. 135.
FUNAKOSHI, M., and KAWAMURA, Y. (1968). *J. physiol. Soc. Japan*, **30**, 282–293.
WIBERG, A. (1969). *Acta oto-lar.*, suppl. 251.

OLFACTORY SURFACE AND CENTRAL OLFACTORY CONNEXIONS IN SOME VERTEBRATES

T. S. REESE AND M. W. BRIGHTMAN

Laboratory of Neuropathology and Neuroanatomical Sciences,[*] National Institute of Neurological Diseases and Stroke, National Institutes of Health, Bethesda, Maryland

THE first-order neurons in the vertebrate olfactory system are bipolar neuroepithelial cells embedded in the olfactory epithelium. Each bipolar cell has a single ciliated dendrite that makes initial contact with odorants and an axon that makes synaptic connexions with second-order neurons in the olfactory bulb. We report here on an electron microscopic study of this dendrite at the surface of the olfactory epithelium in certain vertebrates and on the synaptic organization of the axonal endings in the olfactory bulb of the rat.

The ciliated endings of bipolar cell dendrites, together with adjacent villous apices of supporting cells and the mucus overlying the epithelium, have been termed the olfactory surface (Bannister, 1965; Andres, 1966). The general features of the olfactory surface are similar in a variety of vertebrate olfactory epithelia, including that of mammals (Frisch, 1967; Seifert and Ule, 1967; Okano, Weber and Frommes, 1967; Andres, 1966, 1969), birds (Graziadei and Bannister, 1967), amphibians (Bloom, 1954; Reese, 1965), teleost fishes (Bannister, 1965) and cyclostomes (Thornhill, 1967). Olfactory cilia arise from the bulbous ends of bipolar cell dendrites; these have been misnamed "olfactory vesicles". The cilia have specialized distal segments characterized by a reduction both in diameter and in the number of ciliary tubules. In animals with lungs, parallel arrays of distal segments lie just below the surface of the viscous mucus covering the olfactory epithelium. Marked departures from this general pattern have not yet been observed at the olfactory surface of any vertebrate. In this report it is shown that the olfactory surface in a primate conforms in its general features to the vertebrate pattern and in more specific features to the mammalian pattern. The olfactory surface in the elasmobranch, however, is basically different in that cilia are lacking on the dendritic ends of otherwise typical bipolar cells.

[*] Work on elasmobranchs was done at the Mote Marine Laboratory, Sarasota, Florida.

Previous reports that tight junctions join the apical ends of bipolar and supporting cells (Reese, 1965) raised the possibility that cells in the olfactory epithelium are electrically coupled (Robertson, 1963). More recently, two types of tight junction have been recognized: true tight junctions which occlude extracellular spaces, and gap junctions which appear to be sites of cell-to-cell electrical coupling (e.g. Brightman and Reese, 1969). We have previously shown that true tight junctions but not gap junctions are present in frog olfactory epithelium (Brightman and Reese, 1969). Here we report on the primate and the elasmobranch and in the latter demonstrate that tight junctions prevent protein injected intravenously from reaching the olfactory surface.

A variety of neuronal processes, besides axons from bipolar cells, enter the olfactory glomerulus. Dendrites from mitral, tufted and small periglomerular granule cells as well as axons, probably few in number, from various other sources contribute to the complex glomerular neuropil (Blanes, 1897; Ramon y Cajal, 1911). Thus, besides being a location for synaptic interactions of afferent axons from bipolar neurons with second-order cells in the olfactory bulb, the glomerulus might also be a site for synaptic interactions among these components. We have presented a preliminary report on such "dendro-dendritic" synapses within the glomerulus (Reese and Brightman, 1965) and here give our results in detail. Some unusual morphological features of the numerous, small periglomerular neurons are also discussed.

METHODS

Macaque monkeys (*Macaca mulatta*) three years of age, sexually immature nurse sharks (*Ginglyomostoma cirratum*) weighing 1–5 kg, and adult guitar fish (*Rhinobatus lentiginosus*) weighing 0·5–1 kg were used for studies of olfactory epithelium. The olfactory epithelia were fixed by a two-stage perfusion of buffered aldehydes through the aorta (Reese and Karnovsky, 1967).* For elasmobranchs, the first-stage fixative contained 0·14 M-sodium cacodylate, 0·20 M-sodium chloride, 1·25 per cent glutaraldehyde, 1·0 per cent formaldehyde, and 50 mg per cent calcium chloride. In the second-stage fixative the concentration of glutaraldehyde was increased to 5·0 per cent and that of formaldehyde to 4·0 per cent. Elasmobranchs were injected intravenously with 0·25–1 g per kg body weight of horse-radish peroxidase (Sigma type II) one hour before fixation. Histochemical

* Heads of macaque monkeys perfused in this manner were kindly supplied by Dr Ronald Clark.

demonstration of peroxidase activity by Karnovsky's method (1967) was followed by fixation in osmium, treatment of blocks with aqueous uranyl acetate (Karnovsky, 1967), dehydration in alcohols and acetone, and embedment in Araldite (CIBA Ltd, Duxford, England).

Olfactory bulbs were studied in young adult white rats, some of which were given 150 mg acetazolamide per kg body weight one hour before fixation or dehydrated to 80–90 per cent of their normal body weight by giving them 3 per cent sodium chloride solution to drink. They were then perfused through the heart with balanced salt solution and buffered osmium by techniques that we have described previously (Lenn and Reese, 1966). Slices of olfactory bulb were embedded directly in Maraglas (Marblette Corp., Long Island, New York).

All sections were stained with lead citrate and examined with an RCA 3E or AEI 6B electron microscope.

RESULTS

Olfactory epithelium of the monkey

Flat sheets of fixed olfactory epithelium from the nasal septum, near the cribiform plate, were prepared for electron microscopy. The layer of mucus covering the fixed epithelium was 10–25 μm thick but might be expected to be twice as thick in fresh material (Reese, 1965). Two types of mucus were seen in electron micrographs: patches of fibrillar material extending from the cellular surface up to an outer limiting band, several micrometres thick, of a second, denser type of mucus (Fig. 1). Just below this layer of condensed mucus were bundles of 10–30 distal segments of olfactory cilia containing two or more subfibres, two being the most common number (Fig. 2). Frequent variations in these cilia were redundant folds of ciliary membrane, splits in the array of subfibres, and large vesicles enclosed by the ciliary membrane (Fig. 2, inset). Closer to the cellular surface of the epithelium the pattern of subfibres was the typical "9+2" found in a variety of motile types of cilia and characteristic of the proximal segments of olfactory cilia (Reese, 1965). These proximal segments presumably connect distal segments with basal bodies in the dendritic tips of bipolar cells, as they do in the frog where they have been studied in serial sections (Reese, 1965). The numbers of cross-sections of distal segments in the monkey were markedly fewer than in the frog. This could be due to three factors: fewer cilia per dendrite, lower density of dendrites, and shorter cilia. Of these three factors, the last probably accounts for the largest part of the difference, because the first two seem roughly the same in the frog

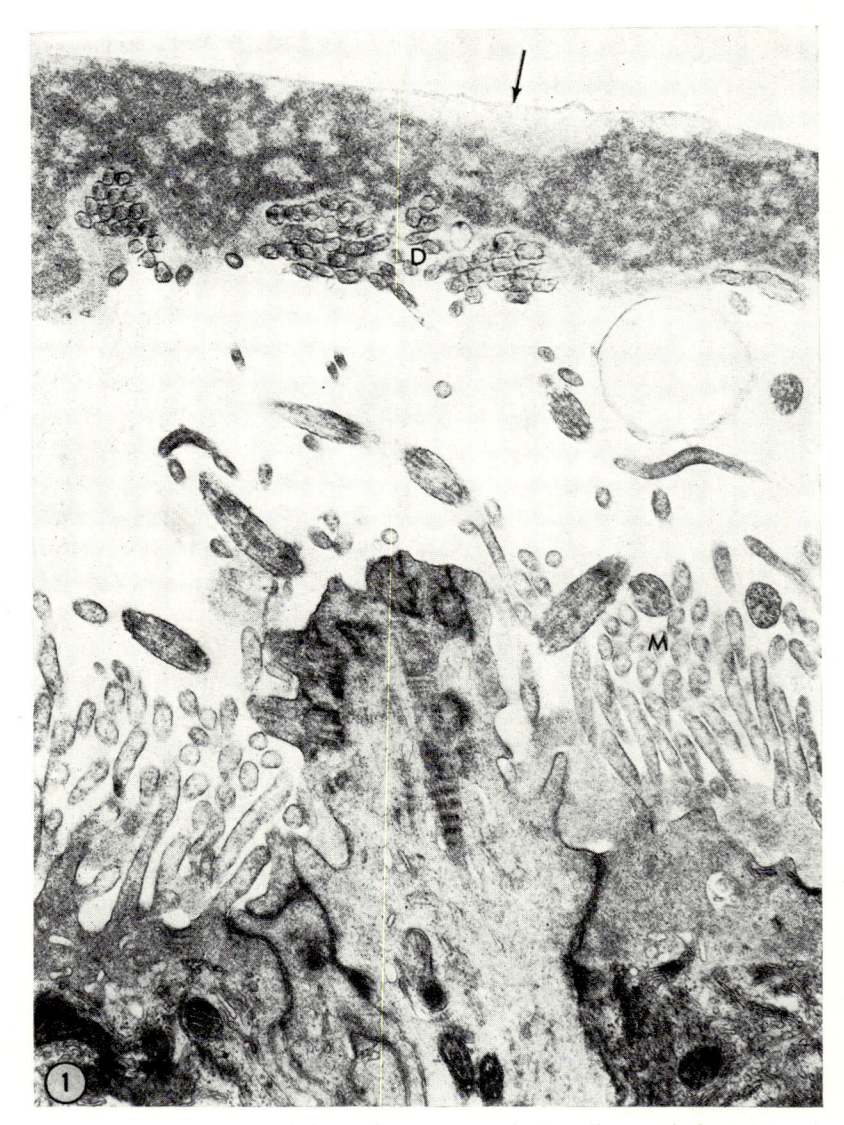

Fig. 1. Olfactory epithelium of macaque monkey. Bulbous end of apical dendrite from an olfactory bipolar cell is flanked by supporting cells. Ciliary rootlets and basal bodies in the dendrite of the bipolar cell are cut in various planes. The distal segments (D) of the olfactory cilia lie in clusters parallel to the surface of the mucus (arrow) covering the epithelium. Rounded profiles lacking central subfibres belong to microvilli (M) which arise from supporting cells. ×23 000

and the monkey. Another feature of the olfactory cilia in the monkey is that they are all cut at approximately the same angle in any plane of section, implying that, as in the frog, their distal segments lie in parallel arrays near the surface of the mucus. Supporting cells possess many villi which may be up to 10 μm long, reaching more than halfway across the overlying

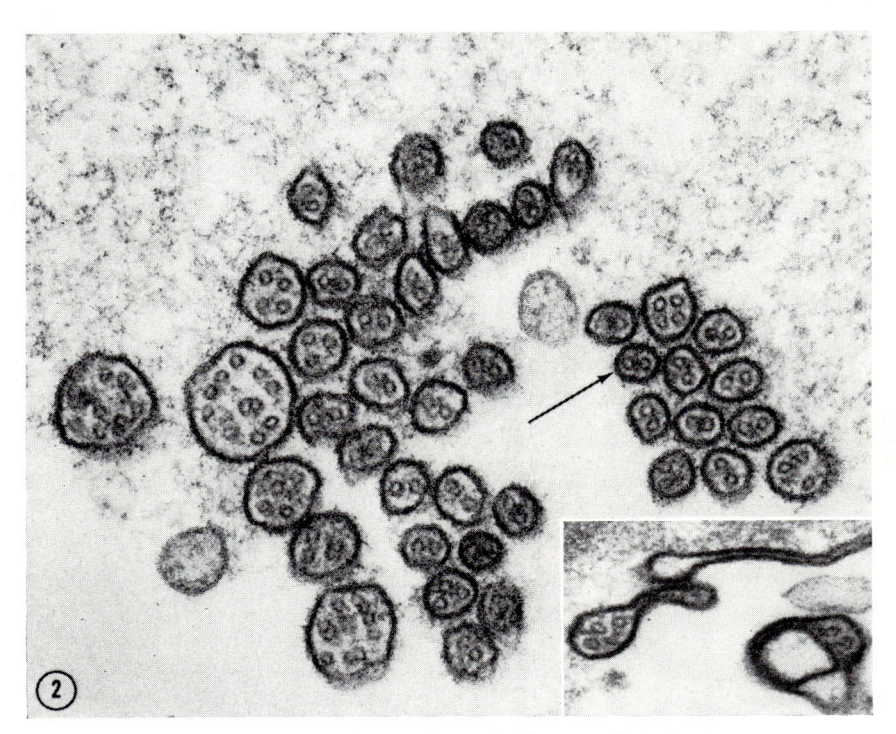

FIG. 2. Cross-section through a cluster of olfactory cilia in the macaque monkey. The predominant pattern, presumably for most of the length of the ciliary distal segments, is a single pair of tubular subfibres (arrow) that are joined to each other and to the cell membrane by a fuzzy material. ×83 000
Inset shows distal segments with an included vesicle (lower right) and with redundant cell membranes (upper left). ×83 000

mucus layer (Fig. 1). A third component of the olfactory surface consists of small villi occasionally found on the dendritic tips of bipolar cells (Fig. 1).

The tight junctions between the apices of bipolar and supporting cells, near the overlying layer of mucus, were examined after treatment with aqueous uranyl acetate to show details of the relationship between outer leaflets of adjacent cell membranes. An extensive system of true tight junctions joined the apical ends of supporting and bipolar cells, occluding all visible extracellular space at these points (see Fig. 5). Because tight junctions were found in every plane of section, it is inferred that they form complete encircling belts branching from one pair of cells to the next. Below we discuss the application of tracers to similar junctions in the elasmobranch (Fig. 5, inset) and give additional evidence that these junctions completely seal off the underlying extracellular spaces from the overlying mucus of the olfactory surface.

Olfactory epithelium of the elasmobranch

The olfactory sac in elasmobranchs contained multiple parallel partitions or septae separated by seawater-filled slits. Three distinct types of epithelium covered these septae. The epithelium capping the apex of the septum contained no neuronal cell bodies. Lining the troughs at the base of the septae were folds of a second type of epithelium, that appeared to be olfactory (see Discussion, p. 138). An unusual feature was the large size of its bipolar cell somata (Gerebtzoff, 1953), approximately 15 μm across and 20 μm deep. The apical "dendrites" of these cells ended as small villous tufts that extended into the ambient seawater. On the sides of each septum were flat-topped ridges oriented parallel to the flow of seawater and these ridges were covered with a third, olfactory type of epithelium composed of olfactory bipolar neurons, supporting cells and basal cells (Fig. 3). The round nuclei of typical bipolar neurons lay deep to a row of supporting cells with oval nuclei. The bipolar cells had a single apical dendrite protruding above the surface and ending as an enlarged bulb. A surprising feature of this dendritic ending in the elasmobranch was that short villous protrusions but no cilia originated from it (Figs. 3 and 4). At the apical border of supporting cells were villous protrusions, similar to those on bipolar cells, and cilia of a typical motile variety with deep rootlets and a "9+2" pattern of subfibres (Figs. 3 and 4). Supporting cells did not contain secretory droplets but were readily distinguished from bipolar neurons because mitochondria in the latter were consistently smaller. No glands lay under the olfactory epithelium and no mucus was observed at the olfactory surface. Here and there in the epithelium were a few goblet cells but their secretions might be expected to be washed away in the flow of seawater directly over the surface of the epithelium. Thus the olfactory surface in the elasmobranchs consisted of cilia from supporting cells and, close to the cellular surface of the epithelium, a mat of microvilli originating from supporting and olfactory bipolar cells.

When horseradish peroxidase was injected into the circulation one hour before fixation and appropriate histochemical techniques were applied, a

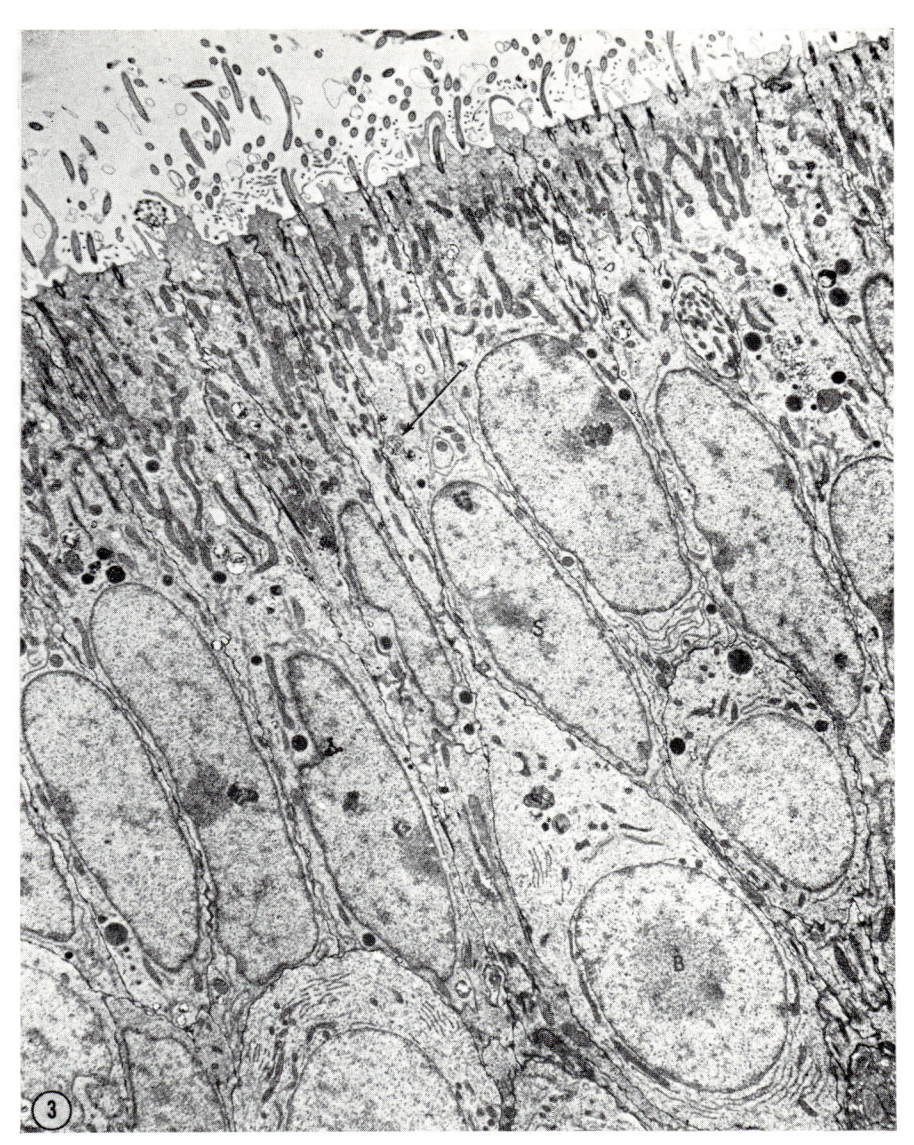

Fig. 3. Nearly the complete thickness of the olfactory epithelium from a guitar fish. Round nuclei at lower right belong to olfactory bipolar cells (B) whereas oval nuclei near surface of the epithelium belong to supporting cells (S). Mitochondria in supporting cells are larger than those in bipolar cells. The apical dendrite of one bipolar cell (B') can be traced to its end at the surface of the epithelium. In the elasmobranch the protruding ends of the apical dendrites of bipolar cells have a few villous extensions instead of cilia. Similar villi originate from supporting cells and together with those on olfactory cells account for the many small processes found at the surface of the epithelium. Typical motile cilia as well as villi originate from the distal ends of supporting cells. Animal injected with horseradish peroxidase; epithelium incubated for demonstration of peroxidase. Reaction product is apparent in large vacuoles (arrow) in supporting cell marked (S). ×3800

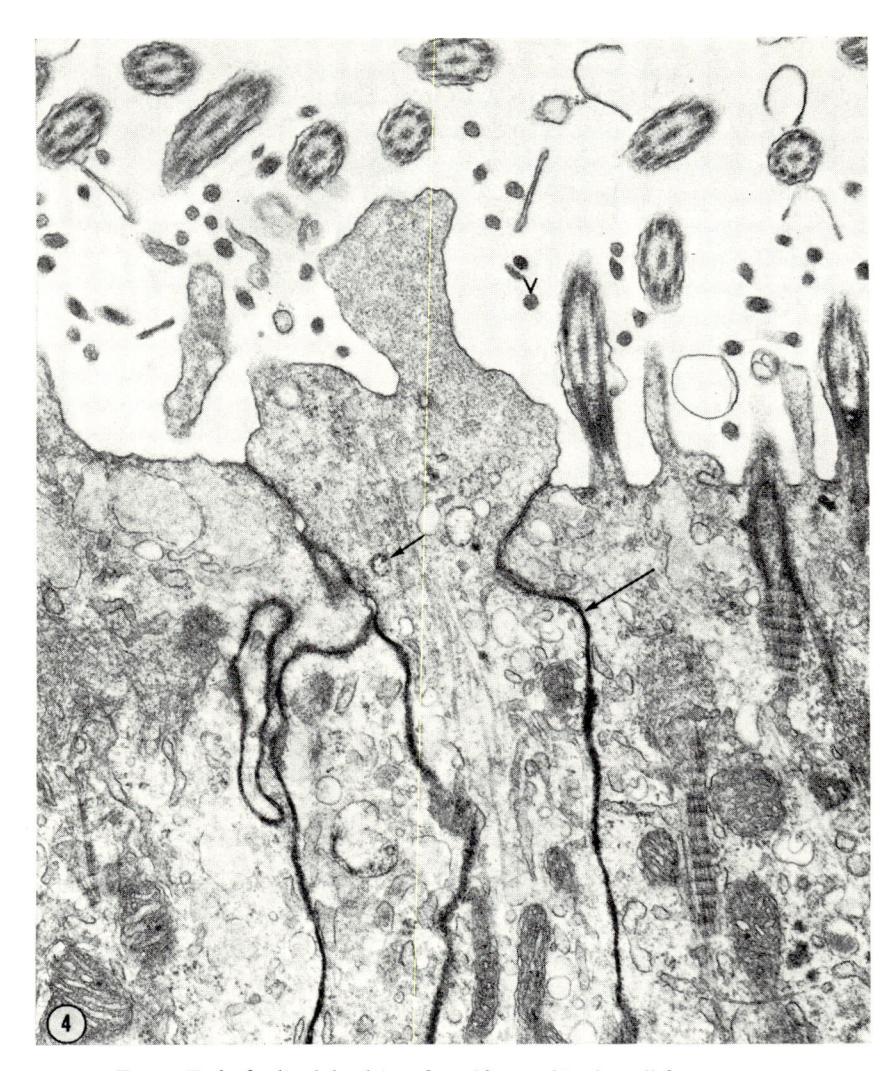

FIG. 4. End of a distal dendrite of an olfactory bipolar cell from a guitar fish that was injected intravenously with peroxidase. Peroxidase occupies the interstitial spaces between the bipolar cell and adjacent supporting cells but is prevented from reaching the surface of epithelium by encircling tight junctions (at longer arrow, shown in more detail in Fig. 5, inset). Peroxidase is also localized in vacuoles in bipolar cells (short arrow). Tips of dendrites of bipolar cells lack cilia and contain numerous small particles but in other respects closely resemble those in other vertebrates. Small villous processes (V) originating from bipolar and supporting cells are presumably the source of the numerous small processes lying near the surface of the epithelium. ×27 000

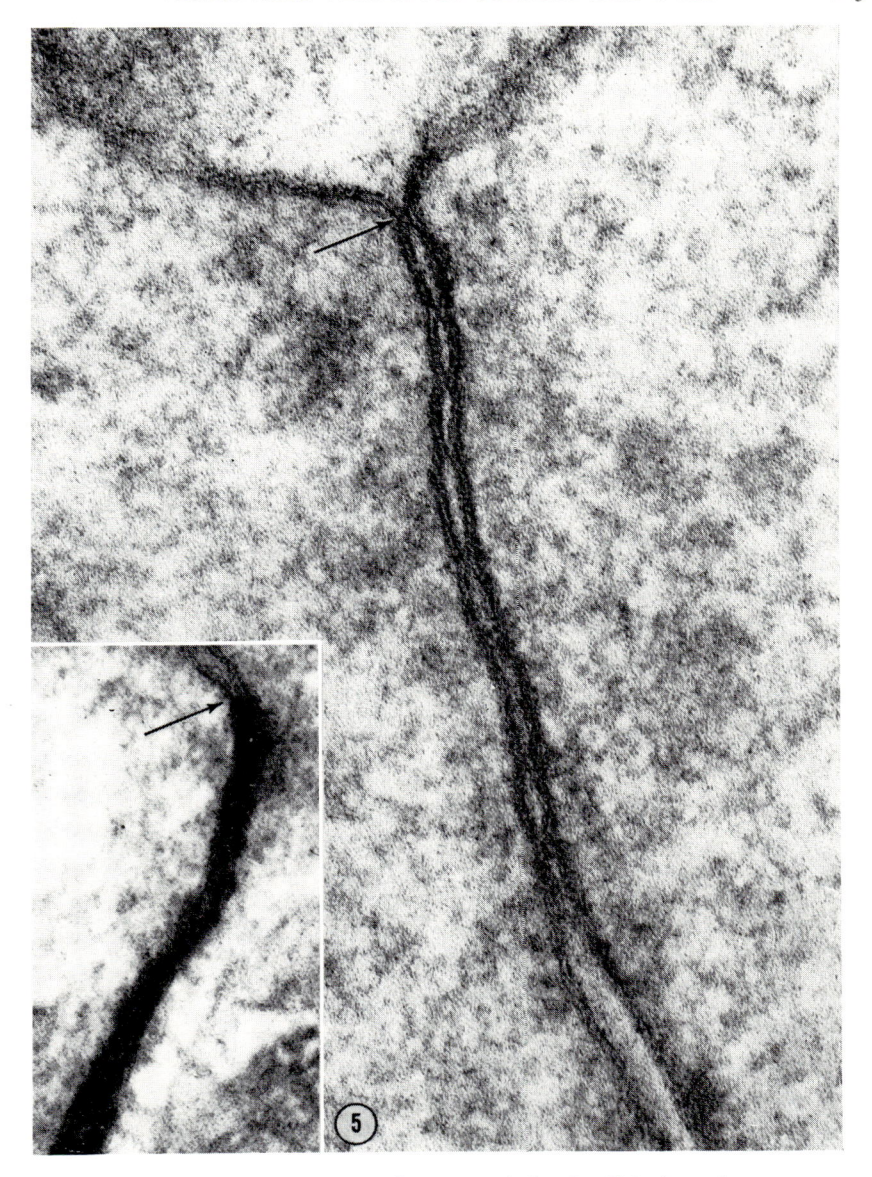

FIG. 5. Row of tight junctions between apical ends of bipolar and supporting cells in the macaque monkey. These junctions are fusions between adjacent cell membranes (arrow) and appear to form continuous belts around supporting and bipolar cells in vertebrate olfactory epithelium. Peroxidase injected into the blood can percolate through the interspaces in the epithelium, stopping only at these apical belts of tight junctions, as shown at the arrow in the inset and in Figs. 3 and 4. ×240 000
Inset is from guitar fish injected with peroxidase. ×160 000

reaction product indicative of peroxidase activity filled all the interspaces in the epithelium below the apical junctions between supporting and bipolar cells (Figs. 3 and 4). Reaction product was also included in vacuoles in supporting cells (Fig. 3) and bipolar cells (Fig. 4). The junctions between supporting cells and bipolar cells in the elasmobranch are true tight junctions, like those in the primate (Fig. 5) and the frog, and at high magnifications are seen to be the sole structure blocking the spread of peroxidase across the apical surface of the epithelium (Fig. 4 and Fig. 5, inset). The implication of this finding is that the tight junctions here as in other epithelia (Farquhar and Palade, 1963) are continuous belts surrounding supporting and bipolar cells, branching from one pair of cells to the next. Substances in the blood would have access to the lateral and basal surfaces of bipolar and supporting cells but not to the olfactory surface, whereas substances in ambient seawater would have access to the olfactory surface only.

Olfactory glomerulus of the rat

The olfactory glomeruli are fields of fine neuropil lying near the surface of the olfactory bulb. Within this neuropil, axons from bipolar neurons in the olfactory epithelium are thought to end on dendrites of mitral and, probably, of tufted and small periglomerular cells (Ramon y Cajal, 1911). The olfactory glomeruli and other layers of the bulb were identified in plastic sections examined with the light microscope. In some rats, the layer of axons covering the bulb was thin and vacuolated and the glomeruli were small. Such animals were eliminated from our series. One or two rows of glomeruli usually lay between the superficial layer of nerve fibres from the olfactory epithelium and the deeper external plexiform layer of the olfactory bulb. Each glomerulus was a roughly circular region of neuropil up to 150 μm in diameter. The glomerular neuropil was of two different types, as distinguished by the intensity of staining with toluidine blue (Richardson, Jarrett and Finke, 1960) (Fig. 6). Within the glomeruli, bundles of axons from the superficial plexus of olfactory nerve fibres were continuous with darkly stained nodules of irregular shape, whereas dendrites entering the glomeruli from the external plexiform layer lay in the lightly stained regions that surrounded the darker glomerular nodules.

Partially encircling each glomerulus was a cluster of small periglomerular cells (Fig. 6). The majority were only 5 μm to 10 μm in diameter and possessed few features to identify them as neurons or glia. Similar small cells, presumably glial, lay among the bundles of nerve fibres entering the glomeruli from the superficial nerve fibre layer. A second, larger but less common type of periglomerular cell was 10–20 μm in diameter and had abundant, lightly stained cytoplasm containing Nissl bodies. Dendrites were frequently seen arising from these cells. On the basis of their resemblance to the tufted cells lying in the external plexiform layer, it seemed likely that these were the periglomerular tufted cells. Occasionally seen were small, darkly stained cells presumed to be oligodendrocytes.

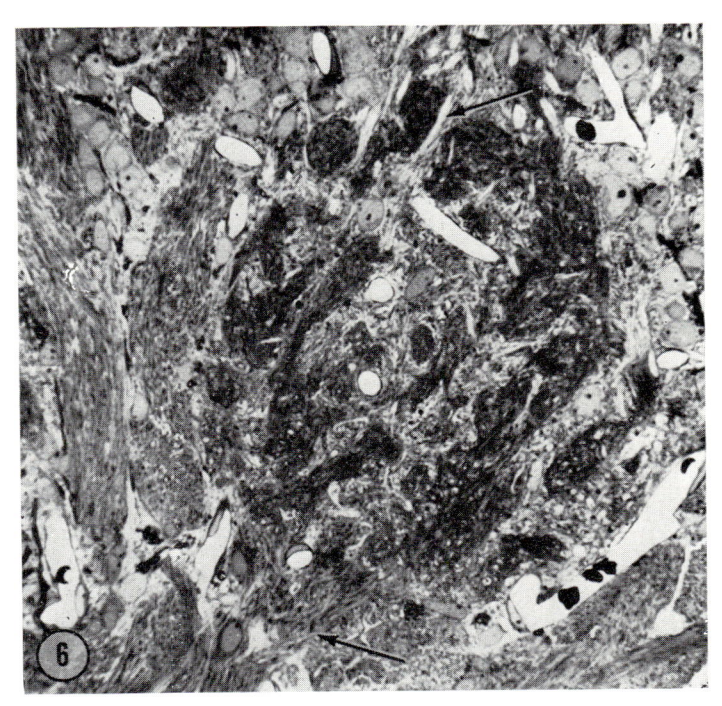

Fig. 6. Light micrograph of an olfactory glomerulus from the rat. Fascicles of olfactory nerve fibres enter from the surface of the bulb below (arrow) whereas dendrites enter from deeper layers above (arrow). Bundles of nerve fibres at left are going to a more deeply situated glomerulus. Neuropil of the glomerulus is of two distinct types: darkly stained and continuous with bundles of incoming fibres, or lightly stained and continuous with incoming dendrites. Cap of periglomerular cells, at top of picture, includes large neurons, small neurons and abundant glia. Osmium perfusion. ×530

The regions of the glomerulus distinguished with the light microscope also have a distinct appearance in electron micrographs. The bundles of incoming fibres from the superficial nerve fibre layer were composed of small, nonmyelinated axons (Fig. 7) averaging about 0·2 μm in diameter. Sheets of astroglia segregated them into fascicles containing approximately

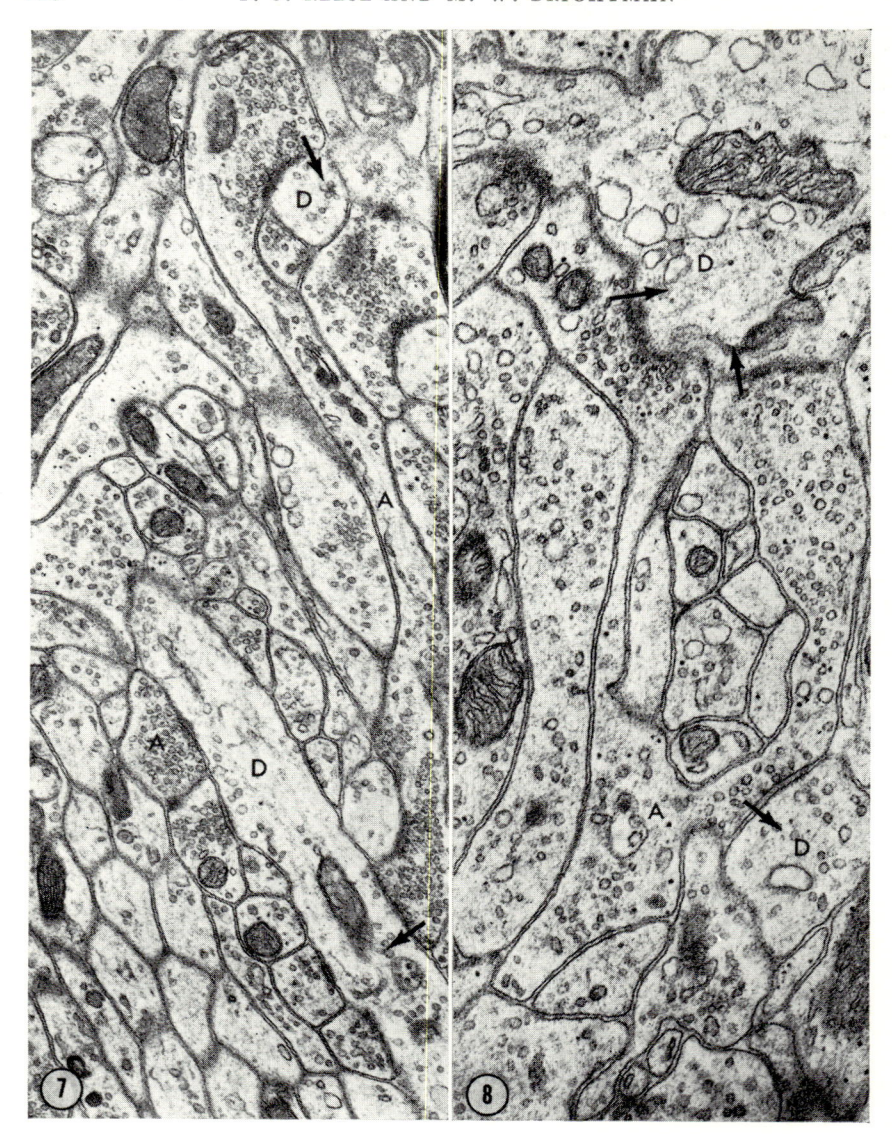

FIGS. 7 and 8. Electron micrographs of synapses in dark areas of the olfactory glomerulus of the rat (see Fig. 6). In Fig. 7 a fascicle of axons (lower A) from the olfactory nerve forms a palisade of synaptic contacts with a small dendrite (D). One axon (upper A) makes synaptic contact with two dendrites, both marked (D). ×22 000

FIG. 8 demonstrates that incoming axons (A) branch and form synaptic contacts *en passant* with dendrites (D) within the glomerulus. Arrows indicate morphological polarity and, presumably, physiological polarity of synaptic contacts. ×32 000

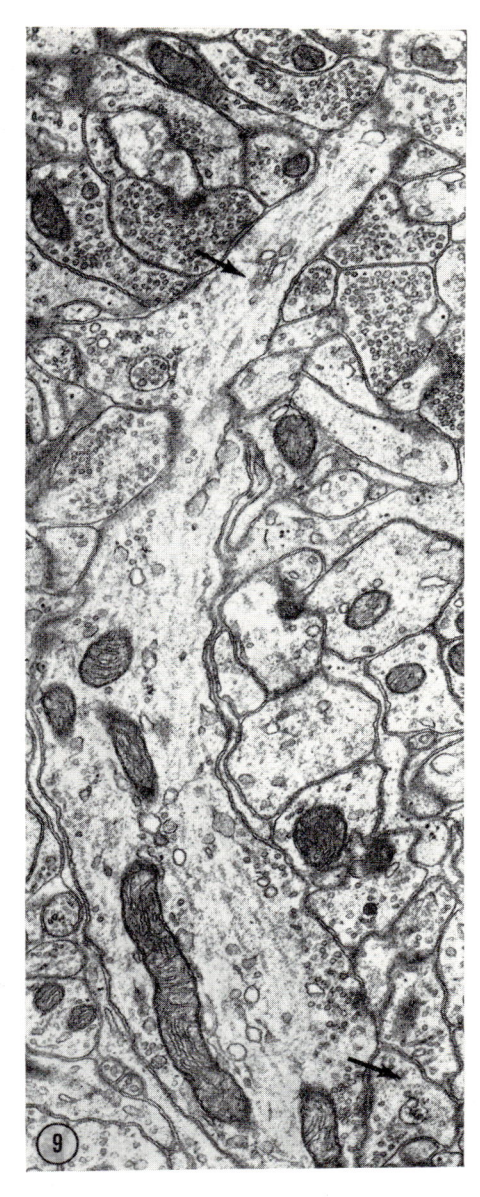

Fig. 9. Axons from the olfactory nerve (at top) make synaptic contact with a dendrite in a dark area (see Fig. 6) of the glomerulus of the rat. The morphological and presumed physiological polarity of these synaptic contacts is indicated by a large arrow. Below, the same dendrite extends into a lighter area of the glomerulus and, at arrow, makes synaptic contact with a lightly stained process. The neuropil in these areas of the glomerulus includes many large, lightly stained processes some of which are thought to be dendrites because of their synaptic relationship with axons from the olfactory nerve. c. ×22 000

25–100 axons. Prominent in their axoplasm were mitochondria, varicosed endoplasmic reticulum and faintly stained, longitudinally oriented neurotubules.

The glomerular nodules stained intensely and consisted principally of vesicle-filled endings of the incoming axons making multiple synaptic contacts with dendrites (Figs. 7–12). The dense appearance of the nodules was attributable to an amorphous but densely stained substance surrounding the vesicles in the axonal endings (Fig. 11). Glia was infrequent within the nodules but sheet-like processes of astrocytes often lay along their borders.

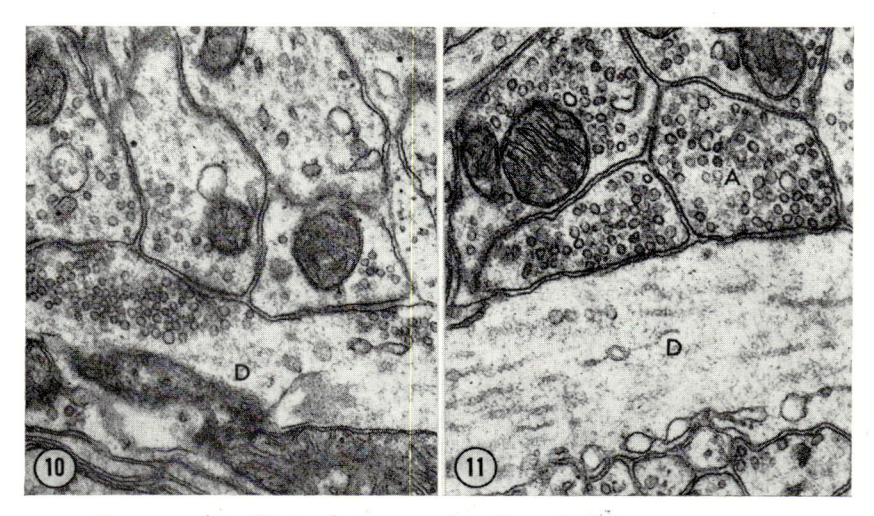

FIGS. 10 and 11. Fig. 11 shows synaptic endings of olfactory nerve axons (A) on a dendrite (D). Fig. 10 shows the same dendrite 10 μm away from the area in Fig. 11. In Fig. 10, the dendrite is presynaptic to a lightly stained process lying adjacent to it. Both ×31 000

The axonal endings lined up in characteristic palisades beside small dendrites (Fig. 7). Many axonal branches and synaptic endings *en passant* were found (Figs. 7 and 8), suggesting that each axon possessed many synaptic contacts and, weaving its way through part of the glomerulus, formed synaptic contacts with many small dendrites. This interpretation is in accord with the appearance of these endings in Golgi preparations, where branches of the incoming axons are seen to undergo tortuous terminal ramifications (Ramon y Cajal, 1911).

The lighter staining of the neuropil around the glomerular nodules was attributed to the high proportion of lightly stained processes (Figs. 9, 10, 13). Many of the lightly stained processes within and around the glomerular nodules were identified as dendrites, on the basis of their large diameter, their postsynaptic relationship with the palisades of afferent axonal endings, or because of their cytological characteristics, particularly the presence of an undulating, irregular contour, ribonucleoprotein granules, and a pale ground substance (Figs. 7–14). Three sizes of dendrite could be distinguished within the glomeruli and in favourable planes of section these sizes were found to correspond approximately to three orders of branching. The

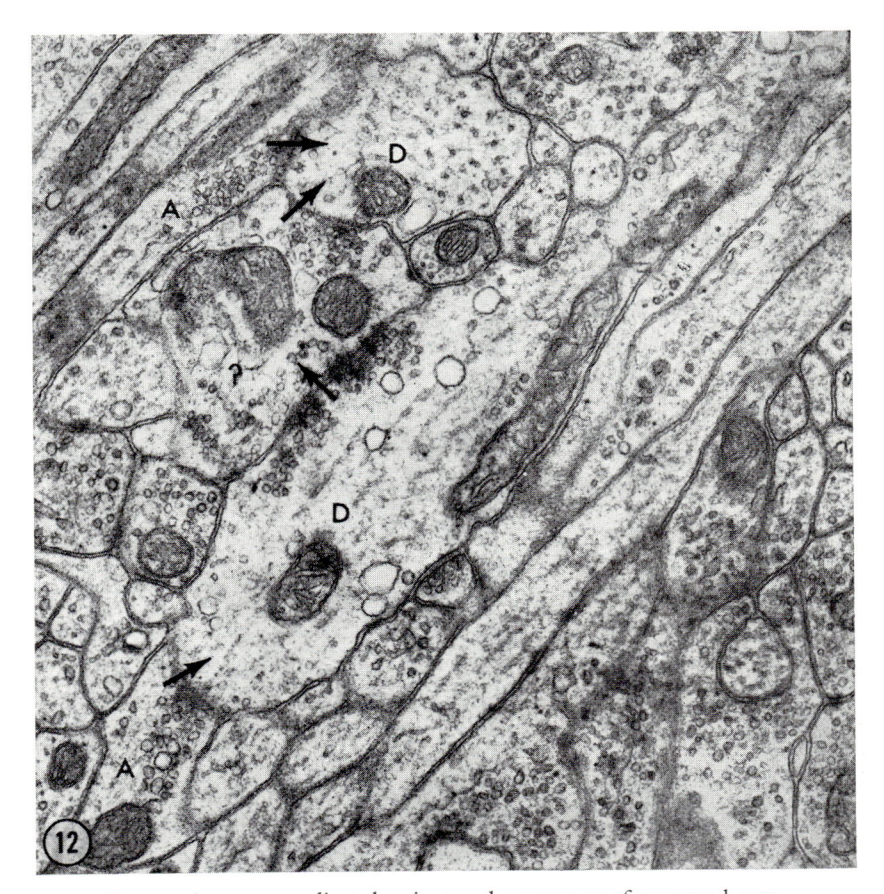

FIG. 12. A more complicated variant on the sequences of synapses shown in Figs. 9–11. Axons from the olfactory nerve (lower A) synapse on a dendrite (lower D) which in turn appears to be presynaptic to another process marked with a (?). This process in turn synapses on another process (upper D) which is thought to be a dendrite because it is postsynaptic to an axon from the olfactory nerve (upper A). The arrows indicate the morphological and presumed physiological polarity of these synaptic contacts. ×25 000

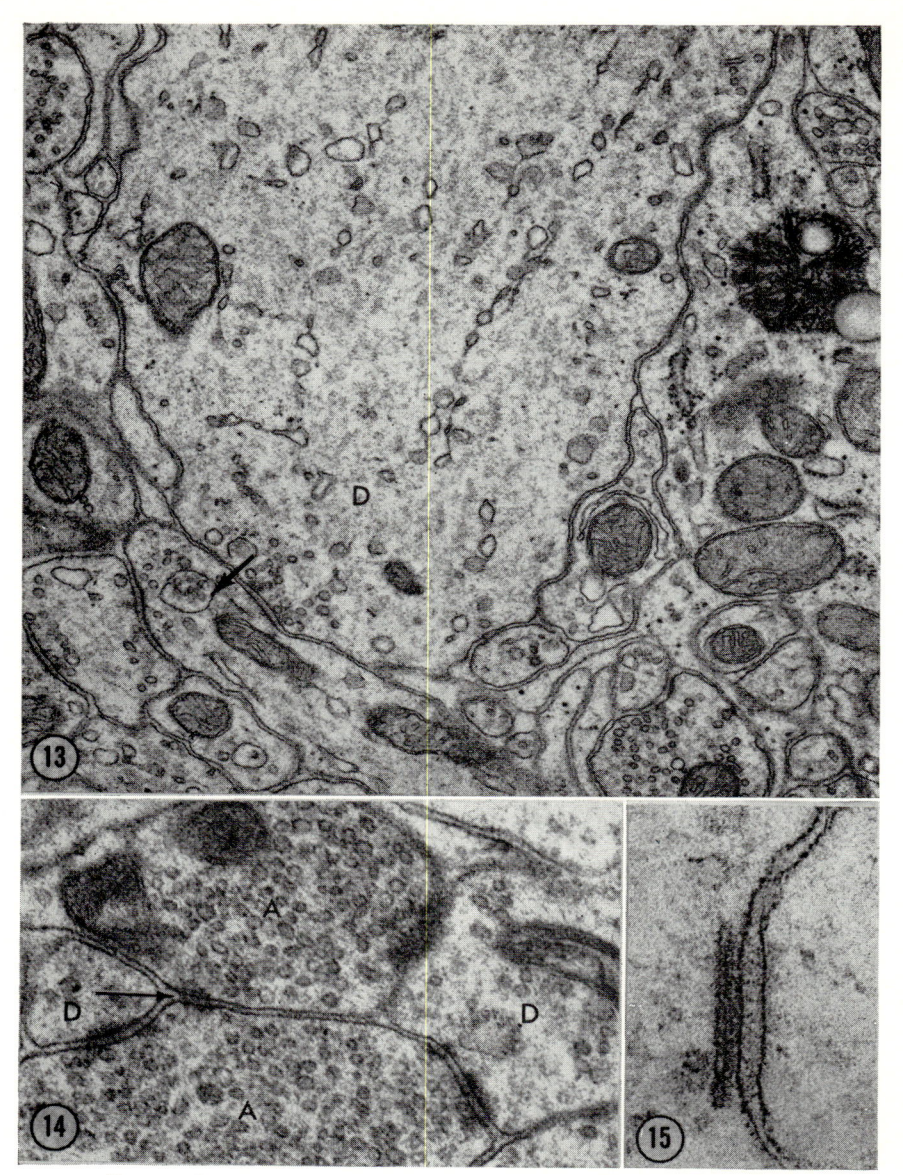

Fig. 13. Large process (D) presynaptic to another smaller process (at arrow) within light area (see Fig. 6) of the olfactory glomerulus. This process is assumed to be a dendrite because of its large size, increasing to 5 μm in diameter somewhat beyond the area illustrated. ×29 000

Fig. 14. Two axonal synaptic endings (A) on two dendrites (D). Small arrow indicates punctate junction between axonal endings. Features which differentiate this from a synaptic contact are its short length, symmetry and lack of associated synaptic vesicles. ×35 000

Fig. 15. Structures resembling synaptic junctions associated with a subsurface cistern are occasionally found in the periglomerular region. These occur between dendrites and between dendrites and periglomerular nerve cell bodies. c. ×100 000

largest dendrites, 2–5 μm in diameter, were never found within glomerular nodules (Fig. 13). Their size and content of rosettes of ribonucleoprotein particles characterized them as trunks or main branches of dendrites entering the glomerulus. Intermediate dendrites 0·5–1·5 μm in diameter and dendritic twigs 0·1–0·4 μm in diameter were found in both the nodular and perinodular regions as well as interdigitated among fascicles of incoming axons (Figs. 7–12 and 14). Rosettes of ribonucleoprotein granules were sometimes present in intermediate dendrites but were lacking in the finer dendritic twigs. The identification of the smallest processes as dendrites rested mainly on their having along their borders palisades of axonal endings. Palisades of axonal endings were thought to correspond to the zones of contact between the axons of the afferent olfactory fibres and the fine terminal branches of dendritic tufts of mitral, tufted or granule cells seen in Golgi preparations (Ramon y Cajal, 1911).

A surprising finding was that, within the glomerulus, dendrites of any size could be presynaptic to other processes. Eight instances were found which combined clear identification of dendrites with accurate transverse sectioning of the dendritic synaptic contacts. In six of these instances characteristic palisades of axonal endings were presynaptic to an intermediate dendrite which, in turn, was presynaptic to another process (Figs. 9–12). The low probability of cutting both the axonal and dendritic synaptic contacts in exact transverse section and a tendency of the axonal and dendritic synaptic contacts to occur in separate regions of the dendrite accounted for the low frequency of clear examples of this synaptic pattern. In two other instances, large dendrites were clearly presynaptic to another process (Fig. 13). While this synaptic pattern appears to be infrequent within the glomerulus, these two instances show unmistakable dendrites in a presynaptic relationship to other structures. Instances of larger dendrites presynaptic to periglomerular cells were also found (Fig. 16).

The postsynaptic member of the dendritic synaptic contacts was either a small, lightly stained process, having the appearance of a dendritic twig, or a dendrite of intermediate size containing rosettes of ribonucleoprotein granules (Figs. 9, 10, 12, 13). However, no absolutely certain instances were found where both dendrites united by a synaptic contact had palisades of characteristic axonal endings along their borders. Only when synaptic contacts blurred by an oblique plane of section were also considered could several instances be found where two dendrites postsynaptic to incoming axons appeared to make synaptic contacts with each other or with an intermediate "dendritic" process (Fig. 12). It is certain, however, that dendrites within the olfactory glomeruli frequently form typical synaptic

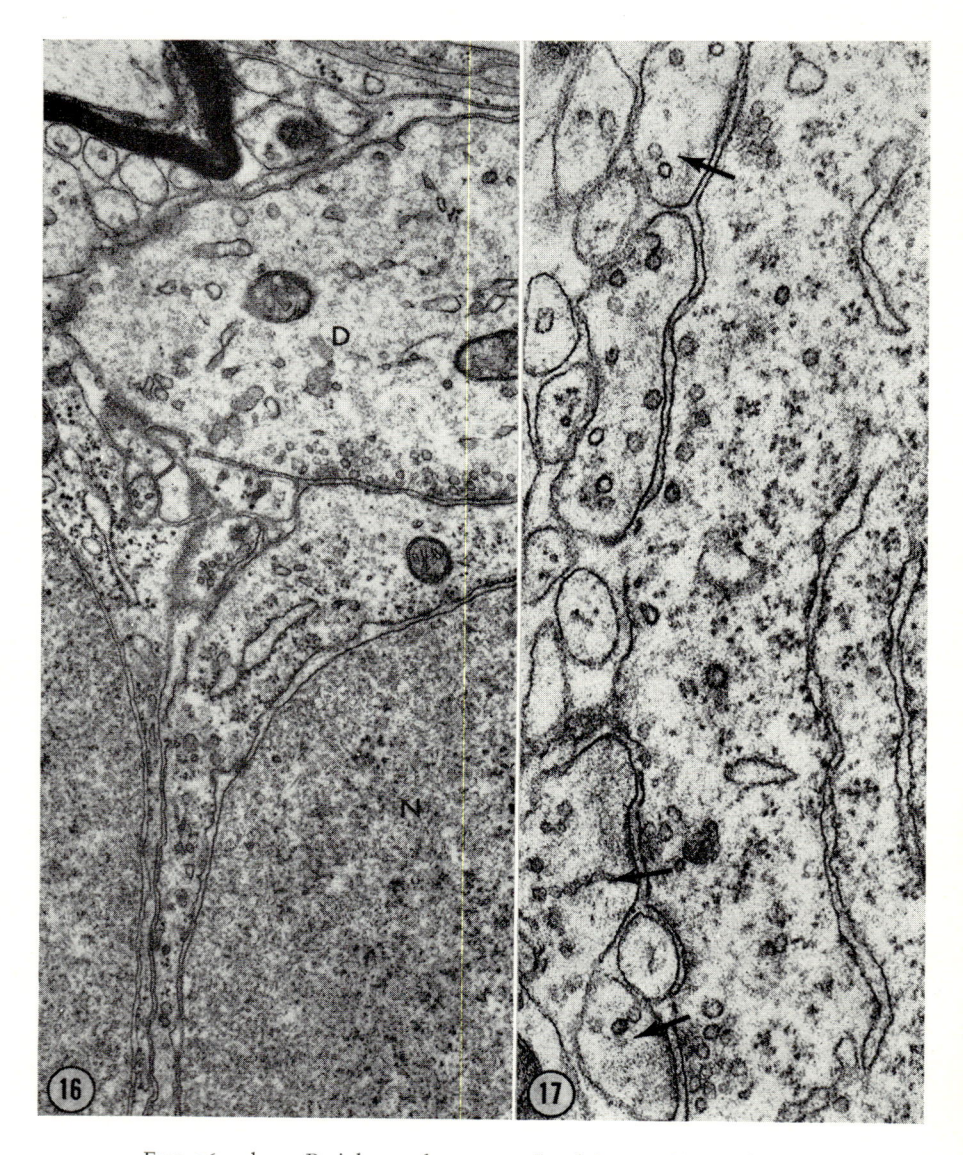

FIGS. 16 and 17. Periglomerular nerve cells of the rat. Fig. 16 shows a small periglomerular neuron (N) and a larger process (D) which forms a synaptic contact with it. This process is presumed to be a dendrite because of its large size. × 10 800

In Fig. 17, a larger cell body, probably belonging to a tufted cell, makes synaptic contact (arrows) with several small processes which also contain vesicles. c. × 54 000

contacts with other processes and that these other processes have the cytological appearance of smaller or medium-sized dendrites. It is likely that these postsynaptic "dendrites" also receive synaptic endings directly or indirectly from afferent axons but analysis of serial sections will probably be required to establish unequivocally that this synaptic pattern is a typical and frequent feature of the olfactory glomerulus.

At the low magnifications used for this study, the dendro-dendritic synaptic contacts resembled synaptic contacts which have been described in other regions of the mammalian central nervous system (Gray, 1959) (Fig. 10). Each synaptic contact consisted of an accumulation of small vesicles at a synaptic junction. In our osmium-fixed material, the vesicles were circular and 30–50 nm (300–500 Å) in diameter and no vesicles within the glomerulus were flattened in shape. Interspersed between them were irregularly shaped patches of dense material that frequently contacted the subjacent cell membrane. Triangular patches of dense material were also found contacting the presynaptic cell membrane. The synaptic junctions consisted of specializations of the cell membranes and the intervening synaptic cleft. At synaptic junctions the pre- and postsynaptic membranes had a denser appearance than the surrounding cell membrane and the intervening cleft was widened and contained a dense material. On its cytoplasmic surface, the postsynaptic cell membrane was intimately associated with a lamina of amorphous dense material. Synaptic junctions were usually 0·2–0·3 μm in length and were often bounded by regions of relative narrowing of the intercellular space. The morphological polarity of these dendro-dendritic synaptic contacts was therefore judged by two factors: a presynaptic accumulation of vesicles and patches of dense material, and a lamina of dense material in the cytoplasm subjacent to the postsynaptic cell membrane. Polarity determined by one of these two criteria was always the same as polarity determined by the other. The axo-dendritic synaptic contacts differed somewhat from the dendro-dendritic synaptic contacts (Fig. 11). The amount of cytoplasmic dense material adjacent to the postsynaptic cell membrane was usually very sparse or absent so that the accumulation of vesicles was the only feature determining morphological polarity. Since the frequency of synaptic contacts was small in comparison to the large number of vesicles in axonal profiles, it is likely that many synaptic vesicles are associated with each synaptic junction.

No instance of synaptic contact between axonal endings was found. Short *maculae adhaerentes* sometimes joined axonal endings and, occasionally, dendrites (Fig. 14) but these junctions were not confused with synaptic contacts because they were shorter, less than 0·2 μm long, had a fuzz symmetrically placed on the cytoplasmic sides of both cell membranes, and were not intimately associated with synaptic vesicles. Particularly in dendrites, which were not packed with vesicles, chance association of vesicles with *maculae adhaerentes* would be unlikely. Another type of junction, found between dendrites and between small periglomerular neurons, resembled a synaptic junction except that a subsurface cistern rather than a cluster of synaptic vesicles was adjacent to it (Fig. 15). Its function is unknown.

Four types of periglomerular cell were distinguished in electron micrographs: astrocytes, large neurons, small neurons and oligodendrocytes. The abundant, lightly stained cytoplasm of astrocytes contained fascicles of fine filaments, numerous glycogen granules, and complex lamellar dense bodies (Fig. 13). The outlines of astrocytes were irregular, conforming to the shapes of surrounding neuronal elements. Processes from astrocytes took the form of flat sheets, in places only 20 nm (200 Å) in total thickness. Astrocytic sheet processes surrounded fascicles of afferent axons in the superficial nerve fibre layer and in the glomerulus, where these fascicles lay adjacent to other regions of the neuropil. One or more layers of astrocytic sheets also surrounded the glomerular nodules and often partially covered the periglomerular neurons.

The large periglomerular neurons, probably tufted cells, were few in number but had easily recognizable features of neurons: subsurface cisterns, stacks of granular endoplasmic reticulum, and synapses on their soma. However, synaptic contacts with their associated synaptic vesicles in the cell body were sometimes found on small processes that also contained synaptic vesicles (Fig. 17). These endings resembled those on mitral dendrites (Rall *et al.*, 1966) where two synaptic contacts are oriented in opposite directions; such a system of dual synapses is thought to occur on tufted cells (Andres, 1965).

Small neurons appeared to be more numerous than astrocytes in the periglomerular region. Their round nucleus was surrounded by a remarkably thin rim of cytoplasm which lacked glycogen granules but contained numerous ribonucleoprotein granules and a dense matrix, absent in the cytoplasm of astrocytes (Fig. 19). In some planes of section, a region with more abundant cytoplasm was found and from this region a thick process often arose. These processes, often pre- as well as postsynaptic to other small processes, were dendrites as judged by their large diameter and content of rosettes of ribonucleoprotein granules. In a few instances, the cell body itself was apparently postsynaptic to a medium-sized dendrite (Fig. 16). A second type of process arose directly from dendrites or the body of

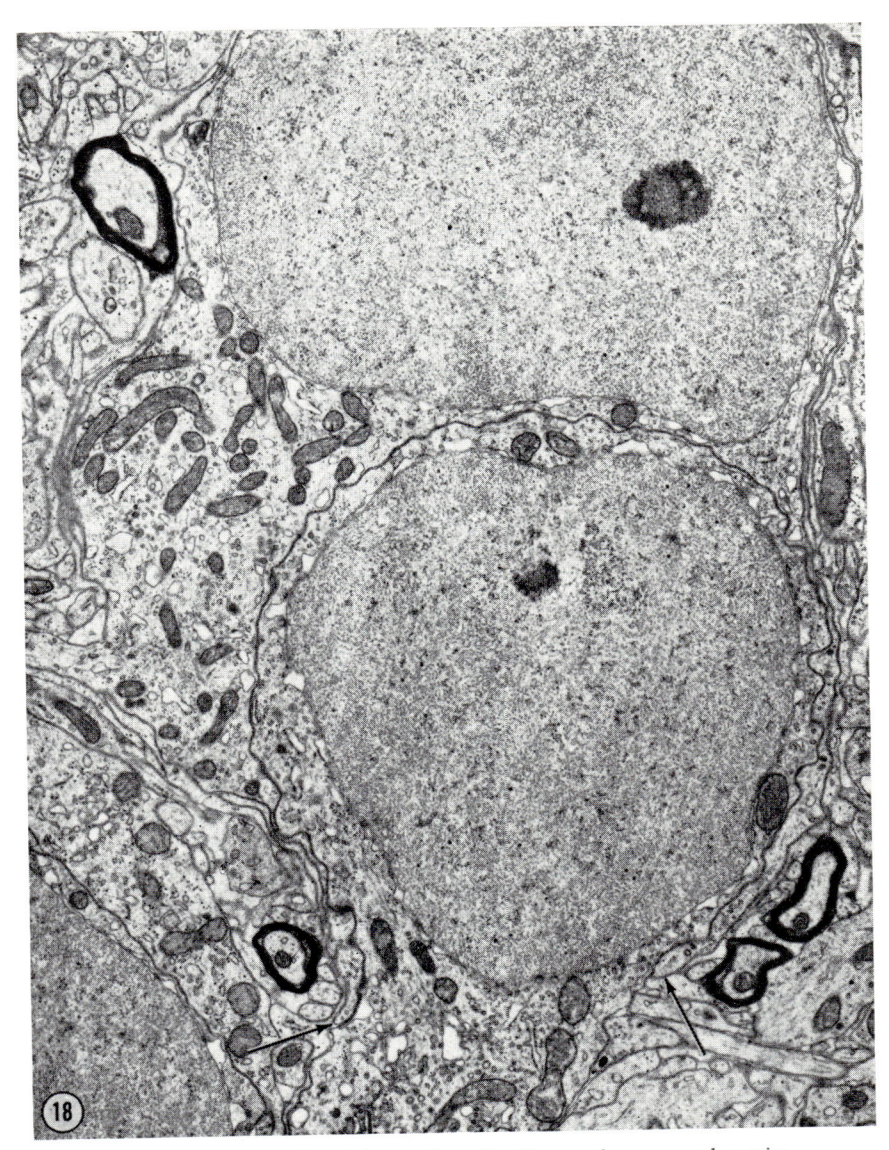

FIG. 18. Two small periglomerular cells. Flattened processes that arise from the upper cell enclose the lower cell and end at arrows. The lower cell contains a subsurface cistern (arrow at left) and has other features which identify it as a small periglomerular neuron. The upper cell closely resembles the lower one and is thought to be a neuron. × 10 000

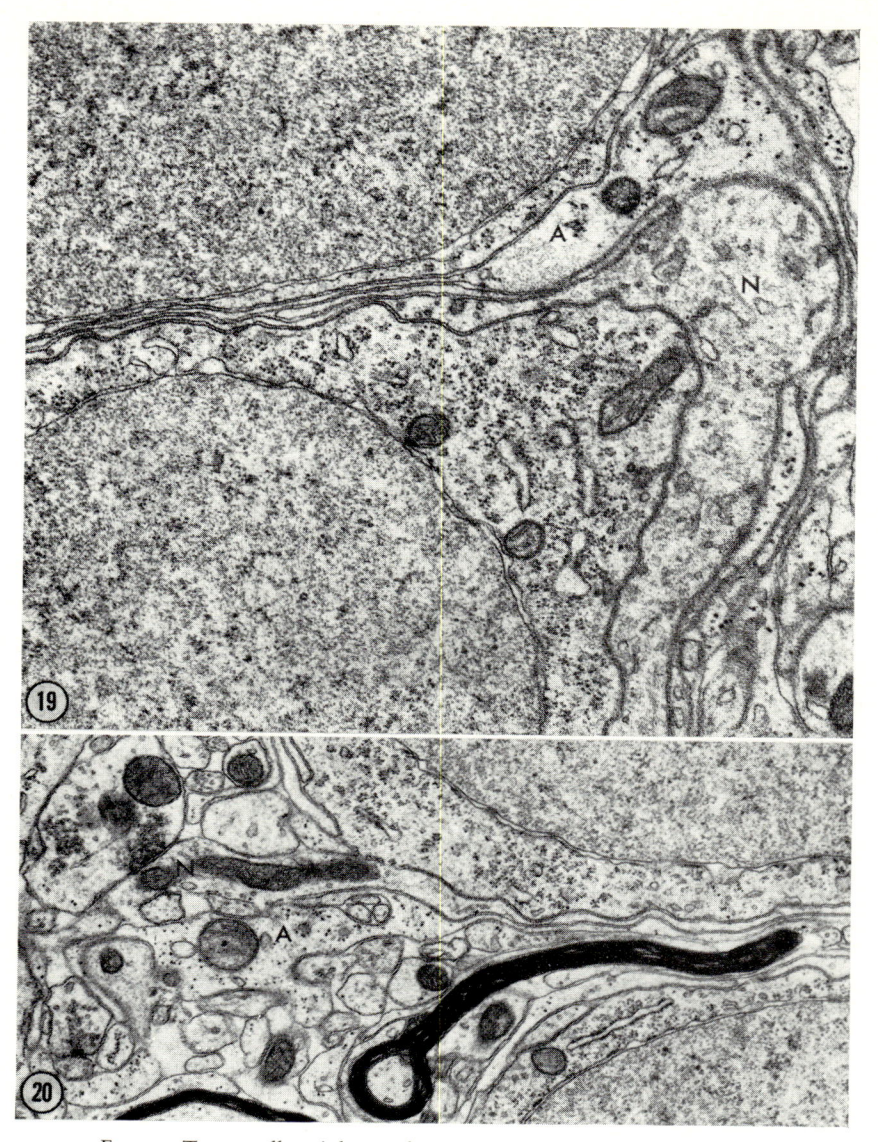

Fig. 19. Two small periglomerular neurons with intervening processes. The upper process (A) contains glycogen granules and bundles of small filaments which identify it as belonging to an astrocyte. The lower process (N) lacks glycogen or filaments and has a dense cytoplasmic matrix which differentiates it from the adjacent astrocytic process. As shown in Fig. 20, such processes can be neuronal and presumably originate from cells like those shown in Fig. 18. ×20 000

Fig. 20. Two small periglomerular neurons with intervening processes (N, A) like those in Fig. 19. Synaptic ending on upper process (N) definitely identifies it as neuronal. In Figs. 18–20 the cell membranes of the neuronal process appear somewhat darker where they contact the adjacent cell. ×17 000

cells presumed, on the basis of their cytoplasmic characteristics, to be small periglomerular neurons. These processes were very thin and resembled astrocytic sheet processes in thickness and extent, forming sheets of flattened fingers rather than having the tubular shape usually associated with neuronal processes (Fig. 18). These neuronal sheets were often wrapped around adjacent small neurons (Fig. 18) and were distinguished from astrocytic sheets by having a darker cytoplasmic matrix lacking glycogen granules (Fig. 19). In these regions of neuronal-neuronal contact, the cell membranes appeared denser but the extracellular space was not bridged by any extracellular fuzz nor was it narrower than is usually found between neurons. The most positive evidence that these perineuronal sheets were also neuronal came from finding three instances of synaptic endings on them or on processes giving rise to them (Fig. 20). The source of these synaptic endings was not determined nor was the relative frequency of neuronal and astrocytic sheet processes, although it was clear that a significant number of small periglomerular neurons are partially wrapped in neuronal sheet processes. At other than perineuronal locations, astrocytic sheet processes are the rule in the neuropil around the olfactory glomerulus.

DISCUSSION

Olfactory epithelium

The olfactory receptor surface in the primate resembles that in other air-breathing vertebrates (Reese, 1965; Andres, 1969). Long cilia originate from dendrites of neuroepithelial cells, cross the covering layer of mucus, and lie in parallel bundles near the air–mucus interface. These cilia have a short proximal segment containing the "9+2" array of subfibres typical of motile cilia (Barnes, 1961) and a longer distal segment, usually containing two subfibres. The presence of similar specialized neurocilia at the surface of the olfactory epithelium in a variety of vertebrates suggests that these cilia have a specific function in the olfactory process.

It has long been supposed that olfactory cilia might be the site where odorous substances initiate events leading to excitation of the olfactory nerve (Hopkins, 1926). Several observations favour this supposition. First is the close proximity of the distal segments of the olfactory hairs to the air–mucus interface, assuring them of rapid contact with air-borne odours (Reese, 1965). Also, cilia are found in other locations which suggest that they initiate excitation in a variety of sense organs (Reese, 1965). A particularly pertinent comparison is to the olfactory hair found in insects (e.g. Slifer and Sekhon, 1964). These hairs are hollow cylinders of cuticle containing neurocilia resembling those in vertebrates. The insect bipolar cells and olfactory cilia are separated from air-borne odours by the cuticle except in the immediate vicinity of the distal segments of the neurocilia where the cuticle is punctured by minute pores. These pores assure rapid contact of odorant materials with the distal segments of olfactory cilia but not with other components of the olfactory organ.

A third consideration is whether a cylinder with the length and diameter of an olfactory cilium could initiate and conduct an effective electrical potential to the bipolar dendrite. From a theoretical consideration of this problem, Ottoson and Shepherd (1967) concluded that such activity in cilia is feasible if it is assumed that they have special properties, like those of olfactory nerve fibres. Certain objections to the supposition that the cilia initiate olfaction have also been discussed by Ottoson and Shepherd (1967).

One of the less common types of olfactory bipolar neuron in teleosts lacks cilia (Bannister, 1965). Bipolar neurons in vertebrate vomeronasal organs also lack cilia (Graziadei and Tucker, 1970). The olfactory epithelium in elasmobranchs appears to be a third example of non-ciliated olfactory bipolar cells and the first example of a primary olfactory epithelium lacking cilia on virtually all of its bipolar cells. It would be interesting to know whether chemical stimulation of such a non-ciliated epithelium would generate a slow potential similar to that generated by ciliated olfactory epithelium (Mac Leod, 1959; Ottoson, 1956), but such data are lacking. There is, however, ample evidence that elasmobranchs are capable of sensitive behavioural responses to water-borne odorants and that these responses are dependent on the integrity of their olfactory epithelium (Hodgson, Mathewson and Gilbert, 1967). The lack of olfactory cilia on this epithelium shows that ciliary basal bodies and axial fibres need not be regarded as necessary features of the olfactory process. In air-breathing vertebrates the association of cilia with olfactory bipolar cells would provide a stiff scaffolding to support, near the air–mucus interface, an expanse of plasma membrane specialized for olfaction. We therefore place the emphasis on the cell membrane as the most likely site for the initiation of olfactory stimulation. Whether this membrane is wrapped around a cilium or a microvillus might be regarded as alternative ways to put sensitive membrane in the vicinity of air- or water-borne odorant materials. Whether, under normal conditions, the ciliary axial fibres produce any ciliary motion is not known but direct observation of frog olfactory epithelium *in vivo* shows that these cilia do not have the beating motion typical of other cilia (Reese, 1965). However, the ciliary fibres in proximal segments might provide for slow undulations or gradual rearrangements of distal segments.

Recent advances in electron microscopic techniques have made it possible to distinguish two types of tight junction (Farquhar and Palade, 1963; Revel and Karnovsky, 1967; Brightman and Reese, 1969). One type, which continues to be called a tight junction, actually occludes spaces between cells. Furthermore, this type of tight junction surrounds cells and branches around neighbouring pairs of cells. In order to cross a sheet of cells joined by tight junctions, a substance must either diffuse through the junction or enter and leave the cells. Some tight junctions have been shown to be impervious to molecules of molecular weight as low as 1800 (Feder, Reese and Brightman, 1969) and, on the basis of indirect evidence, Farquhar and Palade (1965) have suggested that tight junctions might impede movement of even smaller substances. Distinct from this barrier type of tight junction are gap junctions, so named because they are bisected by a narrow gap. Gap junctions do not surround cells and therefore do not prevent diffusion of substances across sheets of cells. Instead, gap junctions appear to be sites of electrical cell-to-cell coupling. Tight, not gap, junctions are now known to be present bordering the apices of bipolar and supporting cells in the primate, frog (Reese, 1965; Brightman and Reese, 1969) and elasmobranch. We have also shown in the elasmobranch that these junctions prevent circulating protein (molecular weight 42 000) from reaching the olfactory surface and that the tight junctions at the apical border of the olfactory epithelium are the only barrier between the olfactory surface and subepithelial interstitial fluid. Flow of extracellular current across the olfactory receptor surface, such as might occur during activation of bipolar cells, would probably encounter an increase in resistance of the extracellular pathway at the olfactory surface so that much of this current would flow through supporting cells, as suggested by Mac Leod (1959). The villi characteristically present on olfactory supporting cells would serve to increase the area of permeable cell membrane available to electrical currents flowing from the olfactory surface to the deeper layers of the epithelium.

Olfactory glomerulus

The principal morphological finding in the glomerulus of the rat olfactory bulb is that dendrites postsynaptic to endings of afferent bipolar cell axons are presynaptic to other neuronal processes, probably dendrites. The terms pre- and postsynaptic are used here in reference to the morphological polarity of synaptic contacts, as discussed below. Synaptic contact between dendrites appears to be a regular feature of the olfactory glomerulus although we cannot, as yet, fit these synaptic relationships into an overall diagram of glomerular synaptic organization. We do not know, for instance, whether dendrites from mitral, tufted and peripheral granule cells receive synaptic endings directly from bipolar cell axons nor do we know the sources of the dendrites joined by synaptic contacts. However, in two instances, dendrites postsynaptic to afferent axons appear to be directly or indirectly postsynaptic to other dendrites receiving synaptic input from afferent axons. If the "dendro-dendritic" synapses in the glomerulus characteristically are such cross-connexions between dendrites receiving afferent synaptic input, two kinds of actions could be postulated for these connexions. If inhibitory, they might be a means of lateral inhibition between different kinds of glomerular input, whereas if excitatory they would amplify the response of discrete regions or of the whole glomerulus to afferent input. Since periglomerular neurons send dendrites to two or more glomeruli (Blanes, 1897), inhibition or excitation might spread along these dendrites from one glomerulus to the next.

In the above discussion, processes are specified as pre- or postsynaptic depending on the morphological polarity of the synaptic contacts joining them to other processes. Synaptic contacts consist of synaptic vesicles clustered on the presumed presynaptic side of a synaptic junction. On the postsynaptic side, clusters of vesicles are lacking near the synaptic junction, but a fuzz (Gray, 1959) may be present on the cell membrane. In the glomerulus, this fuzz is less prominent at primary axo-dendritic synaptic junctions than at dendro-dendritic synaptic junctions. Both types of synaptic junction are distinguished from neuronal–neuronal *maculae adhaerentes* by criteria listed earlier (see p. 133) (Gray, 1961; Ralston and Herman, 1969).

The physiological polarity of synapses in other regions of the vertebrate nervous system has correlated so well with the morphological polarity of the synaptic contacts found at these locations (e.g. Rall *et al.*, 1966) that we feel justified to label processes in the glomerulus pre- or postsynaptic even in the absence of direct physiological data. According to available physiological evidence, afferent axons in the glomerulus can synaptically excite mitral cells (Shepherd, 1963). The synaptic contacts that we and others (Andres, 1965) have presumed to be endings of afferent axons are, as expected, morphologically presynaptic to dendrites in the glomerulus. Whether these are mitral dendrites, or other dendrites which in turn are presynaptic to mitral dendrites, has not been determined.

That dendrites or cell bodies can be presynaptic to other structures was first appreciated in the olfactory bulb. Structures suggesting this possibility were first described by Hirata (1964). More secure morphological identification of the presynaptic structures as dendrites soon followed. Dendritic synapses were found in the plexiform (Andres, 1965) and the glomerular (Reese and Brightman, 1965) layers of the olfactory bulb. Finally Rall and co-workers (1966) correlated various lines of physiological and anatomical evidence and concluded that the large accessory dendrites of mitral cells synaptically excite granule cells which in turn synaptically inhibit mitral cells. Thus it was shown that dendritic trunks of large multipolar neurons could be presynaptic to other dendrites and that such dendro-dendritic synapses could have either an excitatory or inhibitory action.

More recently, morphological studies of other areas of the brain, notably thalamus and midbrain tectum, have shown other instances of medium-sized dendrites in presynaptic positions; the following citations contain complete lists of references to this work (Lund, 1969; Sétalo and Székely, 1967; Ralston and Herman, 1969). In the latter two reports the postsynaptic structures are identified as dendrites, and our results show that processes postsynaptic to dendrites in the glomerulus also appear to be dendrites. Since no conclusive evidence of a "dendro-axonic" synapse has appeared we would suggest that an important feature of dendritic synapses is that the postsynaptic element is either a dendrite or a cell body. The synaptic contacts of amacrine cell dendrites on bipolar cell endings in the retina (Dowling and Boycott, 1965) might appear to be an exception. However, the retinal bipolar cell has no typical axon and might well be regarded as purely dendritic, like the amacrine cell (Shepherd, 1970).

We have presented only a few of the morphological features of the complex periglomerular neuropil. Further discussion of synapses in this region can be found in the work of Andres (1965). The most interesting new finding is the neuronal sheet processes formed by small periglomerular neurons. The neuronal sheet processes are wide, flat fingers that originate from small periglomerular neurons and partially engulf neighbouring neurons; along these zones of contact, the adjacent cell membranes appear denser. Neuronal sheet processes most resemble the calyces on vestibular hair cells that are presumably sites where electrical signals from hair cells are transmitted to the vestibular nerve. Hamilton (1968) has found various membrane specializations, as well as a dense fuzz in the cleft between adjacent membranes, at the apposition of the vestibular calyx with hair cells. We cannot say yet whether such specializations are present at the appositions of neuronal sheet processes with periglomerular cells and therefore cannot speculate whether electrical signals might be transmitted from one cell to another at this site. Finally, the finding of neuronal sheet processes in the olfactory bulb raises the question of whether such processes occur in other areas of the central nervous system. Perhaps it should no longer be assumed that all thin sheet-like processes lying next to neurons originate from glial cells.

SUMMARY

The olfactory surface in the primate, as in other vertebrates with lungs, has parallel arrays of specialized olfactory cilia embedded in a coat of mucus. Olfactory cilia are also present in teleosts but are lacking in elasmobranchs, implying that cilia are not essential for olfaction, at least in animals with gills. Close membrane appositions between olfactory cells and adjacent supporting cells are of the tight variety, implying that some of the current flow during olfactory stimulation might traverse the supporting cell.

Afferent axons from the olfactory epithelium synapse in olfactory glomeruli on dendrites of second-order cells. In the rat these dendrites are, in turn, presynaptic to other processes thought to be dendrites. Evidence is presented that such glomerular dendro-dendritic synaptic connexions would result in synaptic interaction between dendrites receiving synaptic input from afferent axons. Small periglomerular neurons form sheet-like processes, resembling astroglial sheets, which wrap around other small periglomerular neurons. The significance of this unusual neuronal-neuronal relationship is unknown.

REFERENCES

ANDRES, K. H. (1965). *Z. Zellforsch. mikrosk. Anat.*, **65**, 530–561.
ANDRES, K. H. (1966). *Z. Zellforsch. mikrosk. Anat.*, **69**, 140–154.
ANDRES, K. H. (1969). *Z. Zellforsch. mikrosk. Anat.*, **96**, 250–274.
BANNISTER, L. H. (1965). *Q. Jl microsc. Sci.*, **106**, 333–342.
BARNES, B. G. (1961). *J. Ultrastruct. Res.*, **5**, 453–467.
BLANES, T. (1897). *Revta trimest. microgr.*, **3**, 99–127.
BLOOM, G. (1954). *Z. Zellforsch. mikrosk. Anat.*, **41**, 89–100.
BRIGHTMAN, M. W., and REESE, T. S. (1969). *J. Cell Biol.*, **40**, 648–677.
DOWLING, J. E., and BOYCOTT, B. B. (1965). *Cold Spring Harb. Symp. quant. Biol.*, **30**, 393–402.
FARQUHAR, M. G., and PALADE, G. E. (1963). *J. Cell Biol.*, **17**, 375–412.
FARQUHAR, M. G., and PALADE, G. E. (1965). *J. Cell Biol.*, **26**, 263–291.
FEDER, N., REESE, T. S., and BRIGHTMAN, M. W. (1969). *J. Cell Biol.*, **43**, 35a–36a.
FRISCH, D. (1967). *Am. J. Anat.*, **121**, 87–119.
GEREBTZOFF, M. A. (1953). *J. Physiol., Paris*, **45**, 247–283.
GRAY, E. G. (1959). *J. Anat.*, **93**, 420–433.

GRAY, E. G. (1961). *J. Anat.*, **95**, 345-356.
GRAZIADEI, P., and BANNISTER, L. H. (1967). *Z. Zellforsch. mikrosk. Anat.*, **80**, 220-228.
GRAZIADEI, P., and TUCKER, O. (1970). *Z. Zellforsch. mikrosk. Anat.*, in press.
HAMILTON, D. W. (1968). *J. Ultrastruct. Res.*, **23**, 98-114.
HIRATA, Y. (1964). *Archvm histol. jap.*, **24**, 293-302.
HODGSON, E. S., MATHEWSON, R. F., and GILBERT, P. W. (1967). In *Sharks, Skates and Rays*, pp. 491-501, ed. Gilbert, P. W., Mathewson, R. F., and Rall, D. P. Baltimore: Johns Hopkins.
HOPKINS, A. E. (1926). *J. comp. Neurol.*, **41**, 253-289.
KARNOVSKY, M. J. (1967). *J. Cell Biol.*, **35**, 213-236.
LENN, N. J., and REESE, T. S. (1966). *Am. J. Anat.*, **118**, 375-390.
LUND, R. D. (1969). *J. comp. Neurol.*, **135**, 179-208.
MAC LEOD, P. (1959). *J. Physiol., Paris*, **51**, 85-92.
OKANO, M., WEBER, A. F., and FROMMES, S. P. (1967). *J. Ultrastruct. Res.*, **17**, 487-502.
OTTOSON, D. (1956). *Acta physiol. scand.*, **35**, suppl. 122, 1-83.
OTTOSON, D., and SHEPHERD, G. M. (1967). *Prog. Brain Res.*, **23**, 83-138.
RALL, W., SHEPHERD, G. M., REESE, T. S., and BRIGHTMAN, M. W. (1966). *Expl Neurol.*, **14**, 44-56.
RALSTON, H. J., III, and HERMAN, M. M. (1969). *Brain Res.*, **14**, 77-97.
RAMON Y CAJAL, S. (1911). *Histologie du système nerveux de l'homme et des vertébrés*, vol. 2, trans. Azoulay, L. Paris: Maloine.
REESE, T. S. (1965). *J. Cell Biol.*, **25**, 209-230.
REESE, T. S., and BRIGHTMAN, M. W. (1965). *Anat. Rec.*, **151**, 492.
REESE, T. S., and KARNOVSKY, M. J. (1967). *J. Cell Biol.*, **34**, 207-217.
REVEL, J. P., and KARNOVSKY, M. J. (1967). *J. Cell Biol.*, **33**, C7-C12.
RICHARDSON, K. C., JARRETT, L., and FINKE, E. (1960). *Stain Technol.*, **35**, 313-323.
ROBERTSON, J. D. (1963). *J. Cell Biol.*, **19**, 201-221.
SEIFERT, K., and ULE, G. (1967). *Z. Zellforsch. mikrosk. Anat.*, **76**, 147-169.
SÉTALO, G., and SZÉKELY, G. (1967). *Expl Brain Res.*, **4**, 237-242.
SHEPHERD, G. M. (1963). *J. Physiol., Lond.*, **168**, 89-100.
SHEPHERD, G. M. (1970). In *The Neurosciences: Second Study Program*, ed. Quarton, G. C., Melnechuk, T., and Schmitt, F. O. New York: Rockefeller University Press. In press.
SLIFER, E. H., and SEKHON, S. S. (1964). *J. Morph.*, **114**, 185-208.
THORNHILL, R. A. (1967). *J. Cell Sci.*, **2**, 591-602.

DISCUSSION

Lowenstein: People who have studied the retina have an idea of what may happen there functionally; have you any idea for the olfactory system?

Reese: Yes. Synaptic endings of afferent axons excite mitral cells (Shepherd, 1963). Accessory dendrites of mitral cells synaptically excite granule cells and granule cell dendrites, branching extensively, synaptically inhibit large fields of mitral cells (Rall *et al.*, 1966). This mitral–granule system resembles the amacrine–bipolar system in the retina (Dowling and Boycott, 1965). It could provide for "lateral" inhibition of mitral cells, thereby sharpening discrimination between odours. Granule cell excitability is also influenced by centrifugal input from various sources (Price, 1969). Many mysteries remain: the activity of the dendro-dendritic synapses in the glomerulus, the role of the axons of the tufted and the periglomerular granule cells, and the synaptic connexions of the larger, stellate cells in the internal plexiform layer are some examples.

Lowenstein: Is the degree of convergence greater in the olfactory system than in the retina?

Reese: Yes. Most olfactory glomeruli receive axons from bipolar cells scattered diffusely throughout the olfactory epithelium (Le Gros Clark, 1957). About 25 000 axons enter each glomerulus in the rabbit, whereas about 25 mitral cells contribute the same number of dendrites to a glomerulus. Thus 1000 first-order afferent axons enter the olfactory bulb for each second-order axon in the lateral olfactory tract (Allison and Warwick, 1949).

Døving: The ratio in the olfactory bulb of the teleost fish *Lota lota L.* between the receptors and the myelinated secondary neurons in the olfactory tract is about 1000 to one (Gemne and Døving, 1969). There are also a large number of unmyelinated fibres in the olfactory tract in fish and we do not know what their function is and how they are connected to the receptors.

Ottoson: To turn to the receptor mechanism, the cilia of the olfactory epithelium are evidently very fragile, so one would suppose that they are constantly broken off during life, and there must be a renewal process. Do you see any evidence of such outgrowths?

Reese: Centrioles are frequently found abutting the cell membrane of the olfactory bipolar dendrites in the frog. In other tissues, such positioning of centrioles is known to be preliminary to outgrowth of cilia (Roth and Shigenaka, 1964). However, we have never found a clear example of outgrowth in the frog.

Ottoson: I noticed a number of regularly spaced dark spots along the cilia in your light micrographs. Could they be a kind of bulbous expansion?

Reese: The micrograph to which you refer is an aerial view of intact, living olfactory epithelium (see Fig. 1). The finest lines resolved in this picture are probably small bundles of ciliary distal segments. I don't know what ultrastructural feature of these bundles corresponds to the dark (actually bright, in life) spots to which you refer. Since there is on the average one bulbous projection or large vesicle per cilium (Reese, 1965), I don't think that there would be enough bulbs to account for all the dark spots. I have wondered whether twisting of the bundles of cilia would cause these reflections.

Ottoson: In the muscle spindle the sensory endings have numerous

Fig. 1 (Reese). Negative image of the surface of intact olfactory epithelium of the frog, photographed from above with reflected light optics adjusted to give a dark field. Dark spots on this micrograph were light spots in life. The dark lines are parallel bundles of distal segments near the surface of the mucus. No ciliary movement was seen. Exposure time was 30 seconds. ×440

bulbous expansions and it has been suggested that these structures are the actual transducers where the response is generated.

Reese: Vesicles 30–60 nm (300–600 Å) in diameter have been seen in a variety of non-sensory cilia; I reported that such vesicles are present in cilia from the respiratory passages in the frog (Reese, 1965). Vesicles over 0·1 μm in diameter as well as redundant folds of ciliary membrane are found in the frog olfactory cilia. Bulbous swellings of the ciliary membranes, without an associated vesicle, are also seen in the frog but we have regarded these as an artifact of fixation. Before deciding whether vesicles or redundant folds of membrane are related to olfactory reception, we should know whether they are found generally in vertebrate olfactory cilia. Dr Andres has just completed an elegant study of olfactory cilia in the cat (Andres, 1969) and might comment on this question.

Andres: In well-fixed material no true vesicles are present, but only flat, vesicle-like invaginations. When the material is badly fixed the cilia form vesicular protrusions which must not be mistaken for true vesicles.

Moulton: Your evidence from an elasmobranch suggests that cilia are not essential for olfaction; what role do you think they may play when they are present?

Reese: They put cell membrane where the smell is! In the shark there is no layer of mucus, so seawater flows directly over the dendrites of the bipolar cells.

Moulton: Is it not possible that even in species where cilia are present the membrane at the base of the cilium is the essential receptor site and not the cilium itself (as Le Gros Clark once suggested, 1956)?

Reese: Cilia and microvilli provide the means to put large areas of cell membrane into contact with odorous substances. The variety of means that olfactory cells use to achieve this one end suggests to me that these large areas of membrane carry the olfactory receptor sites. In air-breathing vertebrates, the receptor sites are probably on the ciliary membrane of distal segments, lying near the air–mucus interface. In the elasmobranch, the receptor sites would be on the cell membrane covering microvilli or, perhaps, on other parts of the bulbous ends of bipolar dendrites. These data suggested that what supports the cell membrane is not an important factor in the deployment of receptor sites.

Moulton: The internal structure of the cilia doesn't then seem to have any significance, if microvilli can substitute?

Reese: At least the ciliary internal structure may not have any role in the initial interaction between cells and odorous substances. The shark has a good olfactory sense and yet has no neurocilia on its olfactory cells. It would be interesting to know more about the olfactory physiology of the shark.

Moulton: In the rat vomeronasal organ there are microvilli on the receptors instead of cilia (J. Kauer, 1969, personal communication) and the organ appears to be highly chemosensitive—in the rabbit, at least (Tucker, 1963).

Reese: Do air-borne substances have direct access to its surface?

Moulton: We don't know. There are suggestions that air may be getting to the system (e.g. Hamlin, 1929).

Zotterman: The basic problem seems to be whether the cilia are acting as conducting dendrites or not; that is, where is the primary excitatory process taking place?

Reese: If olfactory cilia are assumed to have passive electrical properties, it might be inefficient to have the depolarizing sites at the ciliary tips, so far away from the dendrite. One wonders whether the depolarization would be very effective at the dendrite.

Zotterman: The question is what speed the depolarization could be conducted; how long are the cilia in man, say?

Reese: Let's say 50 μm in man.

Zotterman: At a speed of conduction of a few centimetres per second it would take one or two milliseconds for the depolarization to reach the apex of the cell.

Reese: A thoughtful consideration of the conductive properties of structures the size of olfactory cilia in the frog has been published by Ottoson and Shepherd (1967). They concluded that for passive spread of excitation along an olfactory cilium to be effective, it is necessary to assume special electrical properties for the ciliary membrane or cytoplasm. However, they also suggest that since the olfactory cilia are similar in structure to olfactory axons, the cilia might have some active electrical properties, such as those mentioned by Professor Zotterman.

Good evidence relevant to the role of cilia in chemoreception comes from work on insects. The olfactory structure at the bases of their sensory hairs is quite similar to that of vertebrates. However, the cuticle of the insect covers the bipolar dendrites, supporting cells and proximal segments of the olfactory cilia. Only the distal segments have rapid access to odorous material through contact with pores in the cuticle (Slifer and Sekhon, 1964).

MacLeod: Dr Reese, are the cilia of the supporting cells in the elasmobranch motile and if so, is their movement synchronized?

Reese: They look in the fixed material as if they are moving in synchrony, in a metachronal wave. You need to look for ciliary motion on fresh olfactory epithelium in seawater and I have not done that yet.

MacLeod: So it would be possible to think of them like respiratory cilia?

Reese: Yes, it would.

Lowenstein: Have you looked at the basal feet, and if so, do they point in the same direction? If there is polarization in the direction of the working stroke for instance, you have good evidence for metachronal synchrony.

Reese: In some sections the basal feet point in the same direction. This is evidence for a synchronized ciliary beat.

Beidler: Have you looked at the olfactory epithelium after applying different odours to it? The so-called supporting cell is quite different before and after stimulating it. In one case it may have secretory vesicles and in the other case it is already empty of all its vesicles.

Reese: We have not experimented with stimulating olfactory epithelium before fixation, although some of our fixatives would be quite stimulating. What odours did you use in your experiments?

Beidler: One substance we tested was acrolein, because we were interested in emptying of the sustentacular cells. They must produce these granules at a very high rate, at least in the frog. But of course it is a very irritating substance.

Reese: You can discharge the Bowman glands too, by irritating the epithelium mechanically. They shoot out a stream of granules under these conditions.

Wright: I have a question relating to a much grosser feature in the anatomy. Some years ago I looked at the olfactory end organ of the Pacific salmon (*Oncorhynchus* sp.). Underneath the outer cuticle is a dark coloured structure which I thought would be a rich source of olfactory pigment. However, the dark colour was mainly and perhaps entirely due to what appeared to be melanophores. The surrounding tissue was not coloured that way at all; the melanophores were definitely associated with the olfactory epithelium, and it occurred to me that they might be a light screen, if illumination of the sensitive area in some way interferes with the stimulating process. Are melanophores found here in other fish, and are they absent in creatures where the olfactory epithelium is sufficiently deeply located for light to be unable to reach it?

Reese: Varying numbers of melanophores lie under the olfactory epithelium in amphibia and fishes, but none are present in many mammals. Even in their absence it is not certain that much light would penetrate the skin and overlying bone to illuminate the olfactory epithelium. In some elasmobranchs, such as the guitar fish, the openings of the olfactory sac are very large and much of the olfactory surface is exposed, albeit on the ventral surface of the fish.

Wright: From what you say it would appear that where light can't get at the olfactory organ so much, there aren't so many melanophores; where it can, there seems to be this protection. This might be a clue to the actual process of excitation.

Døving: Melanophores are not present in any great numbers in bony fishes; as a result the olfactory organ is white or yellow. A conspicuous pigmented organ is the osphradium in gastropods, for example the whelk (*Buccinum undatum*), which is dark brown or black. Its colour is due to pigment granules in the supporting cells (Welsch and Storch, 1969).

REFERENCES

ALLISON, A. C., and WARWICK, R. T. T. (1949). *Brain*, **72**, 186–197.
ANDRES, K. H. (1969). *Z. Zellforsch. mikrosk. Anat.*, **96**, 250–274.
DOWLING, J. E., and BOYCOTT, B. B. (1965). *Cold Spring Harb. Symp. quant. Biol.*, **30**, 393–402.
GEMNE G., and DØVING, K. B. (1969). *Am. J. Anat.*, **126**, 457–476.
HAMLIN, H. E. (1929). *Am. J. Physiol.*, **191**, 201–205.
LE GROS CLARK, W. E. (1956). *Yale J. Biol. Med.*, **29**, 83–95.
LE GROS CLARK, W. E. (1957). *Proc. R. Soc. B*, **146**, 299–319.
OTTOSON, D., and SHEPHERD, G. M. (1967). *Prog. Brain Res.*, **23**, 83–138.
PRICE, J. L. (1969). *J. Physiol., Lond.*, **203**, 77P–78P.
RALL, W., SHEPHERD, G. M., REESE, T. S., and BRIGHTMAN, M. W. (1966). *Expl Neurol.*, **14**, 44–56.
REESE, T. S. (1965). *J. Cell Biol.*, **25**, 209–230.
ROTH, L. E., and SHIGENAKA, Y. (1964). *J. Cell Biol.*, **20**, 249–270.
SHEPHERD, G. M. (1963). *J. Physiol., Lond.*, **168**, 89–100.
SLIFER, E. H., and SEKHON, S. S. (1964). *J. Morph.*, **114**, 185–208.
TUCKER, D. (1963). In *Olfaction and Taste* (Proceedings of the First International Symposium, Stockholm, 1962) pp. 45–69, ed. Zotterman, Y. Oxford: Pergamon Press.
WELSCH, V., and STORCH, V. (1969). *Z. Zellforsch. mikrosk. Anat.*, **95**, 317–330.

THE OLFACTORY NEURON AND THE BLOOD-BRAIN BARRIER

A. J. Darin de Lorenzo

The Johns Hopkins University School of Medicine, Baltimore, Maryland

The olfactory mucosa of primates and man extends from the roof of the nasal cavity about 10 mm downwards on either side of the septum and surface of the superior nasal conchae. The total area in the squirrel monkey (*Saimiri sciureus*) is about 300 mm^2. In spite of its remarkably high order of odour specificity and its vast range of sensitivity (Ottoson and Shepherd, 1967) the olfactory mucosa is quite simple in structure. Seen in the light microscope it consists of long columnar cells arranged as a pseudostratified epithelium about 65 μm thick (Allison, 1953; Hopkins, 1926; Parker, 1922). Its surface is covered with a mucous blanket 8–10 μm thick in which cilia can be visualized under ideal conditions. The epithelium consists of three cell types: (*a*) the olfactory receptor, (*b*) the sustentacular or supporting cells and (*c*) basal cells which rest on a basement membrane (Fig. 1). Numerous secretory glands complete the histological description of the olfactory mucosa at this level of resolution.

In recent years numerous descriptions of the fine structure of the olfactory mucosa have appeared. These studies have emphasized the remarkable similarity in fine structure in the frog (Bloom 1954; Reese, 1965), mouse (Frisch, 1964), rabbit (de Lorenzo, 1957, 1960, 1963), primates and man (de Lorenzo, 1968). These observations are summarized and incorporated into a simple drawing (see Fig. 2). There seems little need merely to present further similarities in fine structure in yet another species. Therefore in this study the emphasis is placed upon the structure and modifications of the cell membranes, the olfactory cilia and receptor membrane specializations which may be of functional significance. Finally, the relationship of the olfactory neuron to the blood–brain barrier is examined by the use of colloidal gold tracer particles.

MATERIALS AND METHODS

Young male squirrel monkeys weighing about 1 kilogramme were anaesthetized and perfused through the left ventricle with a fixative consisting of 2 per cent paraformaldehyde, 2 per cent glutaraldehyde, 0·1 M-sodium phosphate buffer and 4·5 per cent sucrose. All fixation was carried out at room temperature.

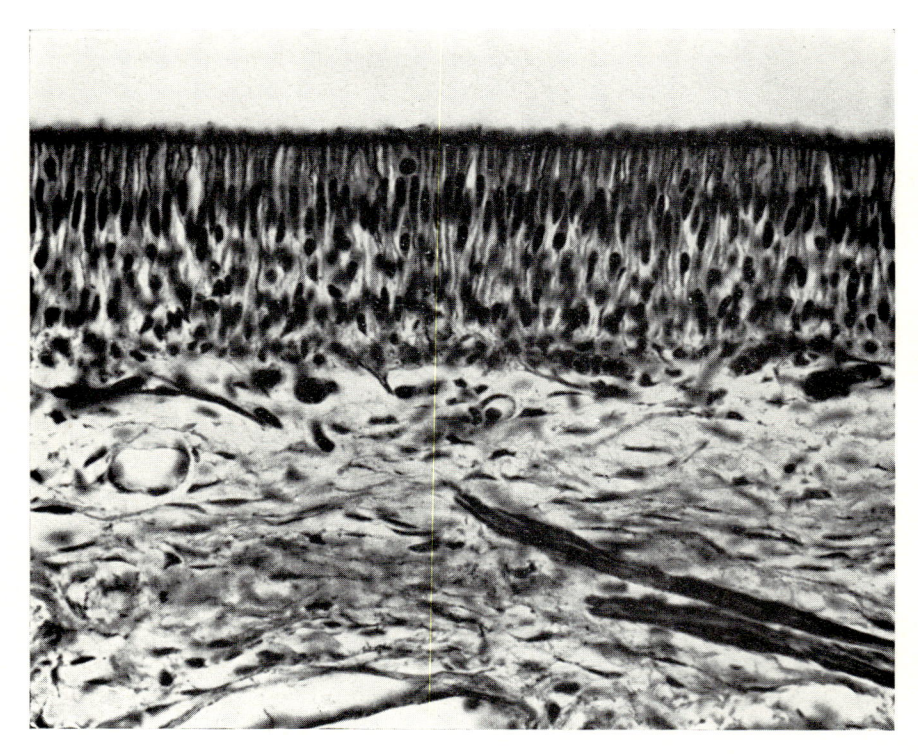

Fig. 1. Section through the olfactory epithelium of the squirrel monkey, stained with Bodian stain to show olfactory receptors and olfactory nerves. × 300

Animals used in the study of the blood–brain barrier received 1 ml of silver-coated colloidal gold (kindly supplied by Abbott Laboratories) intranasally. The animals were killed by perfusion as above, at various times after the application of the colloidal gold (e.g. 15 minutes, 30 minutes, 1 hour and 24 hours). Tissues were processed for electron microscopy in the routine manner.

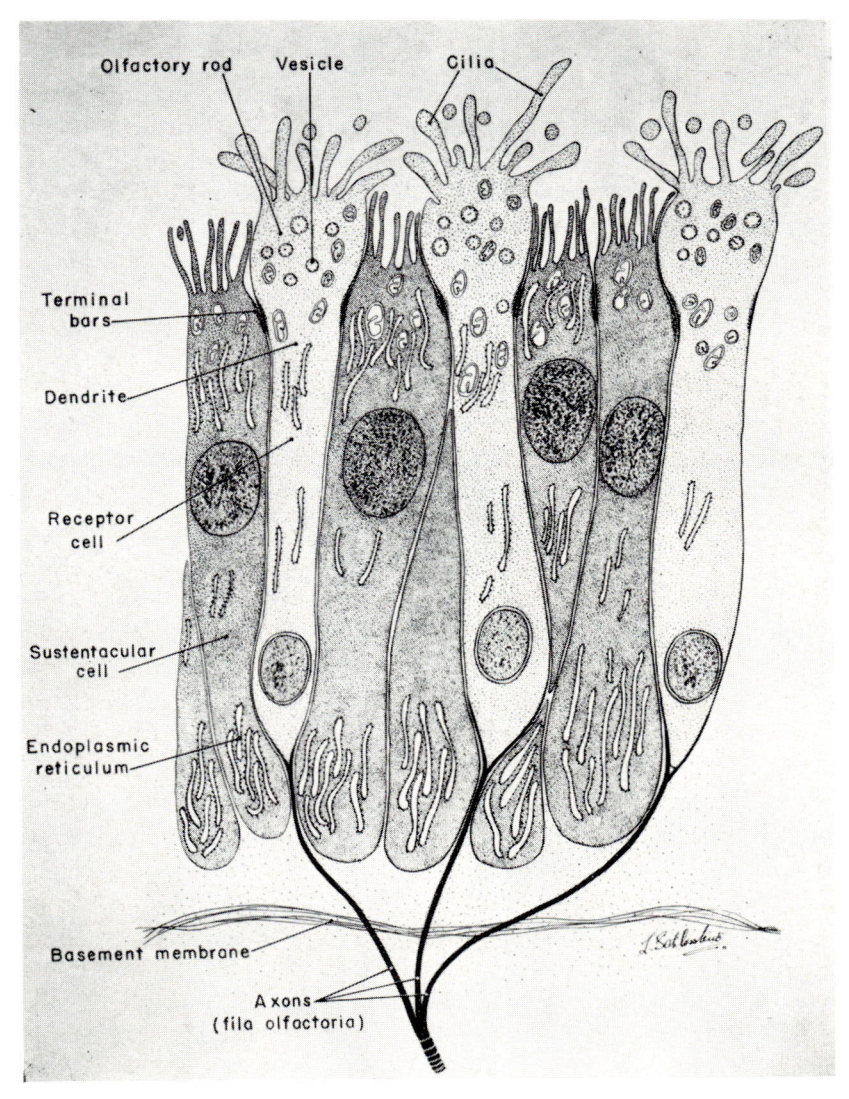

FIG. 2. Schematic representation of the vertebrate olfactory epithelium.

OBSERVATIONS

The olfactory mucosa of the squirrel monkey includes the sensory epithelium consisting of (a) olfactory receptor cells, (b) sustentacular cells and (c) basal cells.

Olfactory cells

The ciliated apical processes of the receptor cells extend above the free surface of the epithelium as the olfactory rod or vesicle. At the region of the epithelial surface, the neck of the receptor is constricted and reduced in size (Figs. 2, 6, 7 and 11). Cilia numbering as few as eight and as many as 20 or more extend from the surface of the olfactory rod and quickly bend to run parallel with the surface of the epithelium (Fig. 3). Receptor cells are usually in contiguity with supporting cells and structural modifications are seen in the contiguous membranes in such regions, and will be described below. Thus the receptor is ensheathed throughout its length except at the apical tip.

Occasionally one sees receptor-to-receptor contiguity. In these cases the contiguous membranes demonstrate the fine structural organization seen at synaptic junctions. Such regions may suggest lateral interaction or inhibition between receptors, as seen in other sensory systems such as the *Limulus* eye.

Olfactory rod

The olfactory rod is the bulbous enlargement described above which comprises the only naked region of the entire olfactory neuron. It is the only cranial nerve whose first-order neuron cell body resides in a distal epithelium. Its cytoplasm contains a number of organelles. Mitochondria are occasionally seen in the olfactory rod but usually they are located in the region of the neck (Fig. 12). They resemble mitochondria seen elsewhere in neural tissues which do not contain granules. Their cristae are arranged longitudinally. Otherwise they are unremarkable except for their conspicuous absence in the typical olfactory rod.

The cell membrane of the receptor consists of the unit membrane configuration. Its uniformity is interrupted by the cilia which protrude from the olfactory rod (Figs. 3, 4, 6 and 12). The outer leaflet of the cell membrane is not smooth but is covered with a fuzzy thread-like surface coating which can best be seen in Fig. 4 at the region indicated by the black arrow. Here the membrane has been sectioned obliquely and the fuzzy coating is easily recognized. In the inset of Fig. 4 one can see the unit membrane characteristics of the cell membrane. In addition, at the region designated by the arrows small globular structures measuring about 2·5–3 nm (25–30 Å) can be resolved adhering to the outer leaflet of the membrane. Similar structures forming rosettes are also seen in the cytoplasm (Fig. 4) and may represent the mechanism for movement of small particles through

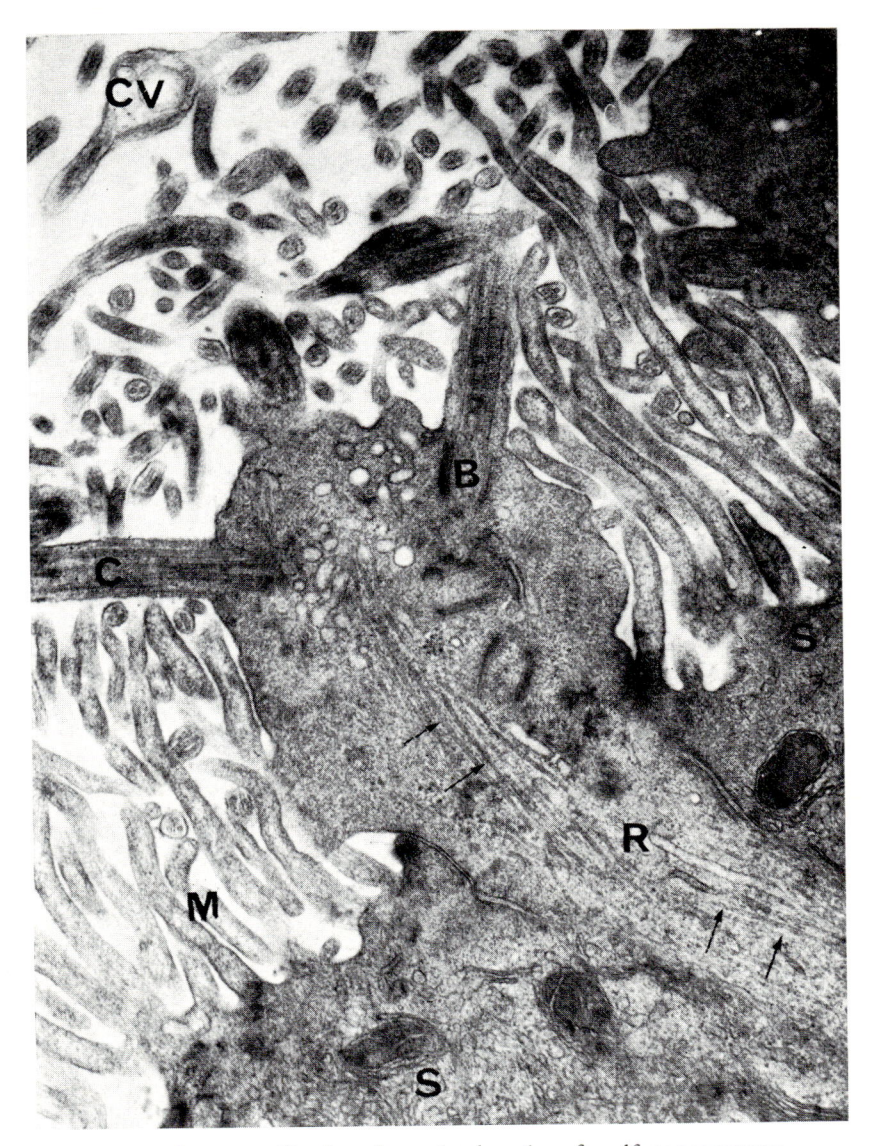

FIG. 3. Low magnification of a restricted portion of an olfactory receptor cell. The olfactory receptor (R) demonstrates microtubules (arrows), basal bodies (B), centrioles, ribosomes and numerous cilia (C) protruding from the olfactory rod. Sustentacular cells (S) and their microvilli (M) are seen. CV, vacuolated ciliary dilatation. × 15000

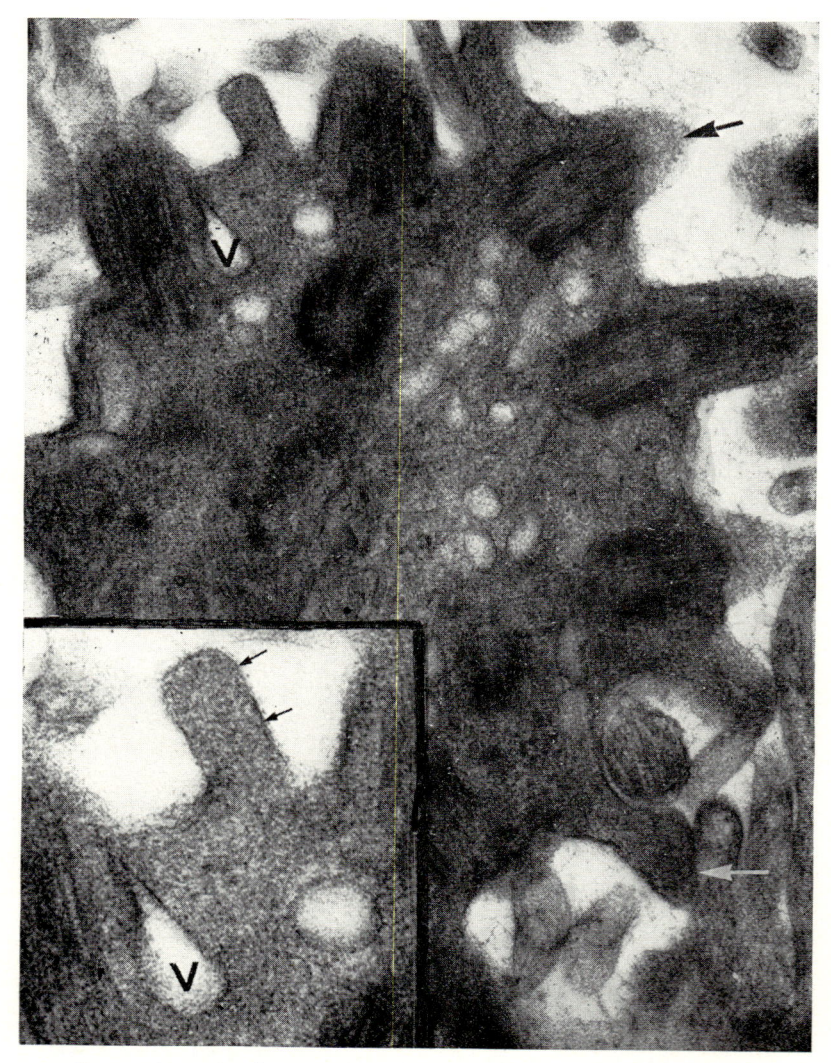

FIG. 4. A restricted portion of an olfactory rod showing fuzzy outer coating on receptor membrane (black arrow) and micropinocytotic vacuoles (V). White arrow indicates junction between receptor cell and microvilli of supporting cell. ×45 000

Inset: High resolution micrograph of fuzzy outer coating of receptor membrane. Note small particles adhering to membrane (arrows) and microglobular matrix in cytoplasm. V, micropinocytotic vacuole. ×90 000

the membrane (Brandt, 1962). An additional modification of the membrane is seen in Fig. 4 where the membrane is being pinched off to form rather typical pinocytotic vacuoles (V). Throughout the olfactory rod but particularly at the apical surface one sees large numbers of intracellular vacuoles measuring 40–120 nm (400–1200 Å) in diameter (see Figs. 3 and 4). Occasionally these vacuoles appear tubular as in Fig. 6, when they measure

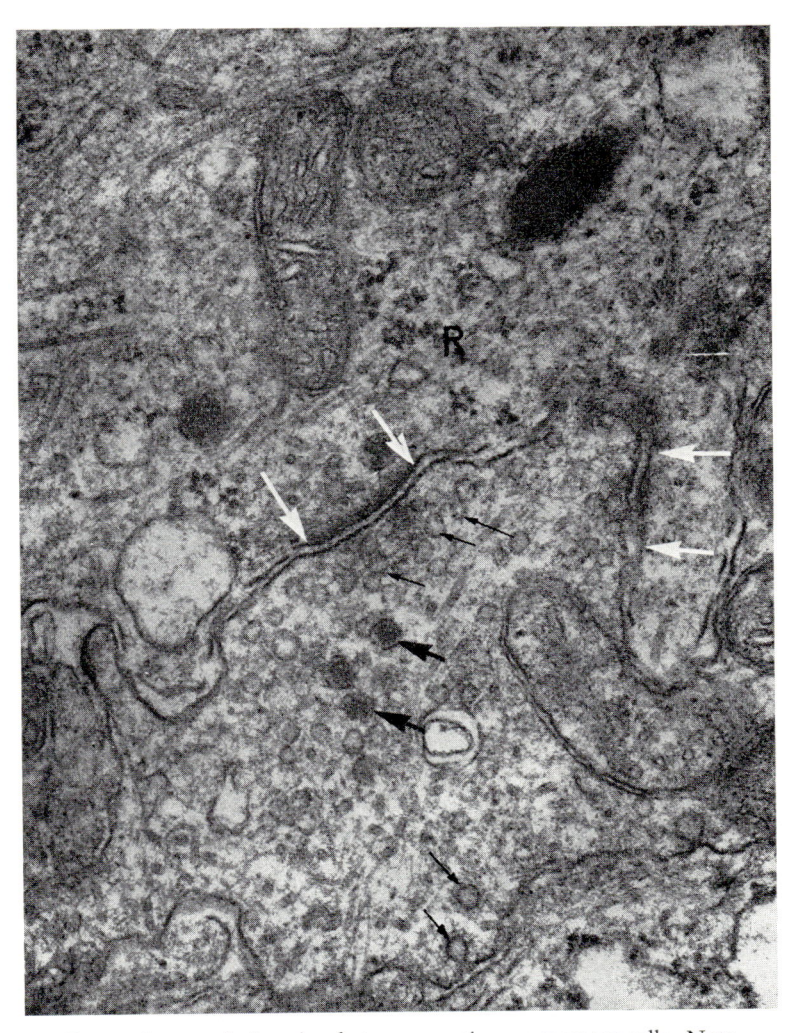

FIG. 5. A synaptic junction between contiguous receptor cells. Note presynaptic vesicles of various sizes (black arrows), rosettes of ribosomes (R), synaptic cleft and postsynaptic thickening. White arrows identify areas of synaptic junctions. × 32 000

about 40 nm (400 Å) in width and can be as long as 0·3 μm. If one examines a large number of such vacuoles one is impressed with the fact that they appear to be continuous with cell membrane invaginations described as pinocytotic vacuoles (V) in Fig. 4. One must conclude that the olfactory rod is busily engaged in "drinking in" materials from the extracellular environment. The cytoplasm also contains a much smaller circular vacuole population, best seen in the inset of Fig. 4. It consists of vacuoles 2·5–15 nm

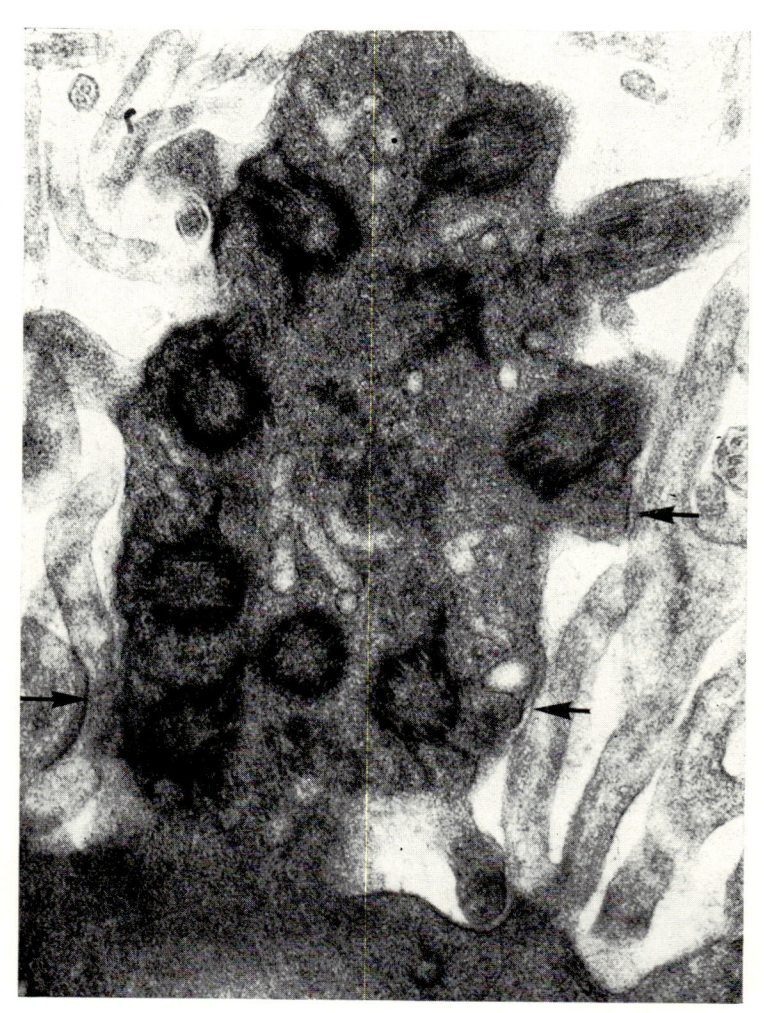

FIG. 6. An olfactory rod with at least ten centrioles and basal bodies in this plane of section. Arrows indicate contacts between microvilli and receptor membrane. × 15000

(25–150 Å) in diameter which have fuzzy edges and seem to fill the cytoplasm, giving it an extremely dense appearance. These structures are preserved only with paraformaldehyde–glutaraldehyde fixation and are not seen with osmium tetroxide fixation (Burton, 1968). Whether they represent a molecular transport mechanism of the membrane remains to be seen.

The cytoplasm of the rod also contains a large number of centrioles, basal bodies, basal feet and rootlets (Figs. 3, 4, 6 and 12). With the fixation techniques used in this study microtubules are consistently seen (Figs. 3–7). They measure approximately 27 nm (270 Å) in diameter and are several micrometres in length. They are seen in the olfactory rod where they seem to end near centrioles and also occur in the region of the perikaryon, where they are less numerous. Microtubules are also seen in the nerve fibres making up the fila olfactoria; thus they appear to occupy the entire neuronal cytoplasm. Unlike microtubules described in the olfactory cells of the mouse (Frisch, 1964), those seen in this study lack a central dark dot or filament. Our observations are in accord with those of Kirkpatrick (1969a, b). Free ribosomes are seen throughout the cytoplasm of the receptor cell.

Olfactory cilia

The olfactory rod contains a variable number of cilia which protrude from the surface for a short distance and then are oriented parallel along the surface of the epithelium in the mucous blanket. In the squirrel monkey the mucous blanket is about 8 μm thick, but the preparative procedures do not preserve the mucus, so only the framework of supporting-cell microvilli and ciliary processes is seen (Fig. 8). Careful counts of selected serial sections reveal that the number of cilia on a given olfactory rod varies from as few as eight to as many as twenty. Like other cilia, they are associated with basal bodies and basal feet, and cross-sections through these regions show the typical nine pairs of peripherally oriented ciliary fibrils (Fig. 6). Within 1 to 2 μm from the rod the diameter of the cilia is reduced by approximately one-half and is then further reduced as it nears its tip. The length of the cilia varies a great deal but this may be a matter of splitting or breaking off during fixation. In this study cilia have been followed as far as 50 μm but this distance probably represents their minimal length. Their parallel orientation and reduced peripheral diameter make their length difficult to measure accurately with the light or electron microscope. However, if one examines cross-sections of cilia at various locations along their length, their fine structure changes dramatically.

Fig. 9 depicts the fine-structural changes in the number and distribution of the ciliary fibrils. Cross-sections near the basal bodies reveal only the peripheral or outer nine pairs of fibrils (Fig. 6). Cross-sections through cilia 0·5 to 1 μm distal to the olfactory rod are seen in Fig. 8. Here they demonstrate the 9+2 organization typical of contractile cilia elsewhere. Cross-sections further along their length are shown in Figs. 7 and 8. The arrowheads in Fig. 8 demonstrate the dramatic reduction in diameter and the loss

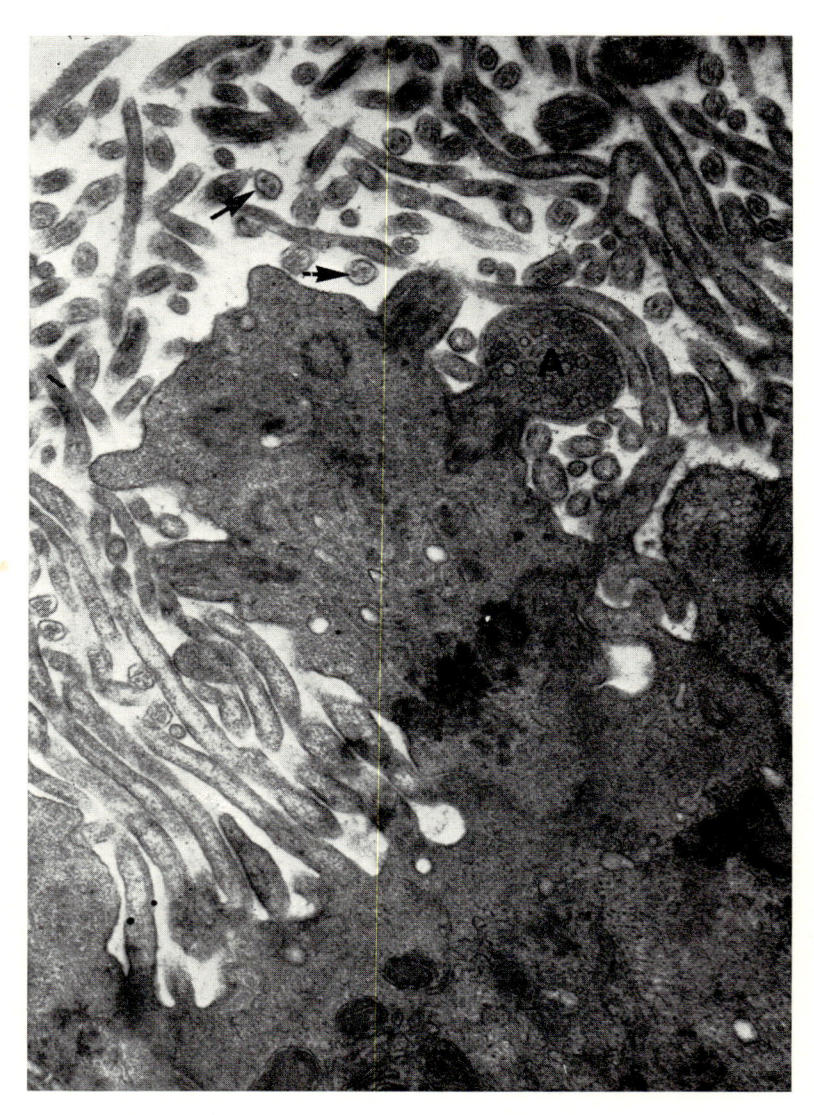

FIG. 7. An olfactory rod with a ciliary vesicle (A) filled with synaptic-like vesicles. Arrows show pairs of ciliary fibrils. × 13 500

of ciliary fibrils. The number has been reduced to four and as the cilium further tapers at its tip the number of ciliary fibrils is reduced to two, as demonstrated in Fig. 7 (upper arrows) and Fig. 8. My impression is that the central fibrils persist to the very terminals and all other fibrils are lost along the way. The significance of this organization remains unclear.

Additional modifications in the cilia are seen in the squirrel monkey and have been described in other cilia. Fig. 7 demonstrates such a structural

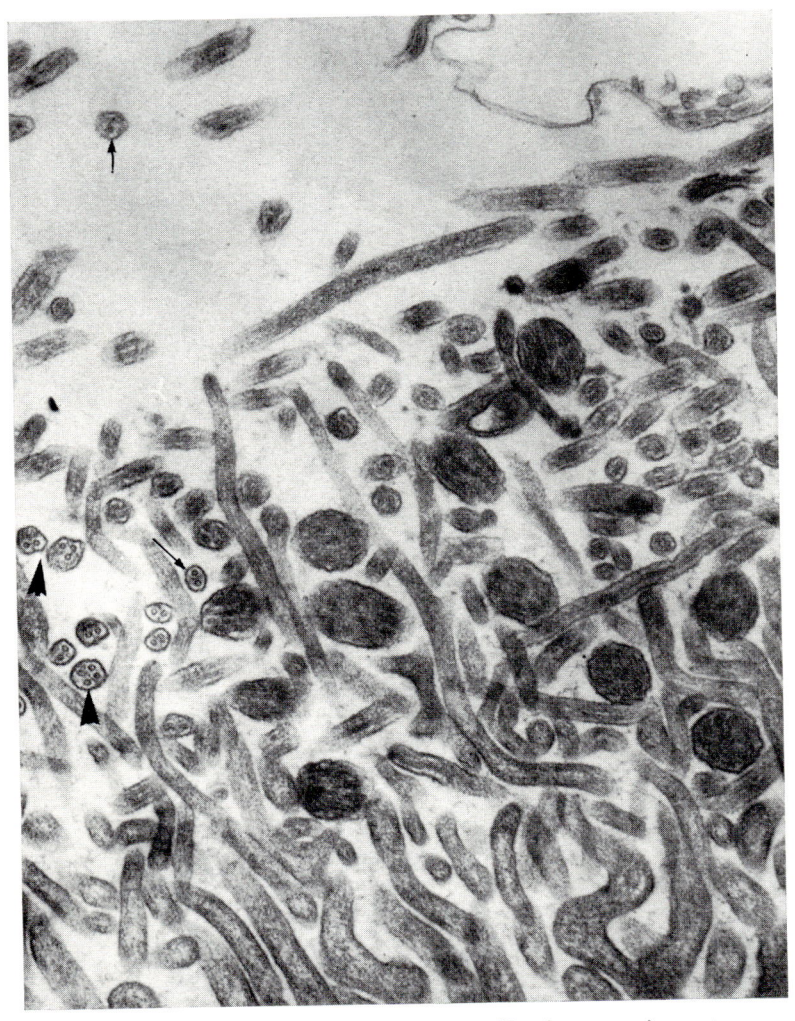

FIG. 8. Cross-sections through olfactory cilia demonstrating 9+2 arrangement of fibrils as well as 4 and 2-fibrillar tips of cilia (arrows).
× 15000

specialization (A). This structure is an ovoid dilatation deriving from a basal body which contains a number of vesicles of the same size and appearance as synaptic vesicles, being 40–75 nm (400–750 Å) in diameter. Some have dense cores and others have light or empty cores. Another kind of dilatation which occurs more frequently is shown in Fig. 4 (CV). These vacuoles occur in larger numbers further from the surface of the epithelium.

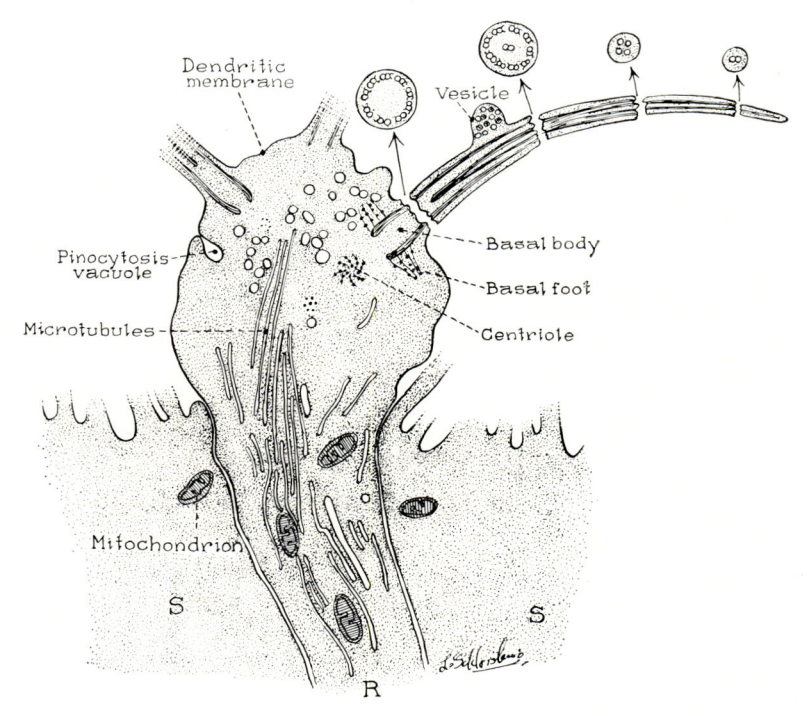

FIG. 9. Schematic drawing of important fine structural components of olfactory receptor.

The cell membrane covering the cilia is similar in fine structure to the receptor cell membrane. The fuzzy outer coating persists, particularly in the regions of the ciliary vacuoles. The matrix of the cilia, however, is devoid of cytoplasmic structures seen in the olfactory rod. Fig. 9 is a diagrammatic representation of the fine structural characteristics of the olfactory receptor and its cilia.

Receptor cell membrane contacts and specializations

As briefly described above, the receptor cell membrane is naked in the

olfactory rod region and in contiguity with other cells throughout the remainder of its length. These cell contacts include:
(1) receptor to supporting cell;
(2) receptor to receptor;
(3) receptor rod to supporting-cell microvilli;
(4) receptor cilia to supporting-cell microvilli;
(5) olfactory nerves and fila olfactoria to supporting cells, basal cells, Schwann cells and neuroglial cells.

The most frequently encountered relationship is *receptor to supporting cell*. This relationship is depicted in Figs. 3, 7 and 12. Here the apposing membranes demonstrate *zonulae adhaerens, zonulae occludens*, desmosomes and tight junctions. These junctions permit the movement of certain proteins between cells and exclude their movement in other regions (see pp. 29–30).

Perhaps the most interesting cellular contact is that shown in Fig. 5, namely *receptor to receptor* contiguity. In Fig. 5 one receptor cell is located to the upper left of the micrograph and another to the lower right. Typical microtubules, mitochondria and other cytoplasmic organelles are easily recognized. The two receptor cell membranes are separated by spaces 20–30 nm (200–300 Å) wide which decrease to as little as 15 nm (150 Å) at their narrowest point. At the regions designated by the white arrows the fine structural characteristics of a typical synapse are seen, namely *postsynaptic membrane web or thickening, synaptic cleft of* 20 nm (200 Å) and *presynaptic vesicles*. Thus the cell on the lower right is presynaptic to the cell at the upper left. The synaptic vesicles are of three kinds: (1) typical vesicles with electron-lucent cores, with a diameter of 40–75 nm (400–750 Å)—these are designated by the upper small arrows in Fig. 5; (2) large vesicles 75–120 nm (750–1200 Å) wide with dense cores, designated with the two large arrows; and (3) large "fuzzy coated" vesicles (lower two small arrows) of the type observed in efferent cochlear endings on the organ of Corti.

The postsynaptic cell demonstrates the typical subsynaptic web seen at synaptic junctions and a few assorted small vesicles. Most conspicuous, however, are the clusters or rosettes of ribosomes (R) located in the postsynaptic processes, as in other synapses. The consistent presence of these fine structural features of a synapse makes it difficult not to accept this region as a synaptic junction. Physiological interaction between cells in the olfactory epithelium is well known. Such receptor-to-receptor cell junctions are few and for the most part lie in the basal regions of the olfactory cells.

Junctions between receptor cell processes and the microvilli of supporting cells are shown in Fig. 4 (white arrow) and Fig. 6 (black arrows). Some membrane specialization seems to exist at such locations but their function, if any, remains unknown. Additional contacts between olfactory cilia and microvilli of supporting cells are constantly seen in the apical mucous blanket. Lack of interaction at the molecular level is difficult to imagine.

The proximal processes of olfactory cells are the axons which compose the olfactory nerves and the fila olfactoria. The size of the fibres and the

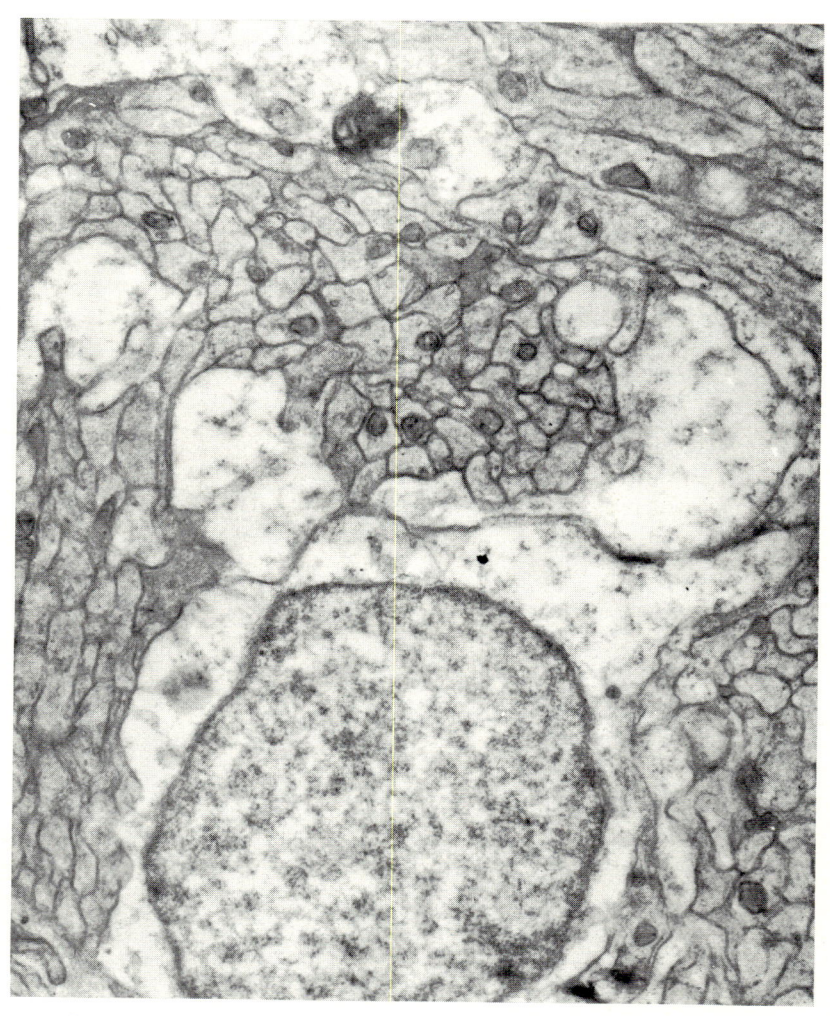

FIG. 10. The organization of the fila olfactoria of the squirrel monkey is identical to that seen in other species, the olfactory nerves being about 0·2 μm thick and organized in fascicles. × 9000

organization of the olfactory fascicles in the squirrel monkey are shown in Fig. 10. Their structure and organization are entirely similar to those described in other species (de Lorenzo, 1957, 1960, 1963; Gasser, 1956). This organization is best described by Fig. 11.

The olfactory neuron and the blood–brain barrier

Foreign substances such as proteins, viruses, particles, dyes and most drugs when injected into the blood vascular system find their way into the

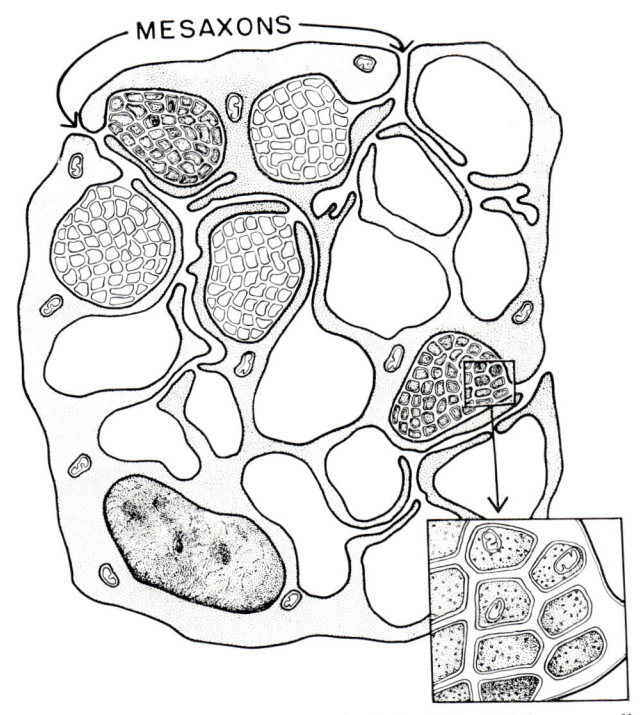

Fig. 11. Schematic representation of relationship of Schwann cell to fascicles of olfactory nerves.

perivascular spaces and the various body tissues. However, they have rather limited access to the central nervous system. The term blood–brain barrier has been applied to this exclusion mechanism and a vast literature exists on this most interesting phenomenon. The olfactory system has been considered to have a unique relationship with the blood–brain barrier since dyes placed on the olfactory mucosa are found to stain the olfactory lobes and brain in short periods of time. Perhaps the most interesting and informative studies are those of Bodian and Howe (1941a, b), who applied polio viruses intranasally in monkeys and studied the rate of progression of the virus in olfactory nerves. They found that polio virus travels "within the axoplasm of axons rather than along sheaths of nerve bundles" (Bodian and Howe, 1941a). Virus was also recovered from olfactory bulbs, thus demonstrating that the olfactory mucosa and nerves provided a portal of entry to the central nervous system. With this concept in mind we began a series of experiments in which we placed herpes simplex and vaccinia viruses as well as particulate tracers such as ferritin and colloidal gold intranasally in squirrel monkeys. The animals were sacrificed 15 minutes, 30 minutes and 1 hour later.

Olfactory mucosa and olfactory bulbs were removed and prepared for electron microscopic examination. Fig. 12 demonstrates an olfactory receptor tip 15 minutes after colloidal gold had been placed on the olfactory mucosa. At the regions designated by the arrows one can see small particles (50 nm, 500 Å) which are the colloidal gold spheres. They are attached to the outer fuzzy membrane as well and appear to be incorporated into the cytoplasm of the receptor cell by the pinocytotic vacuoles previously described. Within 30 minutes one can see aggregates of colloidal gold particles in the axons comprising the fila olfactoria. The upper two micrographs of Fig. 13 demonstrate the particles in olfactory axons (N) which have been cut in cross-section. The bottom micrograph demonstrates the colloidal gold particles in olfactory nerves sectioned obliquely. Clearly the particles reside in the axoplasm. Within 1 hour the particles are seen in the olfactory glomerulus, as shown in Fig. 14. Here one can identify the nerve terminals and the mitral cell dendrites on which they end. Thus the particles have moved from the receptor via the axoplasm and, by a mechanism as yet undetermined, crossed the first-order synapse to reach the mitral cell dendrites. Curiously, however, the particles in the olfactory bulb are no longer found freely distributed in the cytoplasm. They are now all located preferentially in mitochondria (see Figs. 14, 15 and 16). Fig. 15 is from a slightly deeper part of the olfactory bulb and olfactory nerves (N) are seen synapsing on large mitral cell dendrites (D). Note that the intramitochondrial colloidal gold particles reside in both the presynaptic and postsynaptic processes. The actual synaptic junctions are designated by arrows. In Fig. 16 a mitral cell nucleus is located in the upper right corner. A large nerve process near or at its point of synapse on the cell body is seen. In its mitochondria one again sees large aggregates of colloidal gold particles.

It must be emphasized at this point that: (1) intramitochondrial densities or granules are not seen in normal nervous tissues or control studies of olfactory receptor cells, but do occur in mitochondria of liver, kidney and

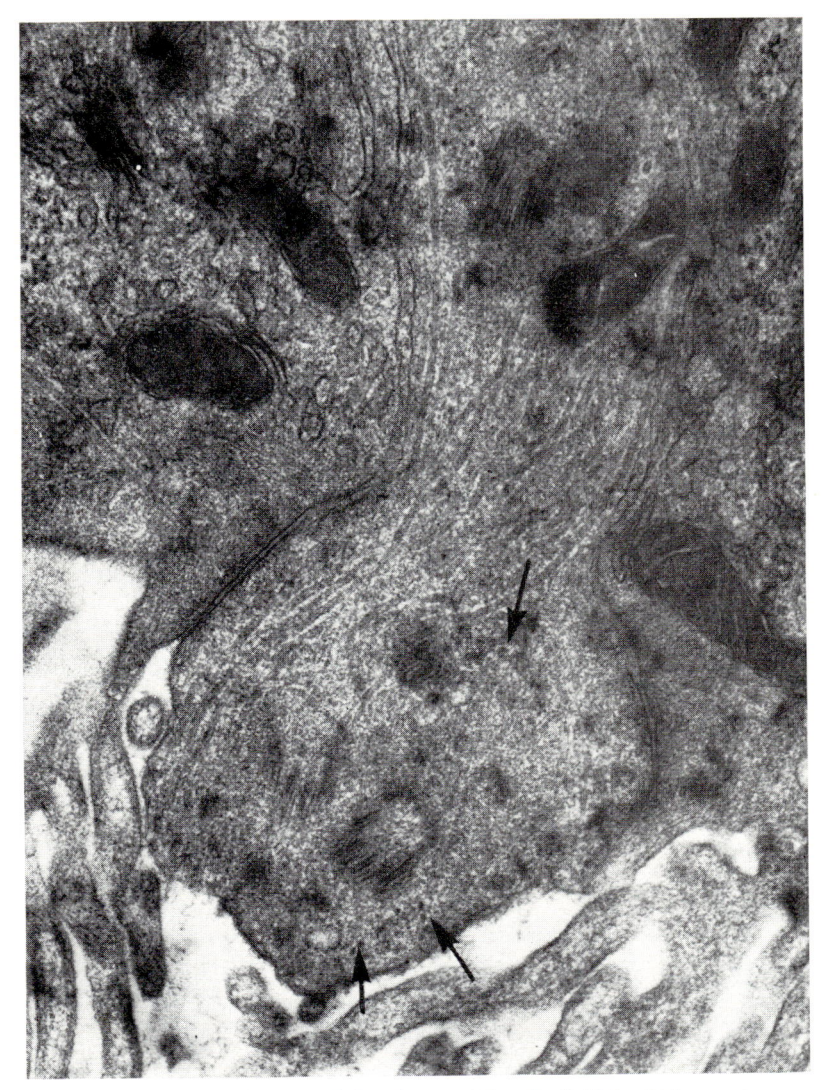

Fig. 12. Tip of olfactory rod showing particles of colloidal gold (arrows) which have crossed the cell membrane and reside in the cytoplasm.
× 24750

other tissue cells; (2) if mammalian Ringer's solution, lacking only the colloidal gold particles, is placed on the mucosa, no such densities are seen in the nerves or olfactory bulbs and (3) the addition or subtraction of calcium in no way affects these observations. We are therefore reasonably

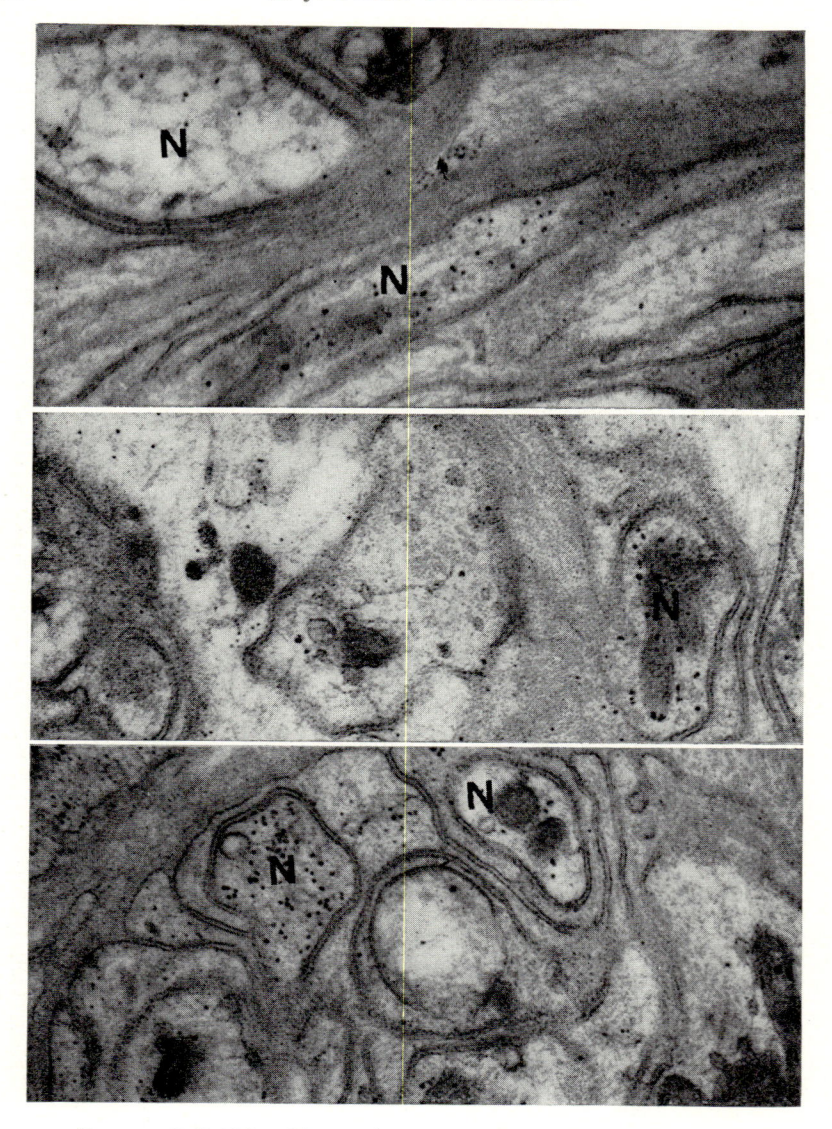

FIG. 13. Colloidal gold particles are seen in axoplasm of olfactory nerves (N). ×24750

confident that we are visualizing the portal of entry and the progression of movement of colloidal gold particles from mucosa to brain.

The movement of these particles is quite rapid since we see them in the olfactory bulbs 30–60 minutes after intranasal inoculation. This time

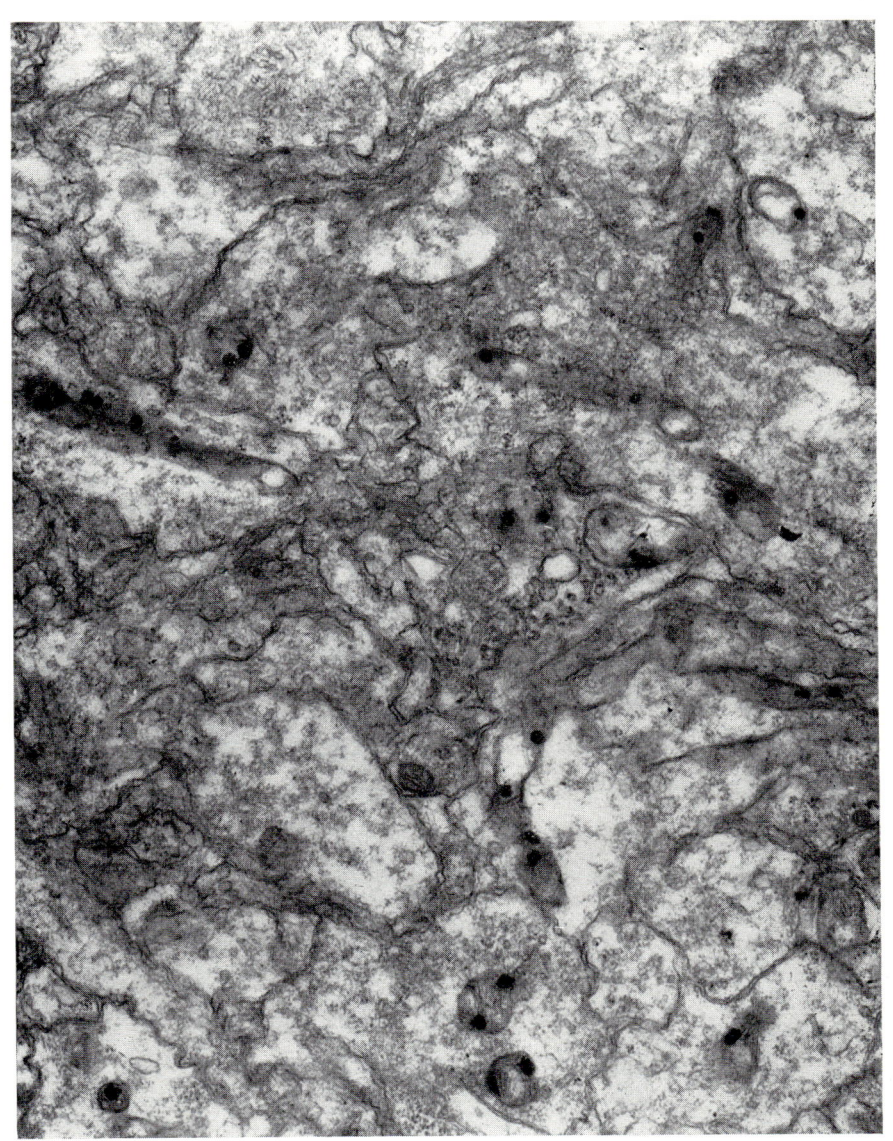

FIG. 14. Cross-section of olfactory glomerulus showing preferential location of gold particles in mitochondria of mitral cell dendrites. ×21 500

corresponds closely with the passage of other materials along nerve fibres. For example, polio virus moves 2·4 mm/hour in the sciatic nerve (Bodian and Howe, 1941a). Herpes virus placed on the cornea reaches the Gasserian ganglion, a distance of about 17 mm in the monkey, in less than 48 hours.

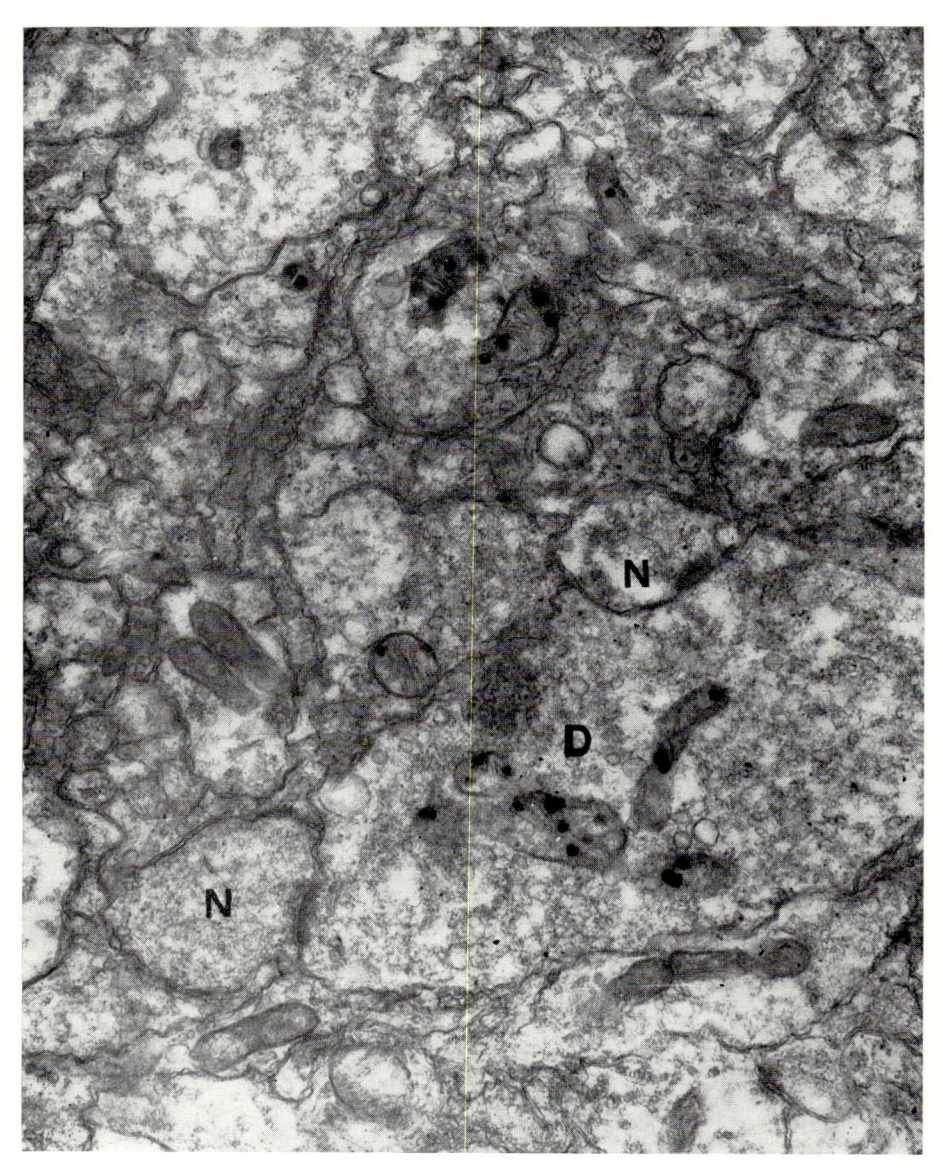

FIG. 15. Further in the olfactory bulb gold particles are seen to have crossed the synapse between olfactory nerves (N) and mitral cell dendrites (D). Note their location in mitochondria. ×24000

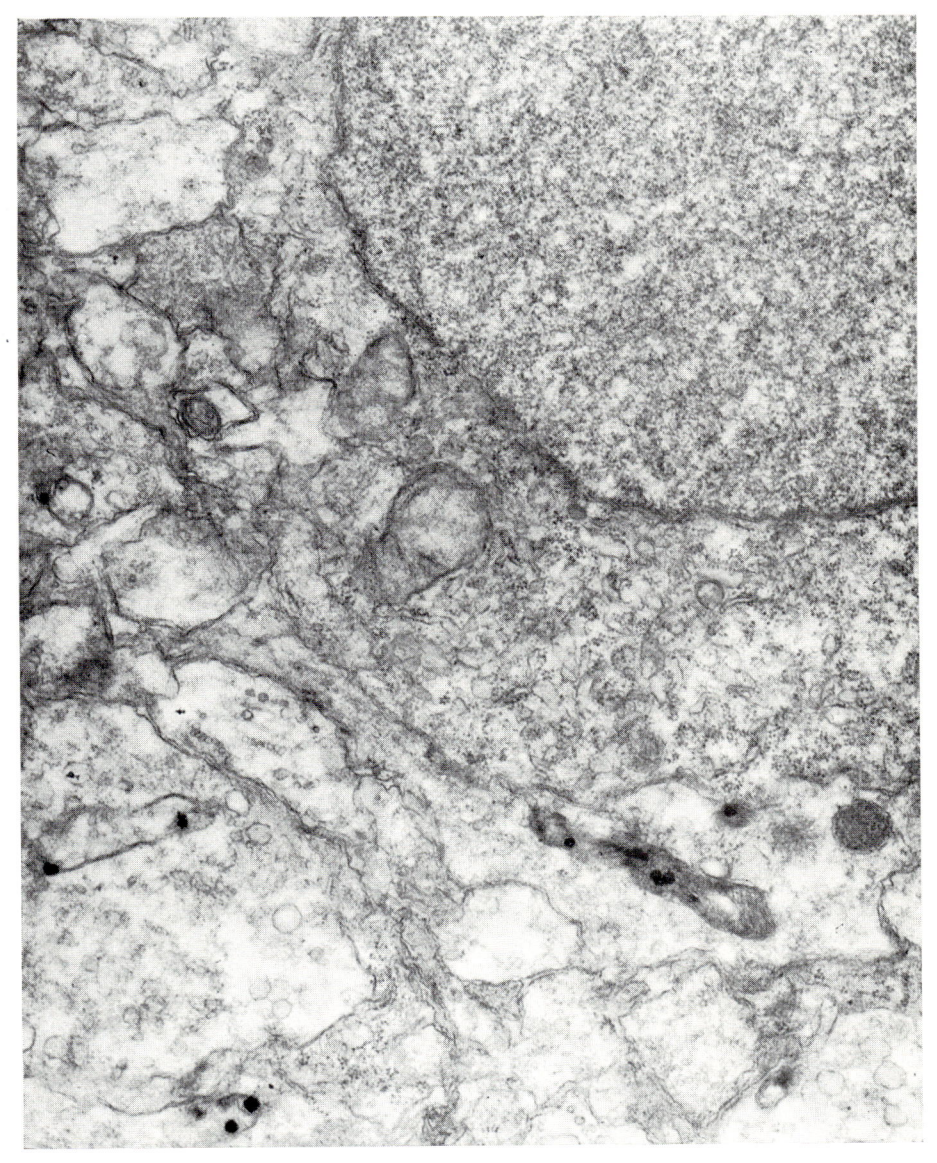

FIG. 16. An axo-somatic synapse on a mitral cell whose nucleus is seen in the upper right corner again shows colloidal gold particles in mitochondria of pre- and postsynaptic processes. × 24000

Our measurements suggest that colloidal gold moves about 2·5 mm/hour in the olfactory nerve.

We can therefore conclude that the olfactory system seems to play a unique role in the blood–brain barrier system. Our observations suggest further studies, both basic and clinical, to resolve the nature and mechanism of the neuronal movement of macromolecules.

SUMMARY

The olfactory epithelium of the squirrel monkey (*Saimiri sciureus*) has been examined with the electron microscope. Emphasis has been placed on the fine structure and histophysiology of the olfactory neuron. The cell membranes of the receptor have a fuzzy outer coating which is continuous with numerous pinocytotic vacuoles. The cytoplasm of the olfactory rod contains ribosomes, mitochondria, microtubules, numerous vesicles, basal bodies, basal feet and centrioles. Each receptor tip or rod is unensheathed and bears 8–20 cilia. These cilia lie parallel to the epithelial surface and are about 50 μm long. They contain the 9+2 ciliary fibrils for only a short distance, when they narrow in diameter and contain only four ciliary fibrils. At their terminals only two ciliary fibrils are seen. Occasional vesicles (dilatations) occur along the length of each cilium and some dilatations contain synaptic-like vesicles. Membrane contacts occur between: (1) receptor cells and supporting cells; (2) receptor cells and other receptor cells; (3) receptor rods and supporting-cell microvilli; (4) receptor cilia and microvilli. Synaptic junctions consisting of subsynaptic web, thickened membrane and presynaptic vesicles are seen between contiguous olfactory receptors, suggesting electrical and chemical interaction at the level of the epithelium.

Colloidal gold particles placed on the nasal mucosa enter the olfactory receptor and are seen in the axoplasm of the fila olfactoria. They reach the olfactory bulb by neuronal transport in mitochondria 30–60 minutes after inoculation. This rapid movement to central nervous tissue suggests a unique participation of the olfactory neuron in the blood–brain barrier system.

Acknowledgement

Research was supported in part by Grant NB RO1 02173-10 from the United States Public Health Service, National Institutes of Health.

REFERENCES

ALLISON, A. C. (1953). *Biol. Rev.*, **28**, 194–244.
BLOOM, G. (1954). *Z. Zellforsch. mikrosk. Anat.*, **41**, 140–154.
BODIAN, D., and HOWE, H. A. (1941a). *Bull. Johns Hopkins Hosp.*, **68**, 248–267.
BODIAN, D., and HOWE, H. A. (1941b). *Bull. Johns Hopkins Hosp.*, **69**, 79–215.
BRANDT, P. W. (1962). *Circulation*, **26**, 1075–1091.
BURTON, P. R. (1968). *Z. Zellforsch. mikrosk. Anat.*, **87**, 226–236.
DE LORENZO, A. J. D. (1957). *J. biophys. biochem. Cytol.*, **3**, 839–850.
DE LORENZO, A. J. D. (1960). *Ann. Otol. Rhinol. Lar.*, **68**, 410–420.
DE LORENZO, A. J. D. (1963). In *Olfaction and Taste* (Proceedings of the First International Symposium, Stockholm, 1962), pp. 5–18, ed. Zotterman, Y. Oxford: Pergamon Press.
DE LORENZO, A. J. D. (1968). In *Medical Physiology*, II, 12th edn., chap. 69, ed. Mountcastle, V. B. St. Louis: Mosby.
FRISCH, D. (1964). *Anat. Rec.*, **148**, 283. (abst.)
GASSER, H. S. (1956). *J. gen. Physiol.*, **39**, 473–498.
HOPKINS, A. E. (1926). *J. comp. Neurol.*, **41**, 253–261.
KIRKPATRICK, J. B. (1969a). *J. Cell Biol.*, **42**, 600–601.
KIRKPATRICK, J. B. (1969b). *Science*, **163**, 187–188.
OTTOSON, D., and SHEPHERD, G. M. (1967). *Prog. Brain Res.*, **23**, 83–190.
PARKER, G. H. (1922). *Smell, Taste and Allied Senses in Vertebrates*. Philadelphia: Lippincott.
REESE, T. S. (1965). *J. Cell Biol.*, **25**, 209–230.

DISCUSSION

Reese: Have you any evidence that the olfactory cilia in the squirrel monkey don't all run in the same direction? It is clear that they all run in the same direction in the frog but they are often figured in other animals as running in all directions. However, can one be sure that they don't turn round before assuming their positions in parallel arrays near the surface of the mucus? If parallel orientation is a consistent property of olfactory cilia, we should discuss its functional implications; for example, in the generation of the slow electrical response.

de Lorenzo: The only generalization I can make is that in sections in which the cilia are displayed longitudinally we never see them turn back on themselves. If they are curving back, we would expect to see this occasionally at least.

Reese: If they are simultaneously curving both vertically and horizontally it would be hard to see it in a vertical or horizontal plane of section.

de Lorenzo: I can't rule this out.

Lowenstein: Is the concept of the cilia carrying out a kind of grasping movement as described by Le Gros Clark something that we have to discard?

Reese: The movement mentioned by Le Gros Clark (1957) was observed by C. Pomerat, who cultured developing mammalian olfactory epithelium.

de Lorenzo: It is very difficult to generalize from tissue culture about movement and migration, however.

Lowenstein: Am I right in saying that you have seen ciliary movement in olfactory epithelium, but only in stumps of cilia, Dr Reese?

Reese: Yes, if olfactory mucosa from a frog is stripped from underlying cartilage and its folded edge examined with phase contrast optics, a variable number of shorter motile cilia are seen among the much longer immobile ones that lie in a parallel array near the surface of the mucus. The number of motile cilia decreases in proportion to the care used in making the preparation and none are present in some areas of some preparations. If a lens is put into the nasal cavity of a living frog and the intact epithelium examined by reflected light optics, no ciliary beat is seen, although a beat is clearly seen in neighbouring non-olfactory regions. My reservations are that reflected light optics (see Fig. 1, p. 145) shows only structures near the surface of the mucus so that ciliary movements might be missed deeper in the mucus, where proximal segments lie. Also, to make these observations one has to light brightly a normally dark cavity. (For fuller discussion of this point see Reese, 1965.)

Wright: We have done some rough calculations, making reasonable assumptions about the physical dimensions of the olfactory cleft in man and the velocity of the air over it. The Reynolds number in the air is extremely high, indicating a very high degree of turbulence. The air passing over the mucus film would therefore take hold of it in a very violent way, so that perhaps the instantaneous pictures of fixed sections don't give a very vivid idea of what is actually happening.

de Lorenzo: D. B. Proctor in our Department of Environmental Medicine is interested in the movement of the nasal mucous blanket. He puts radioactive particles on the nasal mucosa in man and follows their movement, which is very rapid. The movement seems to be in one direction. Such particles placed on the respiratory epithelium are swept down into the lungs very rapidly. The movement of the olfactory cilia doesn't seem to be uniform beating, however, but the cilia seem to be motile. The short jerky movements that Dr Reese described in the frog are more consistent with what Proctor sees in olfactory epithelium in man than the long sweeps of the cilia in the respiratory epithelium.

Reese: Professor de Lorenzo, the structure you showed (Fig. 5, p. 157) in the olfactory epithelium is certainly a synaptic contact, but I wasn't sure of the criteria used for identifying the two cells. If they are both bipolar cells, how frequently do synaptic contacts occur between them?

de Lorenzo: I have seen only two such junctions. Usually receptors are separated from each other by sustentacular cells. In the case described both processes contained microtubules and the cytoplasm was typical of a cell body. Here the two receptor cells were in contiguity with each other. In the perinuclear area or perhaps a little deeper one sees the regular 15 nm (150 Å) spacing of the intercellular space. In the case described here there seems to be a synaptic junction between the two receptors at the level of the nucleus, or just below.

Reese: Did you rule out the possibility that one of the processes is a nerve ending? Nerves have microtubules too. Such nerves might be part of an efferent system, for example.

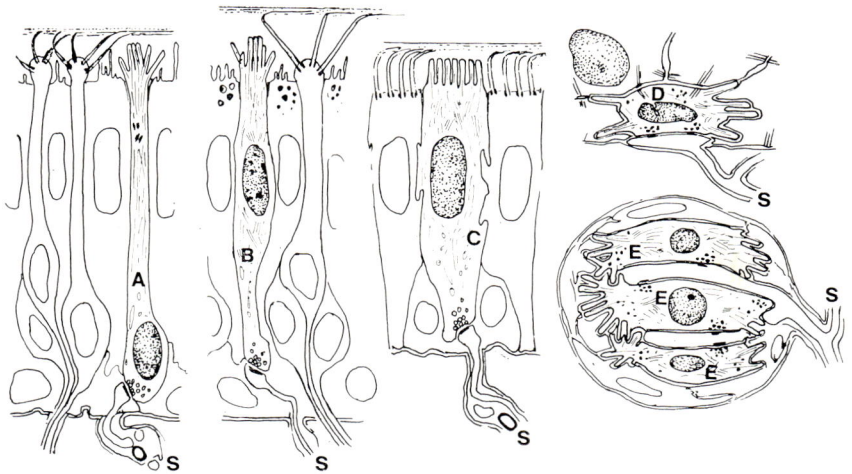

FIG. 1 (Andres). Previously undescribed sensory cells with stiff, finger-like microvilli from the olfactory epithelium of the cat (A), from the olfactory epithelium of the turtle (B), and from the mucous membrane of the trachea of the rat (see Luciano, Reale and Ruska, 1968) (C). Their ultrastructural appearance with a pronounced intracytoplasmic net of filaments resembles that of Merkel's touch cell (D) and the Grandry corpuscle (E). Cells A, B and C have basal synaptic contacts with the intraepithelial sensory nerve endings (S).

de Lorenzo: No, I am sure this was not the case. I have not seen anything structurally that resembles an efferent innervation. There very likely are efferent fibres; such have been described, for example, by Cajal, but I have never seen such endings.

Andres: In the olfactory mucosa of cat, turtle and salamander we have found similar synaptic junctions to those you have demonstrated. These synapses are formed by one particular type of sensory cell and by one sensory axon which approaches the sensory cell, in a similar way to the gustatory cell (see Fig. 1).

Lowenstein: I would like to ask how a granule of gold gets into a mitochondrion.

Reese: Gold granules act as if the cell membrane isn't there; they go in and out of vesicles or from cell to cell apparently indifferent to membranes.

de Lorenzo: The particles are so large, they probably damage the cell membrane, which heals like a Langmuir film which reconstitutes in a few seconds after a hole is made. Ferritin is a better material to use because of this. Ferritin particles adsorb to the membrane and can be clearly identified in the electron microscope.

REFERENCES

LE GROS CLARK, W. E. (1957). *Proc. R. Soc. B*, **146,** 299–319.
LUCIANO, L., REALE, E., and RUSKA, H. (1968). *Z. Zellforsch. mikrosk. Anat.*, **85,** 350–375.
REESE, T. S. (1965). *J. Cell Biol.*, **25,** 209–230.

ANATOMY AND ULTRASTRUCTURE OF THE OLFACTORY BULB IN FISH, AMPHIBIA, REPTILES, BIRDS AND MAMMALS[*]

K. H. ANDRES

Anatomisches Institut der Universität Heidelberg[†]

OVER the last few years electron microscopic study of the olfactory bulb in various mammals has produced important information about its ultrastructure (Hirata, 1964; Andres, 1965a; Rall *et al.*, 1966; Price, 1968). The present study tries to demonstrate that already in the cyclostomes the olfactory system has a basic structure which remains unchanged throughout the vertebrates, right up to primates. The elucidation of the functioning of this pathway will be a task to be shared by physiologists and morphologists.

The distinct layers of the olfactory bulb known from light microscopy make it easier to relate new electron microscopic findings to the whole system. In the lamprey as well as in elasmobranchs and teleost fishes the olfactory bulb is arranged in five layers (Fig. 1): (1) olfactory nerve layer, (2) glomerular layer, (3) mitral cell layer, (4) plexiform layer and (5) periventricular layer. In land-living vertebrates there is a further plexiform layer between the mitral cell layer and the olfactory glomeruli, so that an internal plexiform layer can be distinguished from an external one. Apart from the endings of the olfactory receptor cells in the olfactory glomeruli, and the mitral cells, the granule cells essentially determine the appearance of the olfactory bulb. As the structure of the neuropil is primarily responsible for the characterization of the layers, one should avoid the term "granule cell layer". Granule cells are, rather, embedded in the periventricular and internal plexiform layer and, in terrestrial vertebrates, also in the glomerular layer (Fig. 1).

The light microscopic picture of the olfactory bulb of the lamprey appears very uniform. The relatively sparsely occurring granule cells, together with the mitral cell axons and centripetal fibres, form a dense neuropil. The olfactory glomeruli are very large and only indistinctly defined. The

[*] Supported by the German Research Council.
[†] From April 1970: Anatomisches Institut der Ruhr-Universität Bochum.

mitral cell dendrites are so long that they project into several glomerular branches.

Especially thick dendrites of the mitral cells are particularly striking in *Scyliorhinus canicula* (elasmobranch) and *Tinca vulgaris* (teleost). In these species the large olfactory glomeruli are sharply defined from each other and from the plexiform layer by astrocyte processes (Fig. 2). The plexiform layer is rich in granule cells. The myelinated axons, which can be

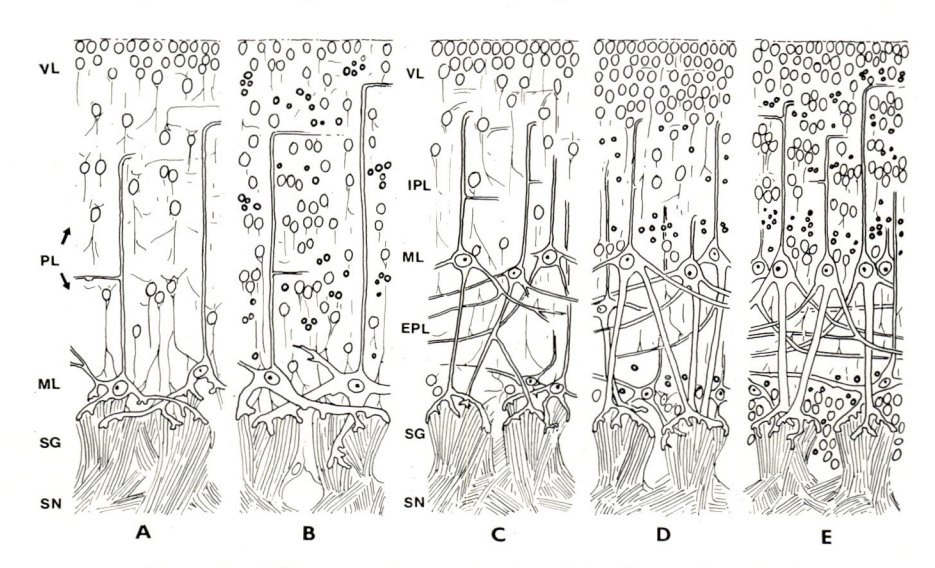

FIG. 1. Semi-diagrammatic representation of the fibres and cytoarchitecture of vertebrate olfactory bulbs. Perikarya, dendrites and axons of the mitral and tufted cells are distinguished. Granule cells are indicated by the oval outlines of their nuclei, and their axons are indicated by fine lines. Small bold circles and dots are cross-sections of myelinated fibres. A: Lamprey, B: Selachian (elasmobranch), C: Amphibian, D: Reptile, E: Mammal, A and B have five layers: olfactory nerve layer (SN), glomerular layer (SG), mitral cell layer (ML), plexiform layer (PL), periventricular layer (VL). Internal plexiform layer (IPL); external plexiform layer (EPL).

associated with the bulbofugal and bulbopetal fibre systems, have a diameter of 2–6 μm in *Scyliorhinus canicula* and of 0·8–3 μm in *Tinca vulgaris*.

In the amphibia (*Salamandra salamandra*, *Rana adspersa*) that we have studied we have found the neuropil to be in a relatively undifferentiated stage, like that in the lamprey. The long accessory dendrites of the mitral cells are a new acquisition, not found in the fishes. These accessory dendrites pass through the external plexiform layer predominantly tangentially to the surface of the bulb. At this stage the accessory and primary dendrites are very fine so that the texture differs very little between the internal and external plexiform layers. At the border of the glomerular layer a few single tufted and granule cells are found. There are no myelinated axons in the neuropil of amphibia.

Of the reptiles (*Terrapene*, *Testudo*, *Varanus niloticus*) that we have studied, *Varanus niloticus* has most myelinated axons in the plexiform layer. The diameter of the fibres is 2–3 μm. The external plexiform layer showed

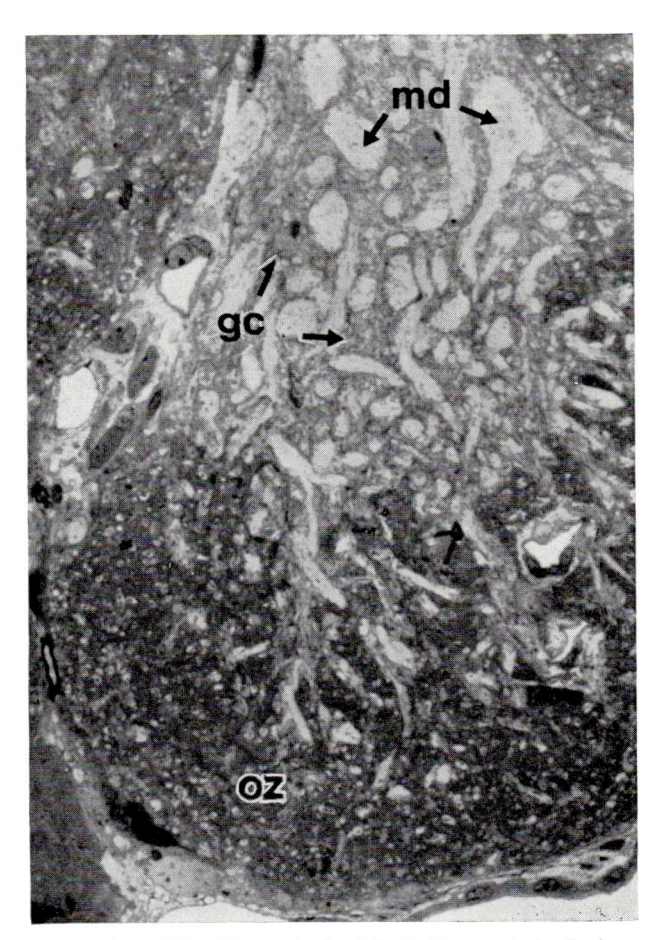

FIG. 2. Section of the olfactory bulb of *Scyliorhinus canicula*. In the outer zone (oz) between the dark mass of fila olfactoria are very fine terminal dendrites of mitral cells. In the inner region are mitral cell dendrites (md) mixed with granule cell axons (gc). Light microscopic picture of a semi-thin section. × 500

relatively broad, light primary and accessory dendrites. In the region of the glomeruli tufted and granule cells are more profuse. The majority of the cell bodies of the inner granule cells lie in the periventricular layer.

The avian olfactory bulb (*Larus argentatus*, *Philomachus pugnax*) resembles the turtle bulb in its cytoarchitecture.

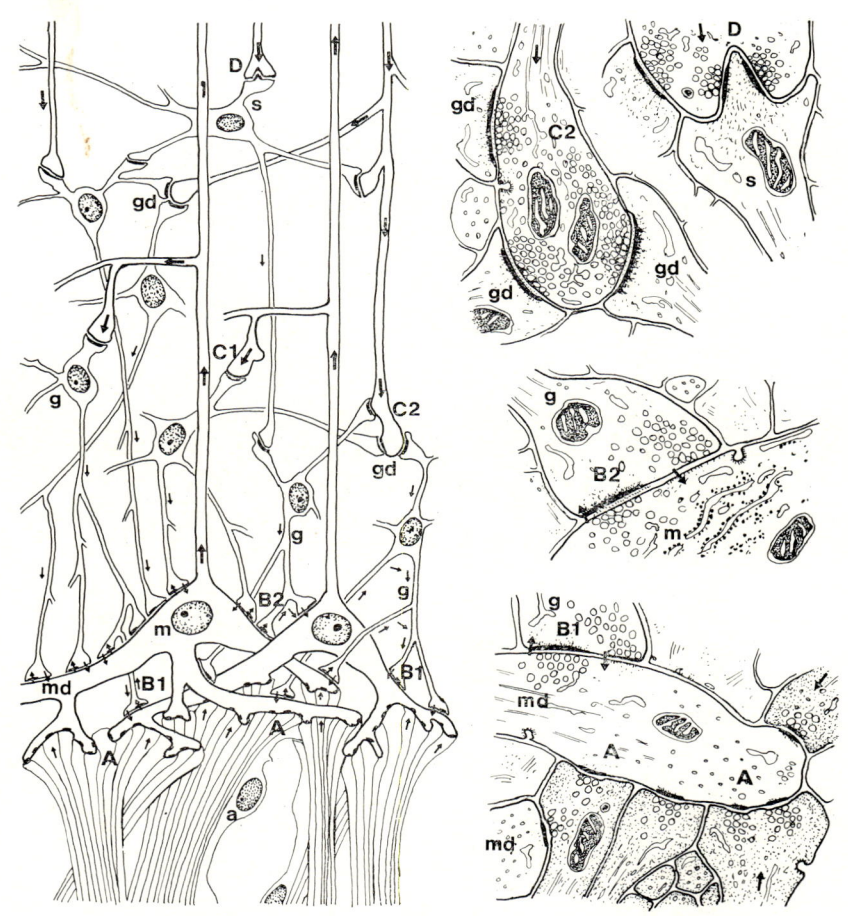

FIG. 3. Diagram of synaptic connexions in the olfactory bulb of fishes. *Left:* hypothetical presentation of neuronal connexions. *Right:* Detailed drawings of types of synapses. Synapses between fila olfactoria and mitral cell dendrites (A), reciprocal granule cell ending on mitral cell dendrite (B1), on the soma of a mitral cell (B2), mitral cell collateral endings (C1) and central bulbopetal fibre endings (C2) on granule cell dendrites; synapses with interdigitated membrane complex in the periventricular zone, probably on stellate cells (D). C1 and C2 form morphologically similar synapses. Granule cell (g), mitral cell (m), stellate cell (s), astrocyte (a), dendrite (d). The arrows indicate the direction of transmission of the stimulus.

Whereas in submammals a mitral cell usually makes contact with several olfactory glomeruli, in mammals the mitral cells are mostly in contact with one glomerulus. Accessory dendrites of the mitral cells undergo their greatest development in length and ramification in the external plexiform layer. The internal plexiform layer is scattered with granule cell groups and interlaced by myelinated axons, of diameter 0·8–2·5 μm. The number of tufted and external granule cells increases considerably in the region of the olfactory glomeruli. The myelinated axons of the centrifugal and centripetal system are also found in this zone.

In amphibia (*Rana adspersa, Xenopus laevis*) and reptiles (*Terrapene, Testudo, Varanus niloticus*) the structure of the accessory olfactory bulb

FIG. 4. Light micrograph of olfactory glomeruli of a young mouse, stained by the Golgi method. The fila olfactoria radiate from below and seem to branch out. It is probable that bundles of fila and astrocyte processes have also been impregnated. × 230

corresponds to the cytoarchitecture of the main olfactory bulb. In mammals (rat, rabbit) the accessory olfactory bulb shows a reduction of cells in the glomerular region as well as thinner primary and accessory dendrites of the mitral cells.

Under the electron microscope the cell types mentioned above are easily recognized by their ultrastructure. Further types of nerve cells known from light microscopy, such as triangular cells, spindle cells, short-axon cells, periglomerular and interglomerular cells, which Nieuwenhuys (1967) mentions in his comparative study of the olfactory bulb in lower vertebrates, have not been studied by us electron microscopically. It is still necessary to establish how far we are dealing with the same functional elements but with different form and nomenclature, in light and electron microscopy.

The considerable advantage of electron microscopy over light microscopy lies in the fact that it is now possible to work out the synaptic contacts between cells and thus their neurophysiological connexions (Fig. 3). Here we are still at the threshold of a completely new and extensive field of research. Our findings already indicate that with the aid of the characteristic types of synapses, we have important criteria which also allow us to draw physiological conclusions from morphological studies.

In the olfactory bulb of vertebrates, including the lamprey, one can demonstrate four basic types of synapses. (1) synaptic contact between the contrast-rich axon terminals of the olfactory receptor cells and either mitral cell dendrites or tufted cell dendrites. In contrast to earlier views which derived from the results of Golgi staining (Fig. 4), no terminal ramification of the receptor axons could be demonstrated in electron microscopic pictures. However, the enlarged terminal may have synaptic contact with two or more mitral cell dendrites. In the lamprey there are also terminals of the fila olfactoria in which two membrane complexes are formed on one dendrite (Fig. 5). The topographical arrangement of the olfactory cell branches on the dendrites appears to be important, since in all vertebrates a dense material can be demonstrated in the intracellular space between the axons. This dense material may help maintain axon-to-axon cohesion (Figs. 6 and 13).

(2) Synaptic contact between granule cell axon and mitral cell, or between granule cell axon and tufted cell. This type of synapse, described by Hirata (1964), Andres (1965a), Rall and co-workers (1966) and Price (1968), is distinguished by the fact that it consists of two different membrane complexes. Here one complex (axo-dendritic or axo-somatic) seems to correspond to type I according to Gray, whereas the other complex (dendro-axonic or somato-axonic) seems to be close to type II according to Gray (1961) (Fig. 7). In the lamprey and the other fish examined this reciprocal type of synapse is found on the dendrites and the somata of mitral cells (Fig. 8). A few of the granule cell terminals penetrate deeply into the glomeruli. In terrestrial vertebrates these reciprocal synapses are predominantly on the accessory dendrites of the mitral and tufted cells; they are less frequent on the primary dendrites and on the mitral cell body. The granule cell terminals of *Philomachus pugnax* contained strikingly abundant granular vesicles (Fig. 9) similar to the so-called catecholamine granules.

(3) Terminals on the dendritic region of the granule cells. The synaptic membrane complexes of these terminals are remarkably large in diameter. Usually one such bouton membrane complex makes synapses with

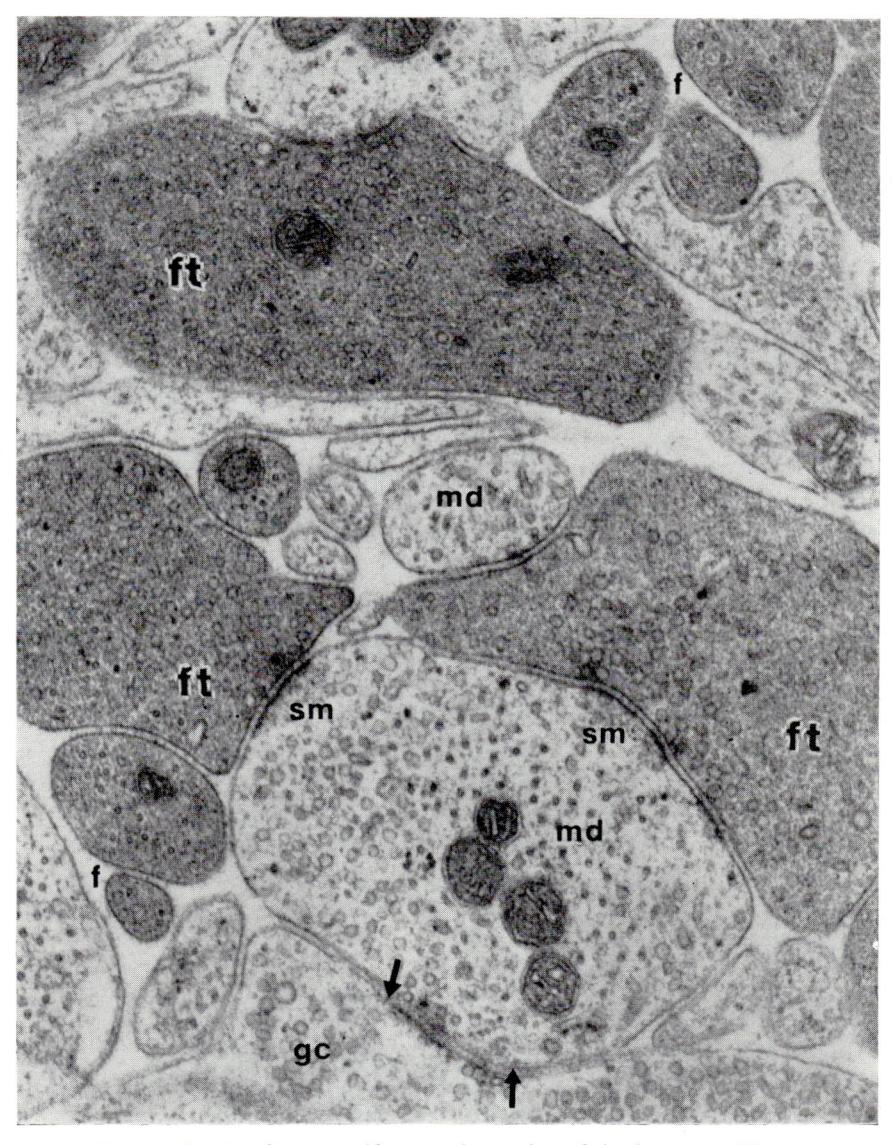

FIG. 5. Section from an olfactory glomerulus of the lamprey. Fila olfactoria (f) and their terminals (ft) with synaptic complexes (sm) are well contrasted. Mitral cell dendrites (md), granule cell endings (gc). Arrows indicate a small reciprocal synapse. Electron micrograph. × 40 000

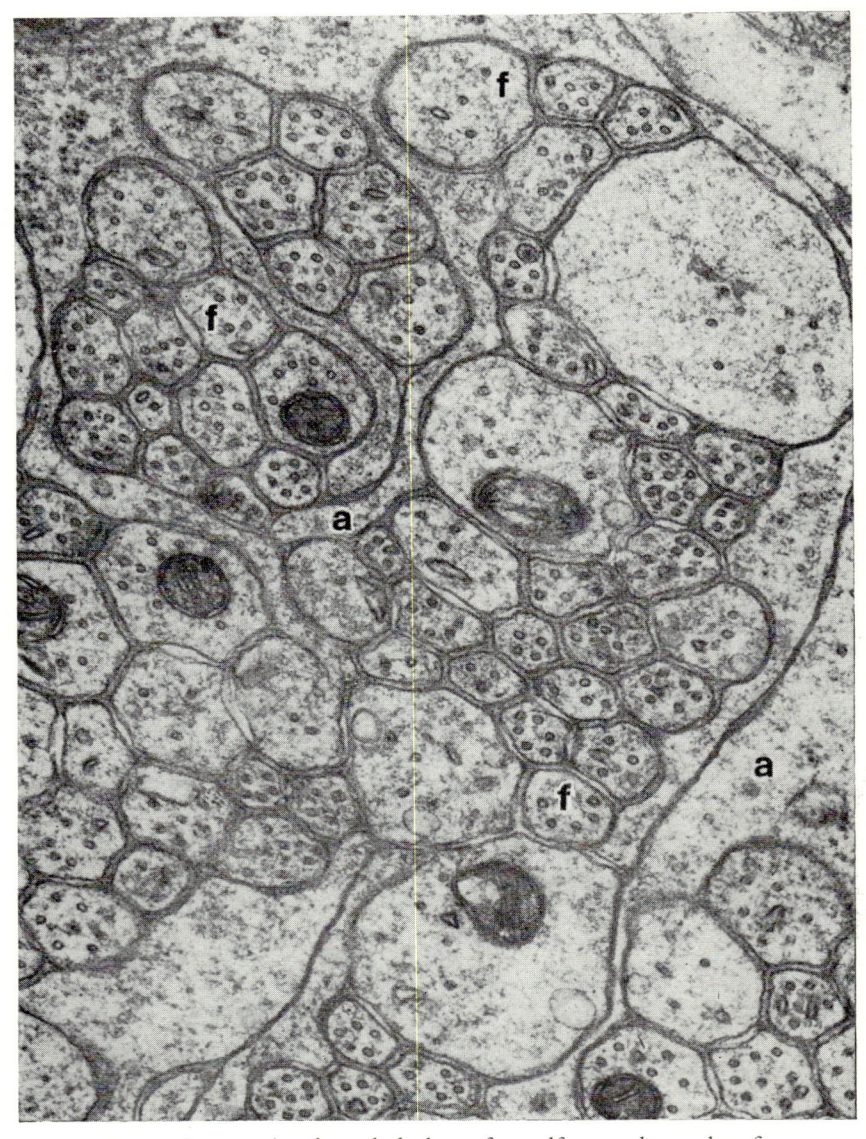

FIG. 6. Cross-section through the base of one olfactory glomerulus of a young cat. The fila olfactoria (f) are embedded in astrocyte (a) endings; between the fila, electron-dense material can be seen. The number of microtubules is remarkably small in the fila olfactoria. Electron micrograph. ×65 000

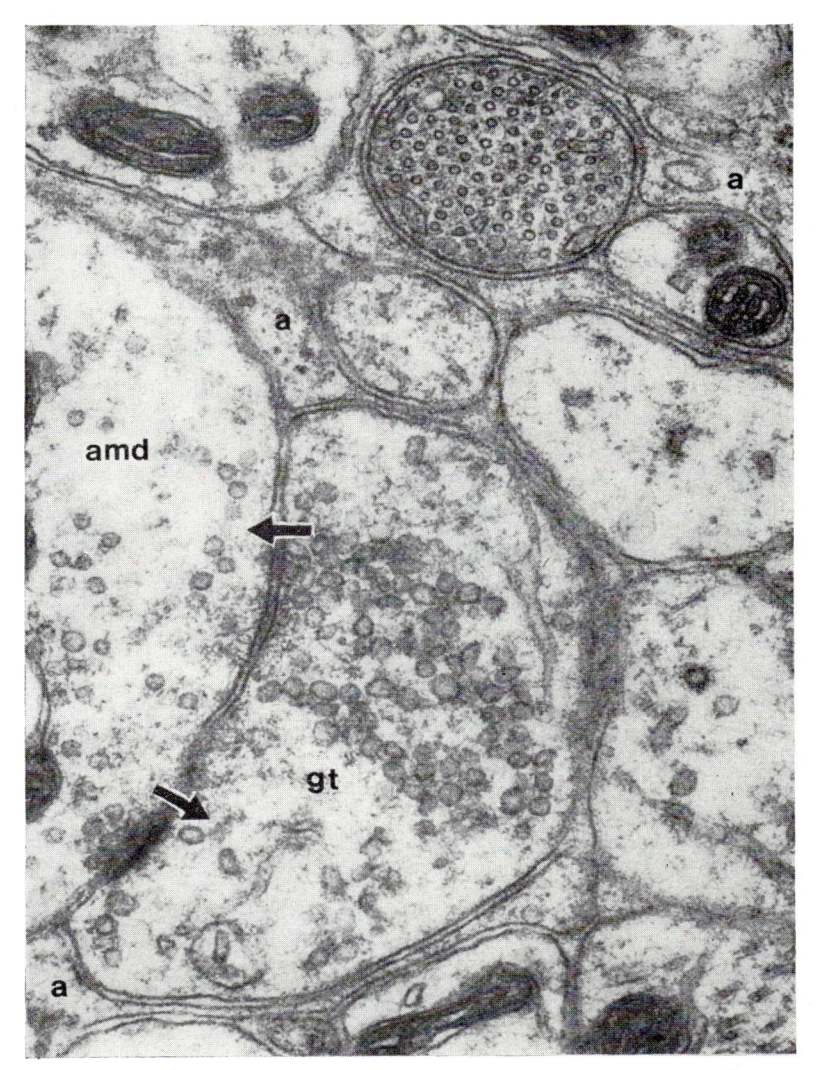

FIG. 7. Section from the neuropil of the external plexiform layer of a young dog. A reciprocal synapse (see arrows) is sited on an enlargement of an accessory mitral cell dendrite (amd). Granule cell terminal (gt), astrocyte processes (a). Electron micrograph. × 70 000

terminals of several granule cells (Fig. 10). Axo-somatic connexions also occur, for example in the rat. In mammals the synaptic vesicles of these terminals have a relatively small diameter of 35–45 nm (350–450 Å). In the lamprey and in higher fishes this type of terminal is distributed throughout the plexiform layer, whereas in terrestrial vertebrates these terminals are more frequent in the internal plexiform layer than in the outer part of the external plexiform layer. The terminals on the outer granule cells usually appear smaller; here axo-axonal contacts are fairly common.

(4) Synapses with interdigitated contact. This axo-dendritic type of synapse is found more deeply, near the periventricular layer (Fig. 11), and in mammals also in the lateral olfactory tract. It is not clear which cell terminals form this type of synapse. Similar types are found in the higher olfactory centres, for example in the nucleus septi (Andres, 1965*b*).

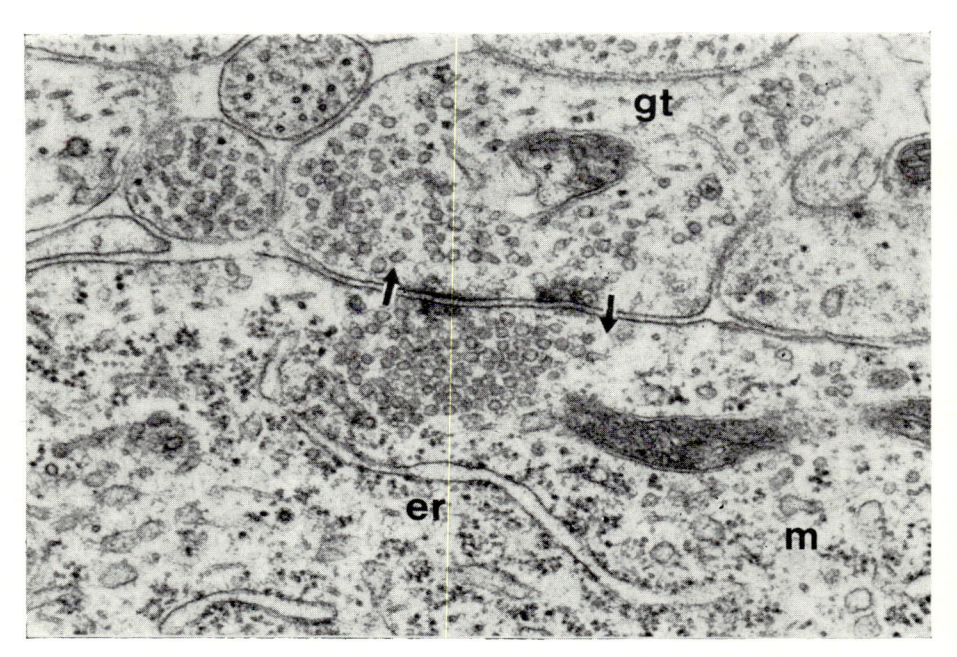

FIG. 8. Reciprocal synapse (see arrows) on a mitral cell perikaryon of the lamprey. Granule cell terminal (gt), mitral cell (m) with Nissl substance (er). Electron micrograph. × 40 000

Up to now morphologists have been unable to produce reliable criteria by which to distinguish olfactory receptor cell types responding to the different primary odours. In this connexion it seems interesting that the olfactory cell axons contain different numbers of microtubules, varying from two to 18. The mitochondria are situated in spindle-shaped extensions of the fila olfactoria. Axons that contain exclusively endoplasmic reticulum could be regenerating fibres. The model shown in Fig. 12 greatly simplifies the actual situation. Allison and Warwick (1949) were able to demonstrate that in the rabbit 26 000 fila olfactoria enter one glomerulus to make contact with 24 mitral cell primary dendrites and 68 tufted cell dendrites.

The results of our study show that the olfactory bulb has essentially the same cytoarchitecture in the lamprey and in all higher vertebrates. This is also true of those cases where the equivalent sensory cells have no flagella or cilia as, for instance, in *Scyliorhinus canicula* or in Jacobson's organ (the

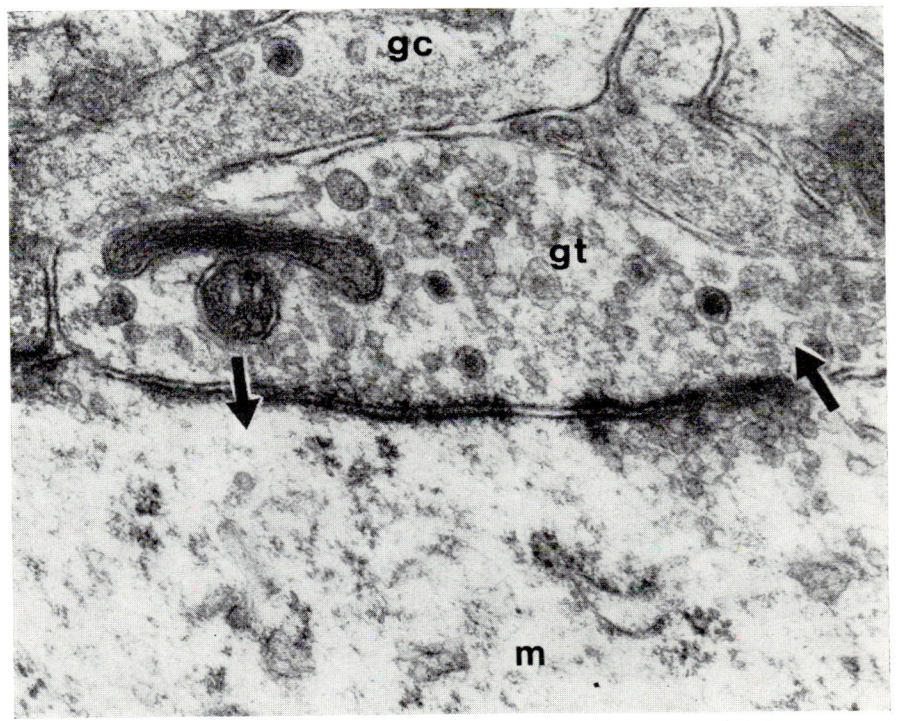

FIG. 9. Reciprocal synapse (see arrows) on mitral cell soma (m) in *Philomachus pugnax*. Besides the ordinary synaptic vesicles the granule cell terminal (gt) contains dense-cored vesicles. Granule cell axon, gc. Electron micrograph. × 60 000

vomeronasal organ) whose fibres lead to the accessory olfactory bulb (Figs. 13 and 14).

DISCUSSION

Herrick already thought in 1924 that the granule cells of the olfactory bulb must be physiologically unpolarized; that is, that their terminals could act transitorily as receptors and also as stimulators of other cells. The morphological substratum of this hypothesis could be the reciprocal

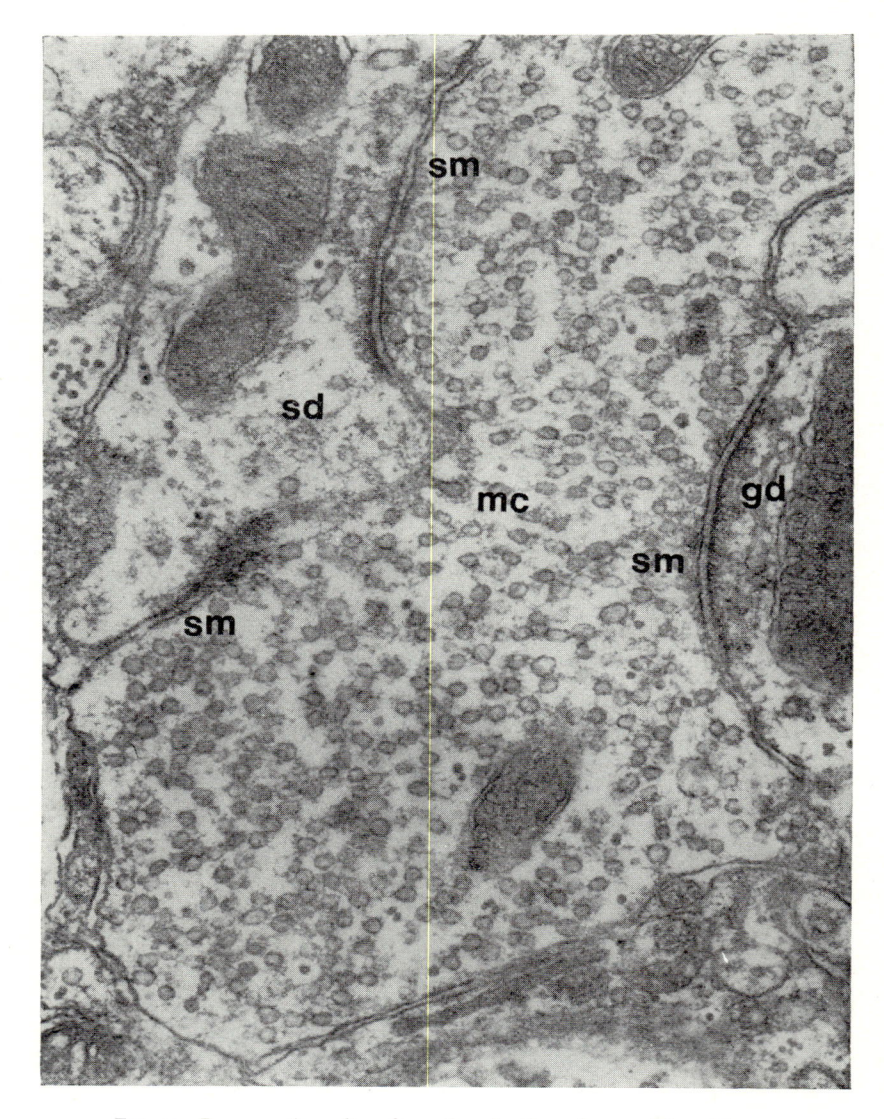

Fig. 10. Presynaptic ending from the plexiform layer of the olfactory bulb of *Scyliorhinus canicula*. This type of ending (sm) with large membrane complexes comes from the myelinated bulbopetal fibres or mitral cell collateral axon. Coated micropinocytotic invagination (mc), granule cell dendrite (gd), stellate cell dendrite (sd). Electron micrograph. × 50 000

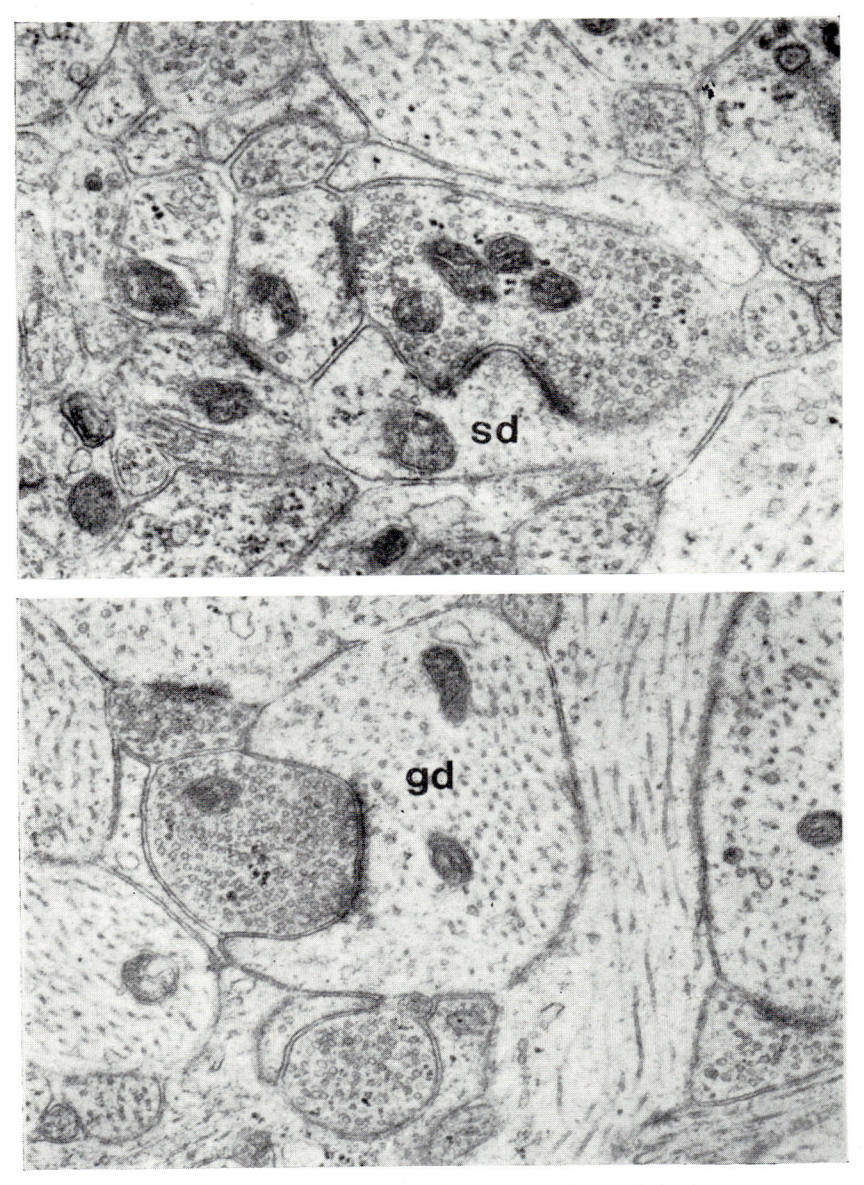

Fig. 11. Types of synapses from the olfactory bulb of the lamprey. *Upper:* From the periventricular layer, with interdigitated contact. *Lower:* Two small and one larger ending from the plexiform layer. Stellate cell dendrite (sd). The large synapses could be formed by the central bulbopetal fibres. Granule cell dendrite (gd). Electron micrograph. × 28 000

synapse described by Hirata (1964) and Andres (1965a). From later physiological studies, Rall and co-workers (1966) suspect that the granule cells have an inhibitory action on the mitral cells. When the lateral olfactory tract is transected (Powell, Cowan and Raisman, 1965; Heimer, 1968; Price, 1968) one can observe degeneration of terminals in the internal and external plexiform layer of the olfactory bulb. With the aid of serial

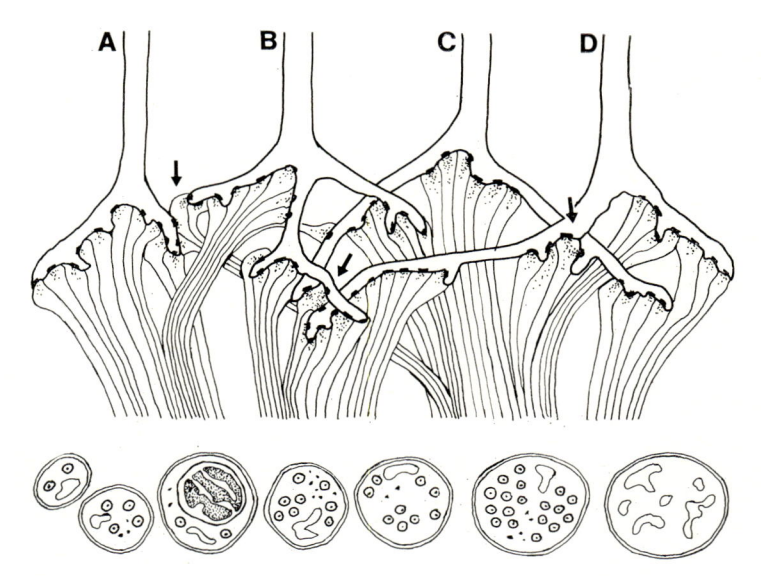

FIG. 12. Schematic drawing of types of olfactory fibres and their interlacing pattern in the olfactory glomerulus. A–D, mitral cells. Several types of fibres can occur in one bundle. Despite this interlacing it is conceivable that single mitral cells are predominantly supplied by one type of fibre. For instance, the primary dendrite B receives mainly fine fibres from different bundles. Single endings form contacts with two or three mitral cells (see arrows).

sections Price succeeded in demonstrating electron microscopically the degenerated terminals on the granule cells.

When one considers the comparative morphological findings in the light of the results cited above, one could hypothesize that the same mechanisms exist in the lamprey and in all higher classes of vertebrates. The inhibition, or modulation, of cell excitation could be produced by the granule cell dendrites in all cases. The input which controls the inhibitory function of the granule cells comes (1) by way of the axo-dendritic synapses between granule cells and the bulbopetal central fibres as well as the accessory dendrites of the mitral cells, and (2) via the dendro-axonic

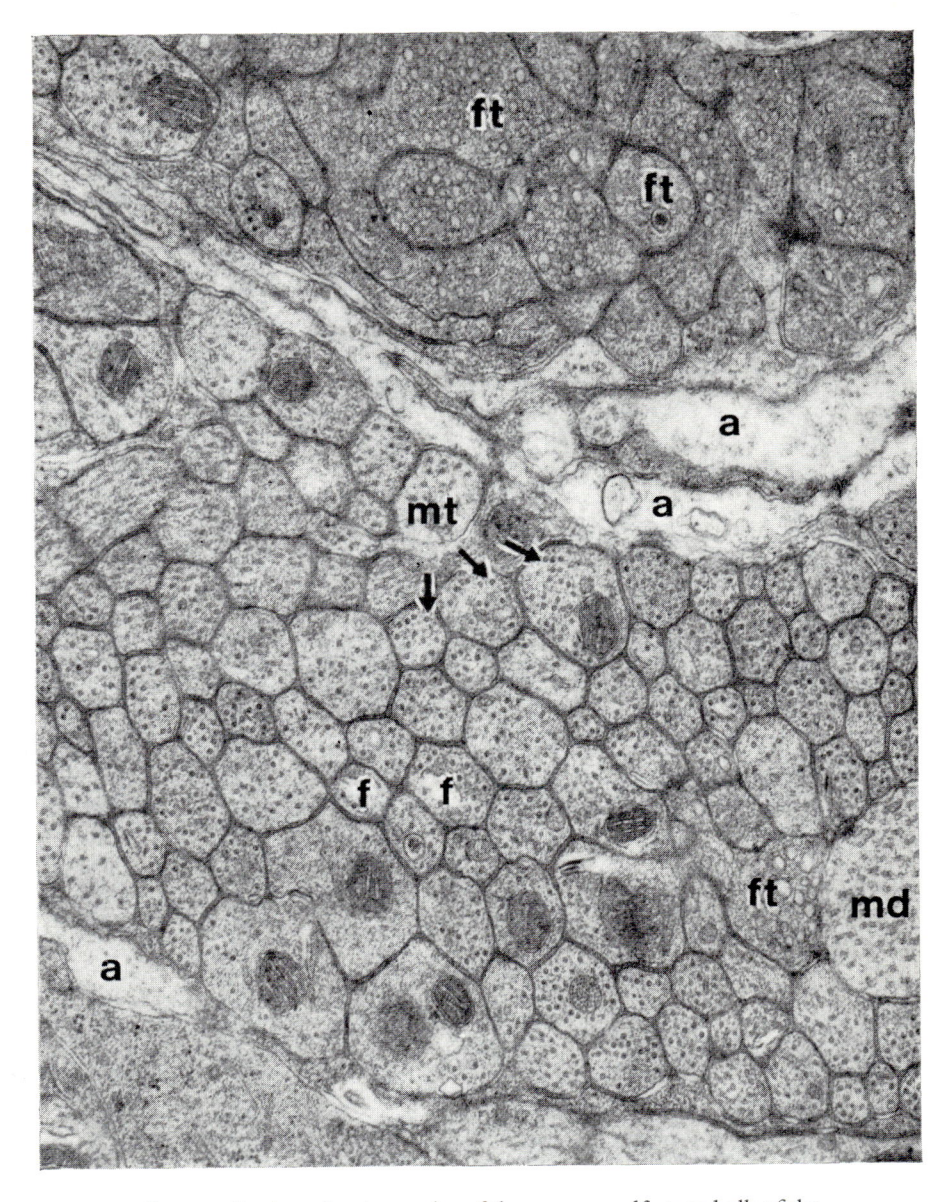

Fig. 13. Section of a glomerulus of the accessory olfactory bulb of the rat. As in the main olfactory bulb the intercellular space contains electron-dense substance between the fila olfactoria (f) and their terminals (ft). The large numbers of microtubules (mt) in the fila olfactoria are conspicuous. Mitral cell dendrite (md), astrocyte processes (a). Electron micrograph. × 30 000

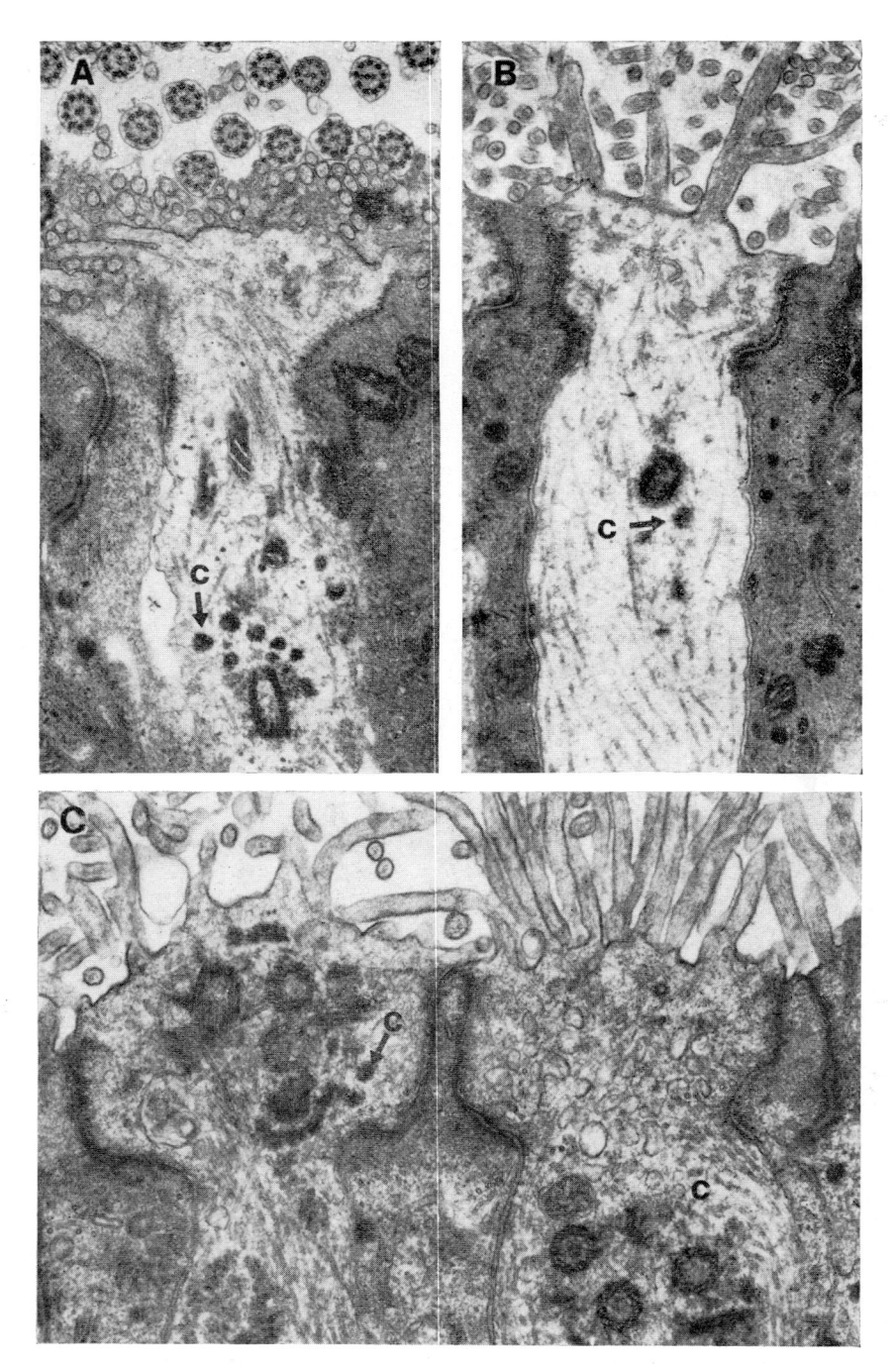

Fig. 14. Examples of olfactory cells without cilia from the vomeronasal organ of *Xenopus laevis* (A), *Varanus niloticus* (B), and the rat (C). The sensory cells have dense clusters of branching microvilli on their surface. Centrosomes (c), centrosome fragments (arrows) and microtubules are regularly found in the sensory end-bulb. The supporting cells of the vomeronasal organ of *Xenopus laevis* have kinocilia. Electron micrographs. A and B × 25 000, C × 32 000

contact between mitral cell dendrites and granule axons. Whereas in the first case probably large regions of the olfactory bulb are influenced, in the second case lateral inhibition could play a more localized role, near the intermittently excited mitral cell, as Rall and co-workers (1966) think. Lateral inhibition will be the more effective, the more mitral cells are excited. On the other hand, the olfactory pathway would be facilitated by weak excitation of a few mitral cells.

The phylogenetic differentiation of the olfactory bulb consists first of an increase in the numbers of granule cells, and an increase in the mitral cell surface, and also of increased numbers of accessory dendrites of mitral cells and intensified feedback from the higher olfactory centres. This higher integration of cell elements leads one to suspect that the olfactory bulb of mammals can receive more highly differentiated odour patterns in comparison to that of other vertebrates.

SUMMARY

Comparative electron microscopic studies reveal identical ultrastructural organization of the olfactory bulb in all classes of vertebrates, including cyclostomes. The size and structure of synapses in the olfactory bulb are specific for each connexion type. The dark endings of the olfactory receptor cells have small axo-dendritic contacts to the bright mitral or tufted cell processes within the glomeruli. Granule cells and mitral cells interact by dendro-dendritic, dendro-axonic and somato-dendritic synaptic complexes which often have "reciprocal" arrangements. Presynaptic endings on the granule cell dendrites and somata contain a large number of small synaptic vesicles and have membrane complexes more than 0·5 μm in diameter. In the central periventricular zone of the olfactory bulb a further type of axo-dendritic synapse is found. As in synapses of the hippocampal region, several small membrane complexes protrude from the axonal ending into the postsynaptic dendrite. There is no fundamental difference between the ultrastructure of the olfactory bulb and the accessory olfactory bulb in anurans, reptiles and rodents.

REFERENCES

ALLISON, A. C., and WARWICK, R. T. T. (1949). *Brain*, **72**, 186–1197.
ANDRES, K. H. (1965a). *Z. Zellforsch. mikrosk. Anat.*, **65**, 530–561.
ANDRES, K. H. (1965b). *Z. Zellforsch. mikrosk. Anat.*, **68**, 445–473.
GRAY, E. G. (1961). *J. Anat.*, **95**, 345–356.
HEIMER, L. (1968). *J. Anat.*, **103**, 413–432.
HERRICK, C. J. (1924). *J. comp. Neurol.*, **37**, 373–396.

HIRATA, Y. (1964). *Archvm histol. jap.*, **24**, 293–302.
NIEUWENHUYS, R. (1967). *Prog. Brain Res.*, **23**, 1–64.
POWELL, T. P. S., COWAN, W. M., and RAISMAN, G. (1965). *J. Anat.*, **99**, 791–813.
PRICE, J. L. (1968). *Brain Res., Amst.* **7**, 483–486.
RALL, W., SHEPHERD, G. M., REESE, T. S., and BRIGHTMAN, M. W. (1966). *Expl. Neurol.*, **14**, 44–56.

DISCUSSION

Mac Leod: It would be interesting to know if the synapses between granule and mitral dendrites are all similar on morphological criteria, such as their synaptic vesicles. Have you any idea, Professor Andres, about the frequency of occurrence of synapses of similar or opposite action between two reciprocally connected cells? That is, can you yet compare quantitatively the frequency of a cell A activating cell B which in turn inhibits cell A with the reciprocal inhibition (or activation) between A and B?

Andres: If you consider the type I synapse of Gray as inhibitory and the type II as excitatory, the reciprocal synapse will inhibit the mitral cell activity if concentrated odours excite many receptor cells. On the other hand excitation of a few receptor cells will not build up lateral inhibition via the granule cell axons. There is therefore a facilitated pathway; of course central bulbopetal fibres may modulate this system. However, before making further theories we have to study the bulb in more detail, including statistically.

Murray: You showed us earlier a cell which you characterized as a sensory cell but not an olfactory cell (see Fig. 1, p. 175). It resembled our type II taste bud cell. Was there a synapse in relation to it, polarized in the proper direction for a sensory synapse?

Andres: A synapse was present which was polarized from the cell to the nerve process, like the synapses of your type II cells of the tongue. I referred to similar cells found in the trachea of the rat by Luciano, Reale and Ruska (1968). The form of such cells is that of a cylindrical epithelial cell but the outer free surfaces show microvilli which contain bundles of filaments which go deep into the cell. In the basal parts of these cells one finds these same somato-dendritic synapses on sensory nerve endings.

Reese: I would like to introduce some unpublished work that gives a more quantitative picture of synaptic organization in the deep third of the external plexiform layer. In order to understand the significance of the synapses between granule cell dendrites and mitral cell dendrites in this layer (Andres, 1963; Rall *et al.*, 1966) it seemed necessary to know what other synapses were present. We therefore attempted to identify all synapses in sets of serial sections from the external plexiform layer near the layer of mitral cell bodies in the cat. We found three types of synaptic endings on mitral dendrites and traced enough terminal processes of these to ascertain that they all might originate from granule dendrites. Two synaptic contacts were associated with most of these endings; one synaptic contact was always oriented so that the granule cell ending was presynaptic

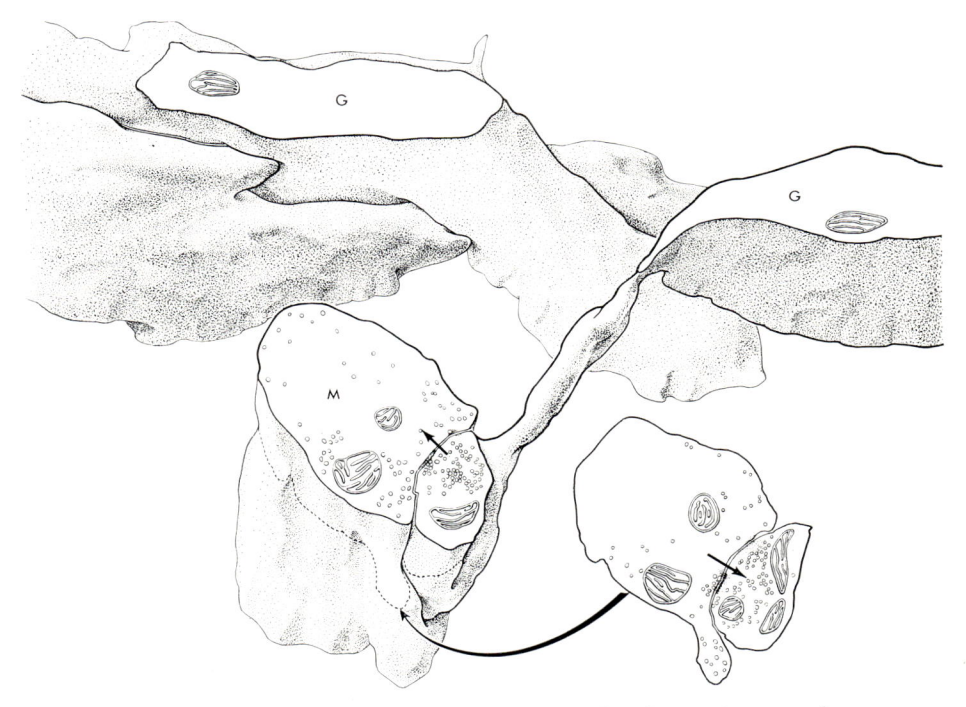

FIG. 1 (Reese). Graphic reconstruction of 24 serial sections cut in a coronal plane through the deep third of the external plexiform layer of a cat. A synaptic ending arising from a granule dendrite (G) terminates on a mitral dendrite (M). Clustering of vesicles indicates that the granule ending is presynaptic to the mitral dendrite, as indicated by short arrow. Cross-section at right is a view of the mitral dendrite at the level of the dotted line at left; at this level the mitral dendrite appears to be presynaptic to the granule ending. Other vesicles in the mitral dendrite are associated with similar mitral-to-granule synaptic contacts.

but the other synaptic contact was always oriented in an opposite manner so that the mitral dendrite appeared to be *presynaptic* to the granule cell ending (Fig. 1). At a few endings the former type of synaptic contact was missing whereas at others there were two instead of one of the latter type of synaptic contact. Thus, nearly all synaptic endings on mitral dendrites, at this level, were of the bidirectional type that originate from granule cell endings; no endings with a single synaptic contact were presynaptic to the mitral dendrites.

Synaptic endings of a very different type were found on granule dendrites or on spinous processes originating from granule dendrites. These endings were the simple bouton type ubiquitous in the central nervous system and their synaptic contact was always oriented so that they appeared presynaptic to the granule dendrite. In our sample of about 100 endings, these boutons accounted for 10–20 per cent of all synaptic endings found, the remaining 80–90 per cent being the endings on mitral dendrites. We did not know the source of the boutons on granule dendrites and the work stopped here. Recently, Price (1968) showed that in the rat some of the endings on granule dendrites degenerate after cutting the lateral olfactory tract and therefore are endings of centrifugal axons originating outside the olfactory bulb. We have now a fairly quantitative picture of the synaptic organization of the deep part of the external plexiform layer in the mammal. Andres' work, presented here, demonstrates that this synaptic pattern is present in olfactory bulbs of other vertebrate classes.

Further development of anatomical data confirms our earlier idea that mitral cell dendrites excite granule cell dendrites which in turn inhibit mitral dendrites. Since each granule cell dendrite contacts dendrites from many mitral cells, these synapses provide a basis for inhibitory interactions between mitral cells. The centrifugal input to the granule cells would modify these interactions.

Andres: I think I have described a similar system (Andres, 1965). Using my morphological criteria I was able to correlate the different types of synapses to the cytoarchitecture of the rat olfactory bulb. In my rat and cat material I have seen the degenerating nerve endings described by Price. Because of the small vesicles and very large synaptic membrane complexes I can distinguish those synapses which are afferent to the bulb from other synapses in the bulb.

REFERENCES

Andres, K. H. (1963). *Z. Zellforsch. mikrosk. Anat.*, **60**, 815–825.
Andres, K. H. (1965). *Z. Zellforsch. mikrosk. Anat.*, **65**, 530–561.
Luciano, L., Reale, E., and Ruska, H. (1968). *Z. Zellforsch. mikrosk. Anat.*, **85**, 350–375.
Price, J. L. (1968). *Brain Res., Amst.*, **7**, 483–486.
Rall, W., Shepherd, G. M., Reese, T. S., and Brightman, M. W. (1966). *Expl Neurol.*, **14**, 44–56.
Reese, T. S. (1968). *Anat. Rec.*, **154**, 408.

EXPERIMENTS IN OLFACTION

Kjell B. Døving

Institute of Zoophysiology, University of Oslo, Blindern, Oslo

The assumption that odours can be arranged in classes according to the similarities in perceived quality was adopted a long time ago. Most of the earlier classification procedures were subjective; frequently the author himself was the experimental subject. Up to the present we find that scientists using precise and quantitative methods in studying physicochemical parameters are reluctant to adopt more sophisticated methods for obtaining biological data. They often rely on their own or their perfumers' judgements. Frequently, however, there are no adequate data available which describe the biological effects of odorous substances. Where biological data are available the odorous substances used appear to be of little or no interest to chemists, as the substances are considered impure and the results therefore unreliable. There is considerable need for cooperation between chemists and biologists, a fact which was recognized officially at the Gordon Research Conference in Enumclaw, Washington, U.S.A. in 1966.

In the present paper I want to discuss some analytical methods that might be fruitful in the approach to an understanding of the mechanisms underlying the primary events in the olfactory system. The paper will not, and it is not intended to, give any clues about what parameters are most important in determining odour quality. The present paper does not indicate the only way to progress, but if it can bridge a gap between physiological thinking and chemical knowledge it will have achieved its aim.

The first part of the paper will describe the electrophysiological events in single receptor cells, in the neurons of the olfactory bulb and in the units of the primary olfactory cortex. The second part will describe the kind of analysis that can be performed on this material, and will show that some methods of analysis are particularly fruitful, in that psychological and physiological data can be correlated. The next part is concerned with visualizing the information in the biological data, and which analytical tools can be used. The fourth part of the paper describes a way of fitting the results of the biological data to the physicochemical parameters. Lastly, the requirements which must be considered in the future when doing these kinds of analyses are discussed.

ELECTROPHYSIOLOGY OF OLFACTORY NEURONS

Receptor cells

The most significant contributions to understanding the physiology of the nervous system have been made by exploring the activity of single nerve cells during controlled experimental conditions. Great progress was therefore expected from studies on single olfactory receptor cells in vertebrates by Gesteland and co-workers (1963). The uncovering of features of the single receptor cells had until then resisted several attempts at exploration, mainly because of the small anatomical size of these cells and their close packing. These authors succeeded in recording activity from single units in the olfactory epithelium and showed that 40 per cent of the cells sampled responded to the 26 chemicals used for stimulation. Most of the receptor neurons increased their spontaneous activity during stimulation. The authors expressed great disappointment in the irregularity of the responses in the receptors, as they were unable to discern the unique properties for which they were searching. The categories that they could construct did not fall into order on the basis of simple chemical properties, or of any psychological grouping of odours. This general disappointment was expressed again later (Lettvin and Gesteland, 1965). However, the chaotic picture given by these authors has not been accepted by others who have since studied the frog's olfactory receptors (Altner and Boeckh, 1967; O'Connell and Mozell, 1969). The first authors studied a small number of receptors with 28 odours and demonstrated that an increased spike frequency was shown to four of these substances by all units tested. They concluded that extreme "generalists", as found among insect olfactory receptors (Boeckh, Kaissling and Schneider, 1965) and described by Gesteland and co-workers (1963), do not exist in the frog. O'Connell and Mozell (1969) reach a similar conclusion from studying the effect of four odours on 101 different receptors; they found that the individual receptors showed a collective tendency towards order.

Olfactory glomeruli

Another approach to the question of coding the olfactory message has been followed by Leveteau and Mac Leod (1969). Although they recorded from the glomeruli in the olfactory bulb they were really investigating the activity of many thousands of receptor axons all ending in one synaptic region. They studied the responses of 128 different glomeruli to 12 carefully selected and qualitatively widely different substances and found that the glomeruli responded to some but seldom to all the different odours, provided that they used low stimulating concentrations. These observations agree with the results of investigations on single receptors in the olfactory epithelium, and confirm a tendency to order. The results of the work of all the authors mentioned above demonstrate that at this level of olfactory coding the excitation of nervous elements is quite common, and inhibition infrequently seen. But not all the odours excite all the receptor cells. The seemingly chaotic picture can be sorted out using the convergence pattern already made by nature, namely the glomeruli. We shall see later how the response patterns in the glomeruli can be further analysed in an attempt to make kosmos out of seeming chaos.

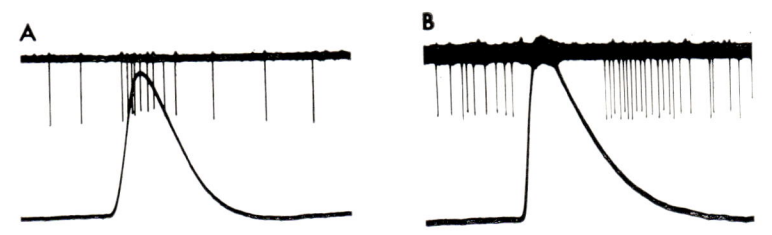

FIG. 1. Recordings from the olfactory bulb in the frog showing excitation (A) and inhibition (B) of two bulbar units in response to stimulation with 1-butanol. The lower traces show the electro-olfactograms. Duration of the trace is about 14 seconds.

Bulbar neurons

The olfactory bulb contains several types of secondary neurons, some of which transmit information from the receptor cells to the brain. These neurons have a distinct spontaneous activity which is changed by olfactory stimulation. Some odours excite the bulbar neurons whereas others inhibit the activity (Døving, 1964). An example of the two types of response is shown in Fig. 1. The upper trace shows the activity of a bulbar unit and the lower trace shows the simultaneously recorded receptor potential of the olfactory epithelium, the electro-olfactogram (EOG) (Ottoson, 1956). The burst of impulse activity starts shortly after the onset of the receptor potential (Fig. 1A) and lasts until the descending phase of the EOG. In another unit nervous activity is inhibited, as shown in Fig. 1B, and the spontaneous activity is abruptly stopped at the beginning of the EOG and is absent until the descending phase of the EOG, followed by increased activity. When double stimuli are applied this post-inhibitory activity is inhibited during the second stimulus, a useful control showing that true inhibition has occurred.

In the olfactory bulb most units—that is, about 60–70 per cent of the population—are inhibited during stimulation. About 10 per cent are unaffected and the rest are excited. Inhibition is thus a dominant feature of neurons in the bulb, in contrast to the properties of the receptors. If one uses a group of odours and records from about a hundred units, a set of data is obtained which may be designated + for excitation, − for inhibition and 0 for unaffected. Such a data sheet seems just as chaotic as the kind of responses described by Gesteland and co-workers (1963).

Neurons in the olfactory cortex

The tertiary structures within the olfactory system have not been systematically studied, but experiments by Haberly (1969) show that inhibition is even more pronounced and conspicuous than in the bulb. He studied the activity of neurons in the prepyriform cortex in the rat and only four units out of 21 were excited by the eight stimuli used. He found no units which were excited by one particular odour and inhibited by another. This last finding might be due to the relatively small number of units studied and the small number of odours used. Units showing excitation usually had small amplitudes. It might be that the units giving larger amplitudes and which were inhibited in his recordings were so specific that the particular odour giving excitation was not found. Presumably one could find differentially affected units by increasing the number of odours and the sample of cortex units.

From the data obtained a change is evident in the appearance of nervous activity at the different levels from the receptors to the brain. The receptors are frequently excited but seldom inhibited by a given stimulus. At the convergence point in the glomeruli the activity is more differentiated, though not conspicuously specific. The neurons in the bulb show a marked tendency to be inhibited by odorous stimuli, and specificity, as indicated by the number of differentially affected units, is prominent (Døving, 1964). In the cortex most units are inhibited, and no specificity like that found in the bulb could be demonstrated, though here the information is as yet incomplete (Haberly, 1969).

The term "specificity" is used in the present context to indicate a differentiated response to a set of odours. For example if a unit is excited by one odour and inhibited by the rest of a set of odours, it will be said to evoke a specific response. The "degree" of specificity can be given a numerical value (Frank and Pfaffmann, 1969; Døving, 1965) to yield various amounts of discriminatory power, but such attributes will not be used in the present study.

A tentative scheme of the integrative processes in the olfactory system is presented in Fig. 2. The receptor cells can only excite the bulbar neurons, but within the bulb the units are either excited or inhibited. The inhibitory effects in the olfactory bulb are presumably due to the intrinsic circuitry within the bulb itself. A scheme for a possible arrangement that might account for electrophysiological events in the olfactory bulb is given by

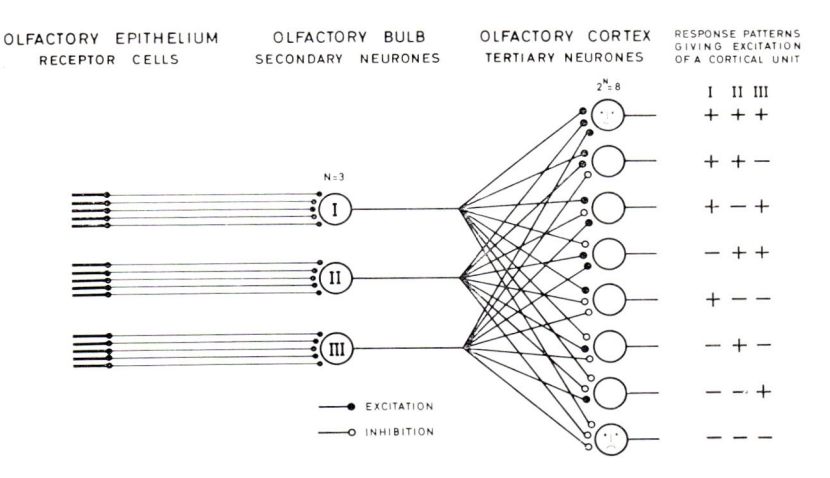

FIG. 2. Tentative scheme of the olfactory system in vertebrates. Three bulbar neurons are shown in synaptic contact with eight different cortical units. The synapses are arranged in such a way that a certain odour will evoke a combination of excitation and inhibition in the secondary neurons, but will excite only one unit in the cortex (see text).

Shepherd (1963). The response patterns observed in the olfactory bulb neurons give rise to excitation and inhibition in cortical units.

The synaptic arrangement is such that when a particular combination of responses in three olfactory neurons of the bulb produces excitation of only one unit in the cortex, a specific response is evoked. The detailed relations between the thresholds at the excitatory and inhibitory synapses which are necessary to account for these effects have not been specified in this diagram. It should also be pointed out that a bulbar neuron is assumed to have both excitatory and inhibitory influences on a cortical cell, which can only take place, according to our present knowledge, through pathways involving interneurons. In Fig. 2, interneurons have not been incorporated, in order to make interpretation easier. The responses of the bulbar units indicated on the right of the figure refer to an activity pattern of three bulbar units and not to a temporal sequence of responses. In the scheme presented here it is also assumed that the cortical cells are spontaneously active, to make it possible to excite a cortical cell by the absence of inhibitory influence. Among the three units in the bulb six combinations can be said to be specific, according to the definition given above, but in the cortex combinations of excitation and inhibition give specific activity in eight cells.

As one goes from a lower level to a higher, fewer neurons are excited in the olfactory nervous system. The features of such a system in the olfactory pathways will be similar to the integrative properties in the visual system, where the stimulus requirements for activation of cells become more refined as one ascends the visual pathways (Hubel and Wiesel, 1962).

STATISTICAL ANALYSIS

The lists of symbols denoting responses in nervous units to different odours are quite chaotic and confusing, as Gesteland and co-workers (1963) point out. These data do however contain a wealth of information and the first step in an analytical procedure will be a statistical treatment of the experimental results. The easily identified types of responses evoked by the olfactory stimuli make the data a convenient subject for chi-square tests. One problem of interest is to test whether any two odours are significantly associated, or in other words, if the odours are treated independently by the olfactory system. If for the sake of simplicity we ignore the unaffected units and concentrate on inhibited and excited units, we can construct a 2×2 contingency table for each pair formed from a set of odours. The entries will be the number of units excited by both odours, inhibited by both, and two entries for those units giving opposite effects. In the glomeruli inhibition has not been observed and at this level responses can be categorized as "response" and "no response" (Leveteau and Mac Leod, 1969).

The results of statistical analysis of electrophysiological data from the olfactory system have been published already (Døving, 1965, 1966a, b) and here I shall therefore only give the matrix obtained by analysing the results from one other electrophysiological investigation.

The data of Leveteau and Mac Leod (1969) are shown in Table I. A Fortran II programme was made to count the number of entries in each category in the 2×2 contingency table and the chi-square values were calculated by the Yates' formula. The matrix gives the degree of association or independence between the $\binom{12}{2}$, or 66 pairs. The higher numbers

TABLE I

CHI-SQUARE VALUES CALCULATED FROM THE RESULTS OF RECORDINGS MADE BY LEVETEAU AND MAC LEOD (1969)

	1	2	3	4	5	6	7	8	9	10	11
(1) Methanol	25·80										
(2) 1-Propanol	0	0·80									
(3) Naphthalene	3·49	1·60	10·22								
(4) Coumarin	6·90	4·11	4·26	19·75							
(5) Pyridine	2·87	14·02	6·64	11·91	9·51						
(6) Nitrobenzene	2·63	4·65	10·57	6·34	4·53	1·51					
(7) Camphor	2·47	0·87	0·07	1·60	1·60	0·53	4·65				
(8) Benzene	1·35	1·05	4·60	0·49	2·10	0·24	1·01	0·41			
(9) 1-Heptanol	2·42	0·36	1·48	2·16	6·59	0·22	3·49	0·28	12·90		
(10) Acetic acid pentyl ester	8·37	2·15	0·17	2·02	8·73	0·27	0	5·08	1·63	7·05	
(11) Decanol	7·53	4·11	0·51	0·83	0·25	2·84	0·07	2·58	0·17	3·91	4·63
(12) β-ionone											

Data were collected from 128 olfactory glomeruli stimulated with twelve odours, the effects being categorized as "response" or "no response".

give a greater probability that the odours are similar. From this table we can see that methanol and 1-propanol were ranked as the most similar odours while camphor and 1-decanol were treated as the least similar odours by the olfactory system at this level.

By going through the different pairs one can distinguish the rank of one pair in relation to that of another. This procedure is convenient for certain tasks where the numbers of odours are small, but it is impractical, and it is impossible to visualize the rank or the relationship between odours when the number increases. We shall return to the problem of visualizing such data below.

In principle the results of Altner and Boeckh (1967) can also be subjected to the same kind of statistical analysis. Their results on the frog, however, are incomplete and the entries in the 2×2 contingency table are too few to perform a meaningful analysis. Gesteland and co-workers (1963) and O'Connell and Mozell (1969) have not given their primary data.

CORRELATIONS BETWEEN BIOLOGICAL DATA

As already mentioned, the information about similarities between odours obtained from electrophysiological data gives rankings between the pairs. In principle it is therefore possible to compare these physiological data with other measurements of similarities, using for instance the Spearman rank correlation coefficient. The psychophysical studies have provided a large number of matrices where different odours have been used. Unfortunately, the various studies can be compared only rarely, because the investigators have used different chemicals. In all cases where it has been possible to compare results, a significant correlation is found. The correlation coefficients calculated from Spearman's formula varied between 0·36 and 0·92, all significant at the 1·0 per cent level or better (see Døving, 1966a,b; Døving and Lange, 1967). The best results have been obtained with homologous aliphatic alcohols. Comparisons have been made between the electrophysiological data (Døving, 1966a) and psychological results (Engen, 1962; Woskow, 1964). It seems that when the odours in general are dissimilar the correspondence is not so good. So far, correlations between the results from different physiological experiments have been impossible to obtain because the same chemicals have not been used.

The experiments by Lange show a good agreement between similarity estimates and confusibility judgements between ten substances (Døving and Lange, 1967). Woskow (1964) finds a good correspondence between successive-interval scaling and mean-category judgements.

MULTIDIMENSIONAL SCALING

All information in Table I and previously published data on odour similarities can be further analysed by multidimensional scaling techniques. The idea with multidimensional scaling is to find the position of n points whose inter-point distances match the original values of the similarities of n objects. The objects in our case are odour substances. The problem is to place the n objects in space in such a way that the distances between the points follow the same order as the original data. How the values of similarities in the original matrices are obtained is of little importance for these methods. A considerable advantage over previously available methods was introduced by Shepard (1962a,b, 1963) when he pointed out the importance of rank between the similarity values, and stressed the importance of a monotonic relationship between the experimental similarities and the distances in the spatial configuration (non-parametric methods).

At present there are three non-metric methods available which were developed for use by electronic computers (Method 1, Kruskal, 1964a,b; method 2, Guttman, 1968; Lingoes, 1965, 1966—"Guttman-Lingoes"; and method 3, Torgerson, 1965; Young, 1968—"Torsca"). In my experience all three methods give essentially the same results on the data I have used. All programmes seek the best solution in the dimension required by the user. If there are many points there may be certain difficulties in fitting the points to the data, especially when the number of dimensions is small. How well the configuration obtained matches the data is given by a factor called *stress* (or in the case of "Guttman-Lingoes", ϕ). It is a positive number and the smaller the better.* Kruskal (1964a) suggests the following verbal evaluation:

Stress	Goodness of fit
20%	poor
10%	fair
5%	good
2·5%	excellent
0%	"perfect"

* Stress, S, is given by the formula

$$S = \sqrt{\frac{(d_{ij} - \hat{d}_{ij})^2}{d_{ij}^2}}$$

where d_{ij} is the original distance between the points i and j and $\hat{d}_{ij}$, the distance given by a solution of the computer programme.

The programme "Guttman-Lingoes" utilizes another parameter named ϕ which is not linearly related to the stress factor used by "Kruskal" and "Torsca".

In all the following computations a Minkowski metric of 2·0 has been used. The distances have thus been calculated as they would be in Euclidean space. Kruskal (1964a) and Wender (1968) found a slightly smaller stress using the exponent $r = 2·5$. A systematic variation of the exponent has not been made in the present material.

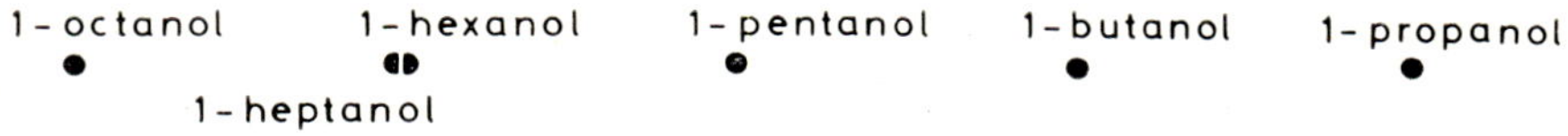

1-octanol   1-hexanol   1-pentanol   1-butanol   1-propanol
            1-heptanol

FIG. 3. A one-dimensional solution given by the "Torsca" programme of data obtained from bulbar units in the frog. Stress=0·05 per cent. (Data from Døving, 1966a.)

• Acetic acid methyl ester

• Acetic acid pentyl ester

                                        • Acetic acid ethyl
        • Acetic acid butyl ester   ester
            • Acetic acid propyl ester

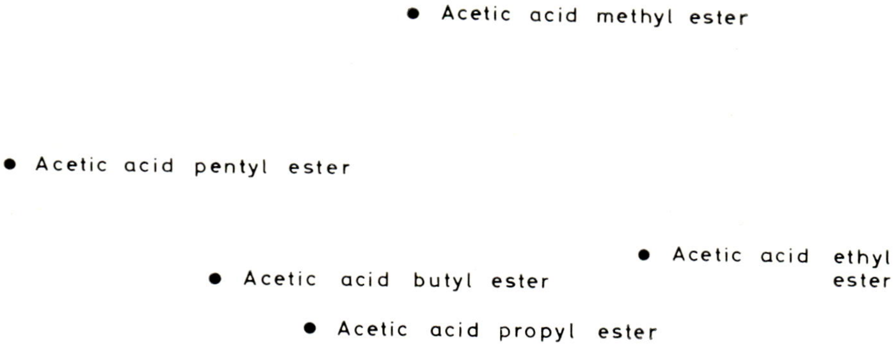

FIG. 4. A two-dimensional solution of the matrix of chi-square values between acetates. Stress=0·0 per cent. (Data from Døving, 1966a.)

The chi-square values of data on the homologous aliphatic alcohols, ketones and acetates obtained in physiological experiments on the frog (Døving, 1966a) are conveniently analysed by these scaling techniques. Fig. 3 shows the results of processing the chi-square values for six alcohols. It was found that an excellent stress value (0.05 per cent) was achieved in one dimension. The substances are arranged according to the number of carbon atoms. 1-Hexanol and 1-heptanol are placed close together, while the distances between the other neighbouring substances are similar. The fact that these substances can be arranged in one dimension with so little stress indicates that only one or possibly two physical parameters determine the relation between these alcohols. Further analysis is made below. Both for the acetates (five odours, Fig. 4) and ketones (four odours, Fig. 5) good solutions were found in one dimension, which is not surprising in relation to the small number of odours for which configurations were sought.

In a fourth series of experiments two odours from each of the three groups were compared (cross-comparison; Døving, 1966a). The odours were 1-pentanol, 1-hexanol, acetic acid butyl ester, acetic acid propyl ester, 2-pentanone and 2-hexanone. In this case a very high stress value (about 20 per cent) was found for the configuration in one dimension, but the two-dimensional solution of Fig. 6 had a stress of only 1·5 per cent. The odours

2-butanone   2-hexanone   2-heptanone
  •   •   •
     2-pentanone

FIG. 5. A one-dimensional solution given by the "Kruskal" programme to the matrix of chi-square values between four ketones. Stress=0·0 per cent. (Data from Døving, 1966a.)

           • 2-pentanone
       • 2-hexanone

• Acetic acid butyl ester

                                    • 1-hexanol
    • Acetic acid propyl ester   • 1-pentanol

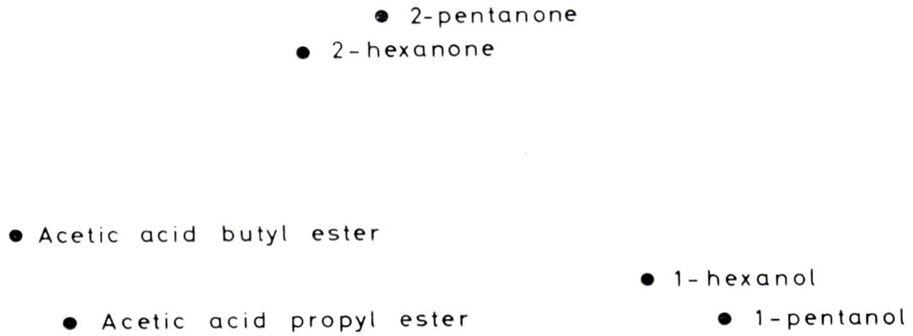

FIG. 6. A two-dimensional solution given by the "Torsca" programme on data from olfactory neurons in the frog. Stress=1·5 per cent. (Data from Døving, 1966a.)

were grouped two by two according to their chemical properties. The fact that a two-dimensional solution was necessary for obtaining a good fit to the experimental data indicates that two or possibly more physicochemical parameters determine the relationship between these six odours. In an earlier study using five quite different chemicals (Døving, 1965) a solution in two dimensions with zero stress was obtained (Fig. 7). The low stress value even for the one-dimensional solution might again be attributed to the small number of odours used in this series. It is surprising to find that in other series, with only five substances, high stress values are obtained for configurations in one dimension (Døving, 1966b). Analysis of the results from the musk series and cross-comparison gave high stress values, as seen in Table II. Configurations had to be sought in three or four dimensions before satisfactory stress values were obtained. The fact that the solutions had to be made in so many dimensions indicates a high degree of complexity even within such a homogeneous group as the musks.

Analysis of the data in Table I from Leveteau and Mac Leod (1969) gives a high stress value in one dimension. Judging from the results of this analysis the authors have been successful in selecting different odours. The two-dimensional configurations have a stress of 18·5 per cent; in four dimensions the stress is about 6 per cent. These high stress values are of significance since they indicate that there are as many factors or physicochemical parameters as there are dimensions.

● salicylaldehyde   ● coumarin

● menthol

                    ● geraniol

● citral

FIG. 7. A two-dimensional solution given by the "Torsca" programme of the chi-square matrix between five odours based upon electrophysiological studies. Stress=0·0 per cent. (Data from Døving, 1965.)

Woskow (1964) and Wright and Michels (1964) have published matrices of similarities obtained by psychophysical experiments on 25 and 50 odours respectively. These authors have treated their data by factor analysis. I have found it of interest to analyse their data with non-metric multidimensional scaling techniques. The two 25 × 25 matrices given by Woskow (1964) are not easily fitted to a spatial configuration even in four dimensions; the stress value is about 8 per cent. However, the large matrix given by Wright and Michels (1964) has an acceptable stress of 5·1 per cent in three dimensions and fair to good stress values for the two-dimensional configuration given in Fig. 8 (stress= 8·0 per cent).

The latter finding indicates a simpler correspondence between the 50 odours selected by Wright and Michels than the 25 odours used by Woskow (1964) or the 12 odours used by Leveteau and Mac Leod (1969).

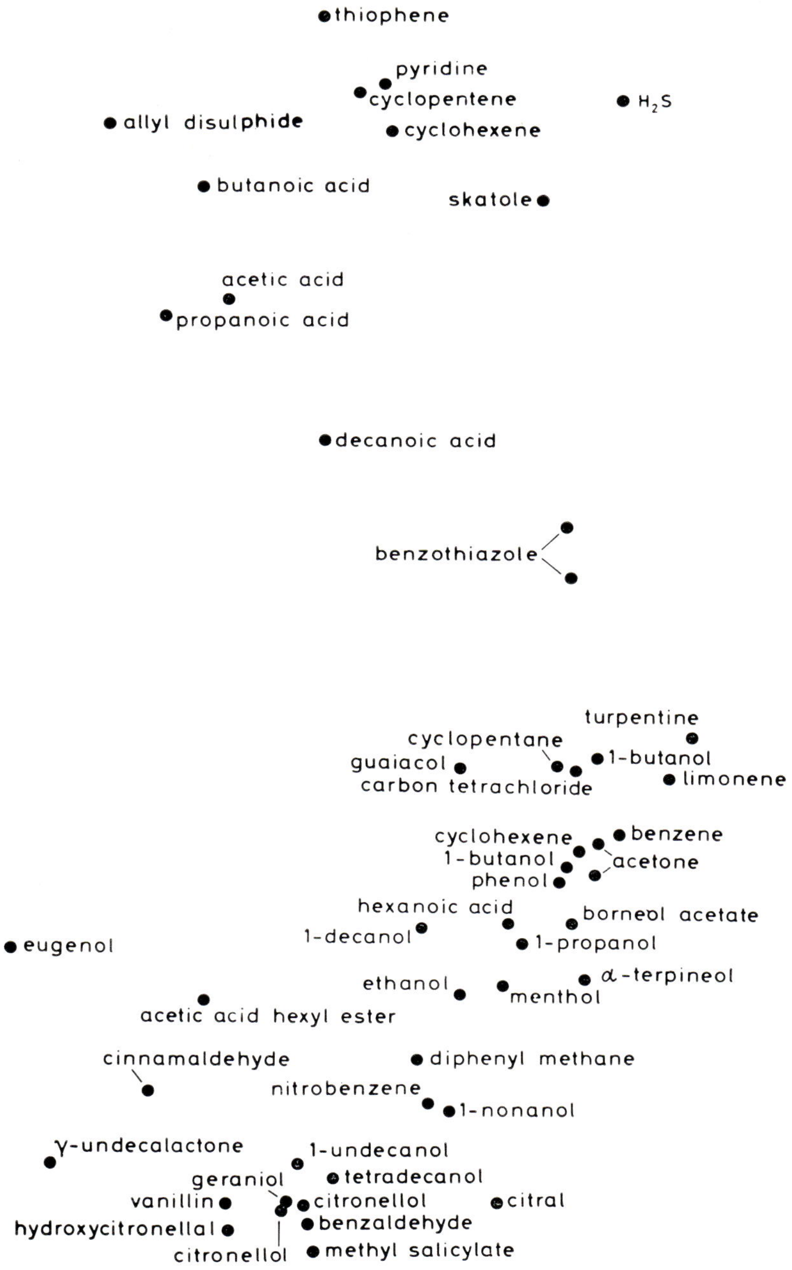

FIG. 8. The two-dimensional solution of the similarity estimates between 50 odours given by the "Kruskal" programme. Stress = 8·0 per cent. (Data from Wright and Michels, 1964.)

TABLE II

STRESS VALUES FOR "KRUSKAL", K, AND "TORSCA", T, PROGRAMMES AND φ VALUES FOR "GUTTMAN-LINGOES", G, FOR SOLUTIONS IN ONE, TWO, THREE AND FOUR DIMENSIONS

Source of data	A*	B†	One dimension K	One dimension T	One dimension G	Two dimensions K	Two dimensions T	Two dimensions G	Three dimensions K	Three dimensions T	Three dimensions G	Four dimensions K	Four dimensions T	Four dimensions G
Doving (1965)	5	60	0·9	2·5	0·1	0·7	0	0						
Doving (1966a) Alcohols	6	95	0·9	0·05	0	0·6	0·2	0						
Doving (1966a) Ketones	4	100	0	0	0	0	0	0						
Doving (1966a) Acetates	5	90	0·4	1·5	0	0·5	0	0						
Doving (1966a) Cross-comparison	6	85	22·8	20·7	5·1	12·6	1·5	0·001						
Leveteau and Mac Leod (1969) 96 units	12	76	45·0	32·4	7·5	14·8	15·3	1·6	6·5	7·7	0·4	6·6	5·6	0·1
Leveteau and Mac Leod (1969) 128 units	12	81	43·3	40·5	10·5	19·7	18·5	2·7	10·9	11·4	1·0	7·2	7·8	—
Doving (1966b) Camphor	5	80	19·0	15·0	6·4	0	0·01	0						
Doving (1966b) Ethereal	5	70	21·2	21·2	3·5	0	1·9	0						
Doving (1966b) Floral	5	80	18·5	18·6	5·5	0·01	1·4	0						
Doving (1966b) Minty	8	80	23·4	28·1	3·2	0·1	1·5	0						
Doving (1966b) Musk	7	43	32·1	26·8	5·9	10·9	10·4	1·6	0·9	0·3	0			
Doving (1966b) Amoore and Venstrom's data‡	8	36	46·8	37·6	12·2	19·7	19·6	3·1	9·8	8·5	0·9	7·9	8·5	0·4
Doving (1966b) Cross-comparison	8	88	40·0	34·6	7·0	14·6	14·7	1·9	5·6	6·3	0·4	3·3	—	—
Woskow (1964)	25		24·6	24·4	3·9	14·5	15·2	1·3	10·2	11·0	0·4			
Wright and Michels (1964)	50		11·9	—	—	8·0	—	—	5·1	—	—			
Ekman (1954a)	20	20	29·4	32·1	6·6	17·2	18·9	1·9	12·1	13·5	0·8	7·0	8·3	0·4
Hsü (1946)	21	7	43·4	44·7	13·3	27·3	25·3	4·1	17·6	17·2	1·7	11·1	11·8	0·9

The stress values are given as percentages.
The solutions have been obtained from data from the references indicated.
* A, the number of odours used in each experimental set.
† B, the percentage of points which violate the triangular inequality given by the "Torsca" programme.
‡ Amoore and Venstrom, 1967, personal communication; see Doving (1966b).

The numbers in Table II give the stress values for "Kruskal" and "Torsca" and the φ values for "Guttman-Lingoes". The values are given as percentages. As seen from the table, all stress factors decrease as the number of dimensions increases.

In all the solutions given in Table II an exponent (Minkowski metric) of $2 \cdot 0$ has been used. The solutions are comparable and in most cases identical, but in some cases, as in the two-dimensional solutions for the cross-comparison between alcohols, ketones and acetates, the "Kruskal" programme has reached a local minimum, giving a stress of $12 \cdot 6$ per cent, while "Torsca" has a $1 \cdot 5$ per cent solution. It should be pointed out that a systematic comparison between the different spatial solutions has not been made.

As seen from Table II, not all methods give the same stress, and in some cases the stress values showed quite large discrepancies. For example, the stress is high for the "Kruskal" solution, indicating that the configuration given by this programme was less good than the other solutions. In such cases the configuration given by the method showing the least stress value has been presented.

HIERARCHICAL CLUSTERING

Solutions based upon multidimensional scaling methods give indications of the grouping of odours, since similar odours will be closer to each other in a spatial configuration than those which are dissimilar. The configurations can give good illustrations of a grouping if the spatial solution is to be found in one, two or three dimensions, but when a solution is found in a higher dimension, any helpful picture is lost. No spatial configuration can define a group or a "cluster" in relation to other clusters. Neither will the spatial configuration give a good solution if clusters occur within other clusters. A hierarchical model tries to overcome these difficulties.

A hierarchical clustering method has been developed by Johnson (1967) which seeks to form a tree structure on which the similar odours are "leaves" on the same "branch". A theoretical model is obtained where any natural arrangement of the objects into homogeneous groups inherent in the data will be visualized. In his programme Johnson (1967) has used both a maximum and a minimum method for measuring distances between clusters. In the version of the hierarchical clustering method used in the Department of Psychology at the University of Oslo, an average method giving the mean of the two extremes is included as an option. All solutions illustrated below are obtained with this average method. By this means it

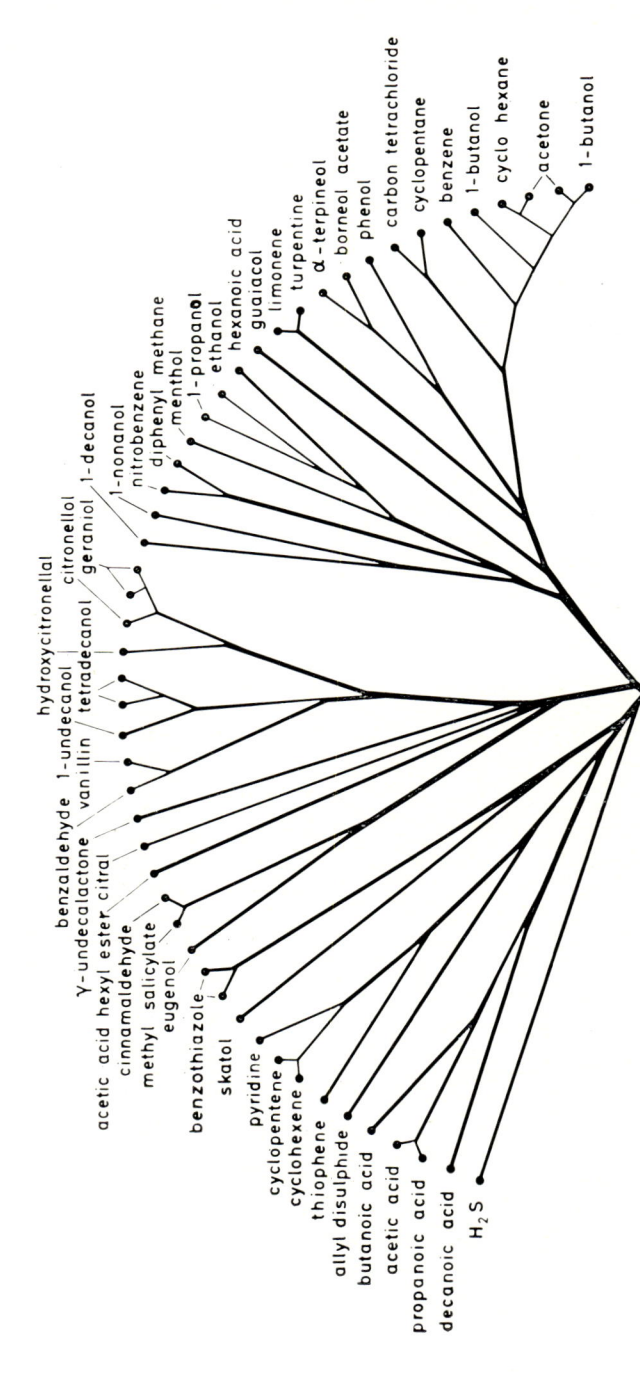

FIG. 9. The hierarchical clustering scheme obtained by analysing the data of Wright and Michels (1964). Average method.

is fairly easy to visualize the odours which are clustered and the order in which the clusters are arranged.

The most interesting results with this method are obtained from the large set of data given by Wright and Michels (1964). Application of the hierarchical clustering method to their data gives a tree structure, shown in Fig. 9. There is a distinct grouping into three different clusters. These authors used the same odours, as different entries, namely acetone, benzothiazole, 1-butanol, geraniol and tetradecanol, and these are naturally linked closely, although some noise is obviously present in the case of acetone and 1-butanol. Hydrogen disulphide is not linked to any other odour

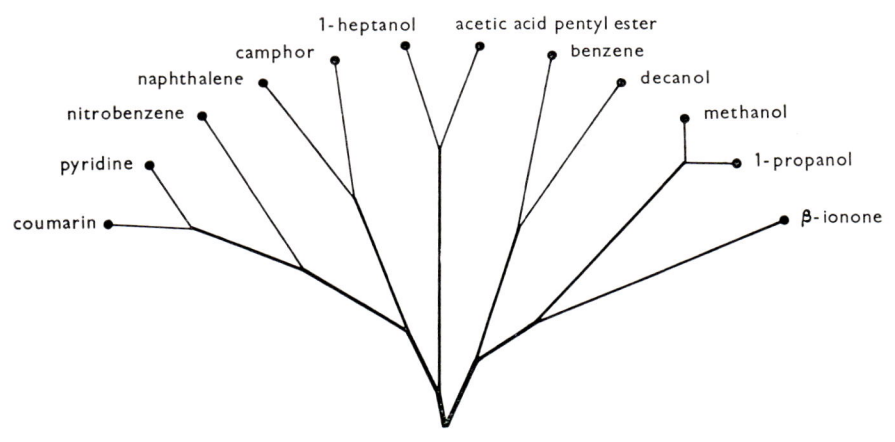

FIG. 10. The hierarchical clustering scheme obtained by analysing the data given in Table I (Leveteau and Mac Leod, 1969). Average method.

and is attached to the "tree" at the very last level, thus forming a "branch" by itself.

Another tree structure can be formed from the data of Leveteau and Mac Leod (1969), and is shown in Fig. 10. In this figure two groups of odours are obtained. One group contained odours from coumarin to acetic acid pentyl ester; the other included benzene, 1-decanol, methanol, 1-propanol and β-ionone.

BIOLOGICAL DATA CORRELATED WITH PHYSICOCHEMICAL PARAMETERS

Where biological data have been obtained, some of the physical constants for the substances used in the experiments are also available. In these cases it should be theoretically possible to correlate the configuration given by the multidimensional scaling techniques with the physicochemical parameters.

The crucial step is to give physical attributes to the axes of the Euclidean solution given by the multidimensional scaling techniques. It is not self-evident that solutions obtained from the biological data with these scaling techniques have physical axes. It may be that they are completely governed by "psychological" factors, like hedonistic factors (see Jones, 1968), though subjective feeling seems to be the main factor in advocating or denying this view.

In the configurations obtained from the biological data only the inter-point distances are of importance. It is therefore possible to perform linear transformations like stretching, shrinking and rotation without distorting the information contained in the original configuration. Since we do not know what the physical axes might be, all known physicochemical vectors should be tested.

TABLE III

CORRELATION BETWEEN BIOLOGICAL DATA ON HOMOLOGOUS ALCOHOLS AND PHYSICOCHEMICAL PARAMETERS, CALCULATED ACCORDING TO THE RANK OF THE 15 DISTANCES BETWEEN THE SIX ODOURS

	Parameter	Reference	Spearman rank correlation coefficient	Significance level
H	Hydrogen binding	1	0·964	<0·001
a	Cross-section	2	0·964	<0·001
$\Delta G_{O/W}$	Adsorption energy	2	0·955	<0·001
M	Molecular weight	3	0·953	<0·001
A	Polar factor	1	0·953	<0·001
$\Delta G_{O/A}$	Adsorption energy	2	0·914	<0·001
$\log K_{O/A}$	Adsorption coefficient	2	0·914	<0·001
p. ol. calc.	Calculated olfactory threshold	1	0·854	<0·001
$\Delta G_{A/W}$	Adsorption energy	2	0·844	<0·001
b	Cross-section	2	0·752	<0·01
$\log K_{37}$	Partition coefficient	1	0·746	<0·01
d	Density	3	0·689	<0·01
p	Critical number of molecules	2	0·275	>0·10
p. ol. exp.	Experimental olfactory threshold	1	0·271	>0·10
$1/p$	Inverse of critical number of molecules	2	0·108	>0·10

References: (1) Laffort (1968); (2) Davies and Taylor (1957); (3) *Handbook of Chemistry and Physics*, Chemical Rubber Company, 46th edn.
A, air; W, water; O, oil.
Biological data on homologous alcohols from Døving (1966a). Number of degrees of freedom is 13. See text for explanation.

In the present study I have compared the one-dimensional solution for the aliphatic alcohols given in Fig. 3 with the physical parameters given by Laffort (1968), by Davies and Taylor (1957) and by the *Handbook of Chemistry and Physics.* All distances between the points have been calculated and compared with the calculated distances between the physical variables. The Spearman rank correlation coefficient was calculated in all cases. The correlation between the biological data in the one-dimensional solution and the seven different physical parameters gives the results shown in Table III.

The terms H, hydrogen binding and A, "polar" factor, have been calculated by Laffort (1968) according to the following formulae:

$$H = \frac{T}{48} + 5 \cdot 1 \left(1 - \sqrt{\frac{R_m}{\rho}}\right)$$

$$A = \frac{15 \cdot 7 V_b}{48 \cdot 6 + V_b}$$

where T is the absolute temperature of the boiling point, R_m is the molar refraction in ml, and $\rho = \sqrt[3]{V_b}$ where V_b is the molecular volume at the boiling point.

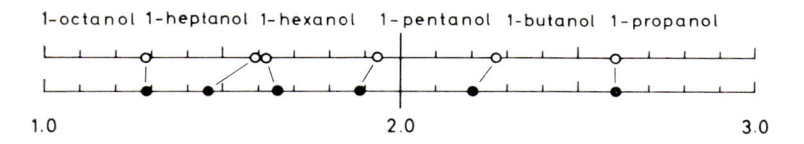

FIG. 11. Hydrogen binding (H, filled circles) (Laffort, 1968) compared with the one-dimensional solution of the biological data (open circles). (Døving, 1966a.)

As seen from Table III the hydrogen binding, adsorption energy, polar factor and molecular weight together with the cross-sections showed the best fit to the experimental data. The threshold values for detection, partition coefficients and densities give correlation coefficients which indicate a less good correspondence with these parameters. The spatial configuration of the six alcohols is compared with the numerical values of hydrogen binding in Fig. 11.

DISCUSSION

In this paper I have tried to give a survey of some analytical methods which can be used to facilitate the correlation between biological data and physicochemical parameters.

The results of physiological recordings can be treated by simple statistical means to give measurements of similarities between the odorous stimuli. The physiological data can thereby be correlated with psychophysical judgements. So far there are no theoretical difficulties in the use of any of these methods; the difficulty arises when introducing multidimensional scaling procedures. These procedures have found wide application in psychophysical research and have also been used by psychologists in olfactory investigations (Hsü, 1946; Ekman, 1954a; Woskow, 1964; Yoshida, 1964a,b). Recently non-metric multidimensional scaling techniques have found applications in analysis of results from studies of the olfactory sense (Døving, 1966b; A. L. Lange, 1969, personal communication; Wender, 1968).

It can be shown that in a space with fewer dimensions than elements, the rank of the distances between the elements locates the position of the elements. The uncertainty of location decreases as the number of dimensions decreases. The elements which are similar or strongly associated are found to be close together, while those which are dissimilar or independent are located far apart. In the present approach the matrices of data from biological studies have a solution in Euclidean space.

The physical constants known for the chemicals used in the biological studies will also be defined in Euclidean space. The final step in attempts to correlate biological and physical data will, therefore, be to make one solution fit the other. Ideally the two solutions should coincide. In the present study only a fit with a rank correlation for a one-dimensional solution for aliphatic alcohols has been obtained.

The data by Leveteau and Mac Leod (1969) are at present the most promising available for analysis, but a Fortran programme for making this comparison is not yet available because of the large number of dimensions in an acceptable solution.

Several methods have been described for fitting spatial solutions. In a study of psychophysical dimensions of similarity among random shapes, Stenson (1968) used a canonical correlation method which should also have an application for present problems, since two different sets of measurements are available on the same set of odours. Further search for the best representation of the biological data and the physical parameters has to wait until feasible methods become available.

Multidimensional scaling techniques have been applied in the present study as a basis for further analysis of the important parameters in the olfactory process and in order, if possible, to connect the biological data with the physical parameters. The biological data are considered to contain the essential information for understanding olfactory discrimination, the results being obtained either from similarity estimates or from electrophysiological recordings. In analyses made previously, authors have sought psychological explanations of their data. For example, Wright and Michels (1964) made a factor analysis of their large matrix and suggested that the first factor was arranged according to affective properties. There might be one or more chemical or physical parameters beyond these affective properties.

An objection to the correlation presented here is connected with the non-parametric multidimensional scaling. In an analysis of similarities between colours, Ekman (1954b) made a factor analysis of his data. Shepard (1962a) made a non-metric multidimensional scaling of the same data and found a two-dimensional solution where the 14 colours were arranged in a ring, very similar to the "colour ring" known for centuries. In such a solution it is difficult to imagine what other factor besides the wavelength of light determines the quality of the colour.

When it comes to the actual correlation between data the first and most conspicuous fact is the lack of cooperation between different authors. In cases when biological results are known the physicochemical parameters are difficult to find or lacking, and *vice versa*.

Theories which seek to bring the quality of odours within the framework of physical parameters have been described by several authors (Amoore, 1962a,b; Beets, 1964; Davies, 1962; Dravnieks, 1968). Davies (1962) and Davies and Taylor (1957) have proposed an explicit theory, the relevance of which to biological data has not yet been fully documented.

Davies (this volume, pp. 265–281) proposes three odorant parameters: desorption rate, molecular size and molecular geometry. He suggests an odour space of at least three dimensions. Findings made with the analytical tools now available seem to indicate that for certain sets of odorous substances the data can be adequately described in two dimensions or even one dimension. But when data on odours from widely separated groups are analysed, as for example the data set obtained by Leveteau and Mac Leod (1969), a solution with acceptable stress is first found in four dimensions.

Most probably the "ideal" or "global" odour space has more than four dimensions. Presumably only restricted parts of the odour space can be visualized in a plane or on a line.

The information presented in matrices from the biological data is difficult to visualize. In order to overcome these difficulties certain analytical steps can be taken. The essential method is that of multidimensional scaling. The methods used in the present study are non-metric, which means that only the ranking between pairs of elements is of importance. Even with so little information from the data, solutions are obtained with all the essential results retained in easily visualized configurations, at least in two-dimensional solutions.

Many authors interested in theories of odour and the factors determining odour quality have used intuitive judgements of similarities. In the early work of Amoore (1962a,b) the grouping was done on subjective assumptions from heterogeneous sources. The lack of adequate experimental data from biological studies has been realized in later studies (e.g. Amoore, 1968). Davies' theory of the quality of odours (1965) was proposed without any biological basis. A later attempt to relate the quality of odours to physical parameters according to this theory seems promising (Theimer and Davies, 1967). The studies by Wright (1964) on the correlation of the quality of olfactory substances with peaks in their infrared spectra are also in need of adequate biological data.

Hierarchical clustering schemes can have useful applications in grouping odours. An advantage of this particular method is the large arrays of data which the programme can handle. The version of hierarchical clustering used at the University of Oslo can take up to 85 points. In future grouping of odours it seems mandatory to explore the possibilities of using this and other analytical methods in the general problem of clustering.

In the present paper a series of methods has been discussed by which biological data may be correlated to the physical parameters of odorous substances. As the final step a configurational fit can be made between the biological data and physicochemical parameters. A conceptual difficulty arises when giving physical attributes to the axes of the spatial solutions from biological data. If these methods are valid the odour quality within a restricted number of odours, and in their own context, can be described by physicochemical parameters. The hope is that these methods of analysing and presenting data will be more frequently used in later studies. These analytical tools may be applied to extract the maximum amount of information and the most easily digested picture of it.

It is hoped that further research along these lines may give indications of solutions to the challenging question of the relation between odour quality and physical parameters.

SUMMARY

A short description is given of various experimental procedures that have led to a fuller understanding of the events in the olfactory system. The mechanisms of coding the quality of the olfactory messages at different levels within the system are discussed.

The electrophysiological data can, if properly evaluated and processed, link physicochemical parameters with sensory perception. Certain ways of handling the electrophysiological data are particularly meaningful in that they can be correlated significantly with the results of psychophysical experiments. Both the electrophysiological and the psychophysical data are the results of the discriminatory mechanisms. These data therefore contain information pertinent to the question of what physicochemical parameters determine odour quality. The data from the biological experiments, based on discriminatory power, give matrices where the relationships between the odours are ranked.

At present three methods are available for transferring such data on the ranked relationship between elements (odours) to spatial coordinates (multi-dimensional scaling). The results of these scaling procedures give "odour spaces" that are obtained exclusively on data from biological experiments. In conjunction with multidimensional scaling methods, a way of grouping the odours by a hierarchical clustering method is described. By this means a "tree structure" is formed where the most similar odours are "leaves" on the same branch. For further processing a method is used by which the "odour spaces" are fitted and correlated to known physicochemical parameters. This method seeks the best orientation of the "odour space" obtained from biological data in a coordinate system where the axes are defined physicochemical variables. The odour qualities are thus described by the physicochemical parameters.

The results of the data processing can be very helpful in understanding basic mechanisms in olfactory processes. For these methods to give optimal insight into the primary olfactory processes, there are certain requirements, discussed here.

Acknowledgement

The data computations have been made on a CDC-3300 machine at the University of Oslo. I want to thank F. Tschudi, Department of Psychology, for his advice and kind help in using the multidimensional scaling programmes adapted to this machine.

REFERENCES

ALTNER, H., and BOECKH, J. (1967). *Z. vergl. Physiol.*, **55**, 299–306.
AMOORE, J. E. (1962a). *Proc. scient. Sect. Toilet Goods Ass.*, suppl. 37, 1–12.
AMOORE, J. E. (1962b). *Proc. scient. Sect. Toilet Goods Ass.*, suppl. 37, 13–23.
AMOORE, J. E. (1968). In *Theories of Odor and Odor Measurement* (Proc. NATO Summer School, Istanbul, 1966), pp. 71–81, ed. Tanyolaç, N. London: Technivision.

AMOORE, J. E., and VENSTROM, D. (1967). In *Olfaction and Taste II* (Proceedings of the Second International Symposium, Tokyo, 1965), pp. 3-17, ed. Hayashi, T. Oxford: Pergamon Press.
BEETS, M. G. J. (1964). *Molec. Pharmac.*, **2**, 3-51.
BOECKH, J., KAISSLING, K. E., and SCHNEIDER, D. (1965). *Cold Spring Harb. Symp. quant. Biol.*, **30**, 263-280.
DAVIES, J. T. (1962). *Symp. Soc. exp. Biol.*, **16**, 170-179.
DAVIES, J. T. (1965). *J. theor. Biol.*, **8**, 1-7.
DAVIES, J. T., and TAYLOR, F. H. (1957). *Proc. II Int. Congr. Surf. Activ.*, 329-340.
DØVING, K. B. (1964). *Acta physiol. scand.*, **60**, 150-163.
DØVING, K. B. (1965). *Revue Lar. Otol. Rhinol.*, **86**, 845-854.
DØVING, K. B. (1966a). *J. Physiol., Lond.*, **186**, 97-109.
DØVING, K. B. (1966b). *Acta physiol. scand.*, **68**, 404-418.
DØVING, K. B., and LANGE, A. L. (1967). *Scand. J. Psychol.*, **8**, 47-51.
DRAVNIEKS, A. (1968). In *Theories of Odor and Odor Measurement* (Proc. NATO Summer School, Istanbul, 1966), pp. 11-31, ed. Tanyolaç, N. London: Technivision.
EKMAN, G. (1954a). *Rep. psychol. Lab. Univ. Stockholm*, **10**, 1-9.
EKMAN, G. (1954b). *J. Psychol.*, **38**, 467-474.
ENGEN, T. (1962). *Rep. psychol. Lab. Univ. Stockholm*, **127**, 1-10.
FRANK, M., and PFAFFMANN, C. (1969). *Science*, **164**, 1183-1185.
GESTELAND, R. C., LETTVIN, J. Y., PITTS, W. H. and ROJAS, A. (1963). In *Olfaction and Taste* (Proceedings of the First International Symposium, Stockholm, 1962), pp. 19-34, ed. Zotterman, Y. Oxford: Pergamon Press.
GUTTMAN, L. (1968). *Psychometrika*, **33**, 469-505.
HABERLY, L. B. (1969). *Brain Res., Amst.*, **12**, 481-484.
HSÜ, E. H. (1946). *Psychometrika*, **11**, 31-42.
HUBEL, D. H., and WIESEL, T. N. (1962). *J. Physiol., Lond.*, **160**, 106-154.
JOHNSON, S. C. (1967). *Psychometrika*, **32**, 241-254.
JONES, F. N. (1968). In *Theories of Odor and Odor Measurement* (Proc. NATO Summer School, Istanbul, 1966), pp. 133-141, ed. Tanyolaç, N. London: Technivision.
KRUSKAL, J. B. (1964a). *Psychometrika*, **29**, 1-27.
KRUSKAL, J. B. (1964b). *Psychometrika*, **29**, 115-129.
LAFFORT, P. (1968). *C.r. Séanc. Soc. Biol.*, **162**, 1704-1713.
LETTVIN, J. Y., and GESTELAND, R. C. (1965). *Cold Spring Harb. Symp. quant. Biol.*, **30**, 217-225.
LEVETEAU, J., and MAC LEOD, P. (1969). *J. Physiol., Paris*, **61**, 5-16.
LINGOES, J. C. (1965). *Comput. behav. Sci.*, **10**, 183-184.
LINGOES, J. C. (1966). In *Uses of Computers in Psychological Research*, pp. 1-22. Paris: Gauthier-Villars.
O'CONNELL, R. J., and MOZELL, M. M. (1969). *J. Neurophysiol.*, **32**, 51-63.
OTTOSON, D. (1956). *Acta physiol. scand.*, **35**, suppl. 122, 1-83.
SHEPARD, R. N. (1962a). *Psychometrika*, **27**, 125-139.
SHEPARD, R. N. (1962b). *Psychometrika*, **27**, 219-246.
SHEPARD, R. N. (1963). *Hum. Factors*, **5**, 33-48.
SHEPHERD, G. M. (1963). *J. Physiol., Lond.*, **168**, 101-117.
STENSON, H. H. (1968). *Percept. Psychophys.*, **3**, 201-214.
THEIMER, E. T., and DAVIES, J. T. (1967). *J. agric. Fd Chem.*, **15**, 6-14.
TORGERSON, W. S. (1965). *Psychometrika*, **30**, 379-393.
WENDER, I. (1968). *Psychol. Forsch.*, **32**, 244-276.
WOSKOW, M. H. (1964). Ph.D. Thesis. University of California, Los Angeles.
WRIGHT, R. H. (1964). *Ann. N.Y. Acad. Sci.*, **116**, 552-558.
WRIGHT, R. H., and MICHELS, K. M. (1964). *Ann. N.Y. Acad. Sci.*, **116**, 535-551.

YOSHIDA, M. (1964a). *Jap. psychol. Res.*, **6**, 115–124.
YOSHIDA, M. (1964b). *Jap. psychol. Res.*, **6**, 145–154.
YOUNG, F. W. (1968). *Rep. Thurstone Psychom. Lab. Univ. North Carolina*, No. 56, 1–27.

DISCUSSION

Wright: The question of multidimensional scaling seems to assume that the odour of a substance like methyl salicylate is resident in a single quality of the molecule, and therefore you can compare its odour as a unique entity with the odour of some other substance. If I can mention insects at a symposium on vertebrates, the results of work by the U.S. Department of Agriculture testing some thousands of compounds for the ability to attract two species of insects, the Mexican fruit fly (*Anastrepha ludens*, Loeu.) and the Oriental fruit fly (*Dacus dorsalis*, Hendel) showed that out of the 2426 compounds tested on both species, 22 per cent attracted the Oriental fruit fly and 8 per cent attracted the Mexican fruit fly. Now if the molecular quality which affects one insect is completely independent of the molecular quality that affects the other, the proportion of compounds which are attractive to both would be expected to be 8 per cent of 22 per cent, which is 1·8 per cent. This is exactly what is observed, indicating conclusively that there are two mutually independent qualities in these compounds, one quality being perceived by one insect and the other quality being perceived by the other species. The coding of the sensation is therefore due not to some unique character but to a unique *combination* of characters. This puts the whole point of multidimensional scaling in a rather precarious position; it is very useful as a psychological exercise, but how much light it throws on the actual transducer mechanism is something else.

Døving: You can use multidimensional scaling on any kind of data where the entries are a correlation coefficient, similarity estimates or chi-square values. Your data are measurements of similarities not between odours but between insects, but probably you can get information pertinent to the question of odour quality by means of the kind of data you describe.

If you have a two-dimensional solution with physical axes, a given position is a single quality but would be described by two different parameters. The "question" of multidimensional scaling does not assume that the odour resides in a single quality of the molecule.

Wright: My point is that it depends which intellectual level, or which physical level, you are working on. If you are thinking about the process of olfactory stimulation as it occurs in the epithelium, you have to recognize the multiplicity of stimulus characteristics in the stimulating molecule. If you are thinking about the subjective evaluation of "does this substance smell like that one?" or "does it produce a similar sensation in the central nervous system?", then multidimensional scaling is useful for expressing the results. I question how far multidimensional scaling will throw light on the actual process of stimulation at the end organ.

Døving: We shall be able to answer this when we get data from the receptor cells and make the same analysis of the data from this level as on data from the bulb. We shall then correlate the observations made from the receptor epithelium, from the olfactory bulb and from psychophysical judgements, and should be able to throw light on this problem. And so far it seems that psychophysical judgements and the handling of data from electrophysiological studies give similar results. It seems too early to reject some analytical methods because of intellectual difficulties of understanding the stimulating mechanisms.

Beets: The correlation techniques worked out by Dr Døving have already revealed very useful information and promise more for the future. My only question about the present paper refers to the choice of materials. In the series of alcohols with 3 to 8 carbons most chemical, physical and structural properties change without showing a break. This also holds good for the olfactory response up to the C_7 alcohol. Since olfactory response shows a break practically at the end of your series (Ottoson, 1958), one can hardly expect any significant stress. Wouldn't it be better to extend the series beyond C_8, up to C_{10} or C_{11}?

Døving: When I did these experiments in 1964 and 1965 I didn't know that I would be here today! But I do intend to select odours with special questions in mind and I would welcome suggestions of a good set of odours which might be useful from the biochemical and chemical point of view. There are two different attitudes to this part of the problem. One is that one ought to work on a rather restricted chemical group, like a homologous series. The other approach, taken by Dr Mac Leod, is to select odours with widely different qualities.

Beets: In a homologous series you know that there is a structural relationship. When you say one should select widely, what you really mean is randomly. I am not in favour of that, when it is so easy to bring into the selection some structural criteria which may make some sense. I don't mean that you should use only homologues, but you can introduce structural details, in the effect of which you are interested. For that reason I wouldn't randomize, but I would select intelligently.

Døving: If you select 12 odours, which are very widely separated in quality, like those used by Mac Leod, you may have to go to an 11-dimensional solution to describe the relationship between the odours.

Beidler: If you change the concentrations of the stimuli you will change the distribution of cells responding and thus upset your classifications.

Døving: When I did these experiments I recorded the receptor potential, and could thereby judge the intensities relative to each other. What I tried to do was firstly to keep the amplitude of the receptor potential the same for all odours within a stimulation series, and secondly I kept it below half the maximum value. This might be too high, but I don't think this would alter the results appreciably.

Beidler: I looked at some data of Dr Don Matthews and he showed that when you increase the concentration an olfactory cell that is inhibitory might become excitatory and *vice versa*. Thus your analysis would change with concentrations. When you say propanol, it is really a given concentration of propanol; if you had chosen another concentration your clusters would have been different.

Døving: I agree that one should take the concentrations into account. I have shown that the type of response may be altered by increasing the intensity in some cases (Døving, 1964).

Pfaffmann: Dr Døving, is there enough information to indicate a difference between the periphery and the bulb, in terms of these clusterings?

Døving: Not yet. It would probably be possible to get such information if one had the raw data from the work of Gesteland and co-workers (1963) or O'Connell and Mozell (1969). Only in one case (Altner and Boeckh, 1967) did I have the raw data, but the number of units studied was too small to make an analysis.

MacLeod: I would like to make another comment about intensity at the bulbar unit level—that is, the mitral cells. There is not much information about intensity, because there is a very important inhibition from granular cells which makes the mitral cell respond by only one spike to any amount of peripheral stimulation. I have seen in the rabbit that the amplitude of the response of the glomerulus is very quickly saturated as one ascends the concentration scale, much more quickly than the peripheral receptor potential. It seems therefore likely that provided a suprathreshold level of excitation is reached, intensity is no longer a dimension of the message from mitral cells. Of course it remains a dimension for the central nervous system, where probably it is transmitted by tufted cells, but that is another question. But at the level where Dr Døving has done his multidimensional scaling, probably it is not of great importance.

Reese: What is meant by a glomerular unit?

MacLeod: One glomerulus is considered to be functionally a unit because it behaves like a unit when you try different olfactory stimuli, giving an all-or-none response profile.

Reese: It would be very important to discover now if the glomerulus really is a unit, signalling a certain odour quality or set of qualities.

Andres: There is great variation in the branching of the mitral cells among the vertebrates. In the lamprey (cyclostomes), in *Scyliorhinus canicula* (elasmobranchs) and in *Tinca vulgaris* (teleosts) there are no real collateral dendrites, only main dendrites. In higher vertebrates we find more and more collateral dendrites from the mitral cells. There appear to be increasing numbers of connexions between olfactory glomeruli.

Døving: But dendrites of mitral cells in the teleosts are connected to several glomeruli?

Andres: They go to several glomeruli, but the total receptive area is very small compared with mammals.

Beidler: It would be very useful to have some way of storing the raw data from single-unit analysis so that later workers would have it available for analysis. It would also be most helpful if Dr Beets would suggest the type of odours to investigate, and also the type of odours we should *not* use!

Beets: The problem is complicated because every research project requires its own selection of odorants. Rather than making a standard series of odorants available to everybody, I think it would be preferable to publish a short list of people who are thoroughly familiar with the problems involved in the relation between olfactory response and odorant structure, for instance two in the United States and two in Europe.* Anyone planning a research project in olfaction can write to one of these people, giving a concise idea of his project and asking for advice on the selection of odorants. The advice given will include the following information: (1) A series of odorants. (2) Molecular structures and stereochemical implications. (3) Analytical requirements (chemical, stereochemical and, eventually, optical purity). (4) Indications where to obtain these materials and, if necessary, how to purify them. (5) The advice *may* in some cases be completed by the supply of samples but this is not an intrinsic part of the service.

Davies: Dr Døving, have you any new data on the different series of

---

* Since the symposium, the following people have been approached by Dr Beets and will be available for advice:

United States: Dr H. U. Daeniker—Givaudan Corporation, 125, Delawanna Avenue, Clifton, N.J. 07014, U.S.A.
Dr E. T. Theimer—International Flavors and Fragrances Inc., R and D Center, 1515 State Highway 36, Union Beach, N.J. 07735, U.S.A.

Europe: Dr G. Ohloff—Firmenich et Cie, P.O.Box 39 Jonction, Geneva, Switzerland.
Dr M. G. J. Beets—International Flavors and Fragrances (Europe), Liebergerweg 72, Hilversum, Holland.

musks: nitro musks, steroid musks and macrocyclic musks? How many dimensions in space do you need to plot these without too much stress?

Døving: I used eight musks. The spatial solution of the results on these substances had to be found in three or four dimensions.

REFERENCES

ALTNER, H., and BOECKH, J. (1967). *Z. vergl. Physiol.*, **55**, 299–306.
DØVING, K. B. (1964). *Acta physiol. scand.*, **60**, 150–163.
GESTELAND, R. C., LETTVIN, J. Y., PITTS, W. H., and ROJAS, A. (1963). In *Olfaction and Taste* (Proceedings of the First International Symposium, Stockholm, 1962), pp. 19–34, ed. Zotterman, Y. Oxford: Pergamon Press.
O'CONNELL, R. J., and MOZELL, M. M. (1969). *J. Neurophysiol.*, **32**, 51–63.
OTTOSON, D. (1958). *Acta physiol. scand.*, **43**, 167–181.

OLFACTION IN MAMMALS—TWO ASPECTS: PROLIFERATION OF CELLS IN THE OLFACTORY EPITHELIUM AND SENSITIVITY TO ODOURS

D. G. Moulton, G. Çelebi and R. P. Fink

Monell Chemical Senses Center, University of Pennsylvania, Philadelphia

Two areas in the study of olfaction that are particularly rich in controversy, speculation and conflicting evidence concern the regenerative capacity of the olfactory organ, and the degree of sensitivity—absolute and differential—of the olfactory system. Knowledge in both these areas is important to an understanding of certain basic functional properties of the olfactory organ, such as the interrelation between different cell types in the sensory epithelium, and the minimum number of molecules necessary to stimulate a single receptor. To provide a focus for discussion we shall briefly review some previous evidence and present some new findings.

CELL PROLIFERATION AND MIGRATION IN OLFACTORY EPITHELIUM OF THE MOUSE

Much work has centred on the question of whether the post-embryonic olfactory epithelium can regenerate, and in mammals, as in amphibians, both negative and positive results have been reported. For example, Nagahara (1940), using mice, reported regeneration of primary olfactory neurons and epithelium 90 days after sectioning the neurons (although the vomeronasal epithelium apparently failed to regenerate after section of the vomeronasal nerve). He described two types of cells: an "active" cell with a small round nucleus, near the surface of the epithelium, and a "resting" cell with a larger, elliptical nucleus, lying at the base of the epithelium. Most of the active cells degenerated after section of the primary neurons, but the number of resting cells increased by mitosis.

Takata (1929), on the other hand, found no regeneration of olfactory cells and no mitosis following transection of the primary olfactory neurons in the rat. Similar conclusions were reached by Smith (1938), who found no regeneration in the rat epithelium after destruction of the epithelium with zinc sulphate; and by Le Gros Clark (1956, 1957), who found no regeneration in the rabbit epithelium—even six months later—after ablation of the olfactory bulb.

The only recent evidence to suggest that the epithelium can regenerate in the adult mammal is a study by Schultz (1960) on the rhesus monkey. He used zinc sulphate to induce coagulation necrosis of the epithelium and found that it gradually reappeared over a period of 6 months to a year. (Investigations on non-mammalian forms are reviewed by Moulton and Beidler, 1967.)

These studies concern the ability of the olfactory epithelium to regenerate after an experimentally induced lesion, or destruction of the olfactory epithelium. Thus—although relevant—they do not directly meet the question which concerns us here: is there a continuous renewal of one or more cell types within the olfactory epithelium of the adult mammal?

The classical answer to this appears to have been that the basal cells give rise to supporting cells. This, however, was not the view of Bimes and Planel (1952). In studies in the guinea pig and other rodents they were able to confirm an earlier observation of Lams (1940) that mitosis occurs—although rarely—at the level of the receptor and supporting cells. They further claimed that as supporting cells age they come together—often in groups of two or three—and, in time, are sloughed off into the nasal cavity. On the basis of these findings the authors concluded that the rate of mitosis within the supporting cell layer was sufficient to account for renewal of supporting cells. They specifically excluded the possibility that the receptors are continually renewed, since they found that regeneration of these cells occurs only after sectioning the olfactory bulb.

On the other hand, they found mitoses among the basal cells, but argued that these cells are destined to renew the cells of Bowman's glands.

Andres (1965, 1966), applying electron microscopy to the olfactory epithelia of dogs, rats and cats, found that mitoses were relatively abundant among the supporting and basal cells. He also described a fourth cell type, often lying below the basal cells, which he concluded was a reservoir for the differentiation and renewal of the sensory receptors. He believes that the entire epithelium undergoes continuous renewal.

More recently, Thornhill (1967), during the study of the ultrastructure of the olfactory epithelium of the lamprey (*Lampetra fluviatilis*), occasionally saw axons darker than surrounding ones, suggesting degeneration. He also noted that the basal cells resemble neurons since they contain microtubules, while tonofilaments are absent. Apparently well-developed cilia were found in some basal cells.

These various studies suggest that in the adult mammal, cell division occurs continuously among basal cells and at the level of the receptor and supporting cells. There is no general agreement, however, about the destiny of the products of cell division—whether, for example, new receptor or supporting cells are formed, or both, or new Bowman's gland cells. Nor is it clear whether the rate of division would be sufficient to account for cell renewal rather than the gradual addition of cells to a slowly expanding population. Other questions concern the relative distribution of dividing cells both along and between the basal and more peripheral zones of the epithelium, and the presence of dividing cells lying below the basal cells.

As a first step towards the solution of these questions we have sought data on the statistical and kinetic properties of cell populations within the olfactory epithelium of the mouse.

Methods

In an initial series of experiments, seven male mice (BUB strain) were used. Their mean weight was 30 g and they were 149 days old. Tritiated thymidine, with a specific activity of about 6·7 curies/millimole, was obtained from the New England Nuclear Corporation. It was dissolved in sterile water and injected intraperitoneally (10 microcuries/g body weight). The mice were killed at intervals of 1 hour, 12 hours, 1, 4, 16 and 30 days after injection.

The anaesthetized mice were killed by perfusing the heart with 10 per cent formalin. The animal's head was removed, skinned and left in 10 per cent formalin for one to two weeks. It was then decalcified in formic acid–sodium citrate solution for 7–10 days, dehydrated in ethanol and dioxane and embedded in paraffin. Sections were cut at 8 μm and every twentieth section selected for further investigation. These sections were then assigned alternately to one of two groups. The first group was stained for 48 hours by Einarson's gallocyanin chromalum staining method. The second group was left untreated. Both groups of slides were then dipped in Kodak nuclear track emulsion NTB3. Slides were stored at 4°C during their exposure time of 60–90 days and developed. Sections not prestained were then stained with gallocyanin chromalum. Counting was done under oil immersion.

To provide further data—particularly for time-intervals not represented in the initial series—a second series of experiments was run. This used six male mice (of the BUB strain) of which two were 141 days old and four were 136 days old. The methods of injection were as outlined above.

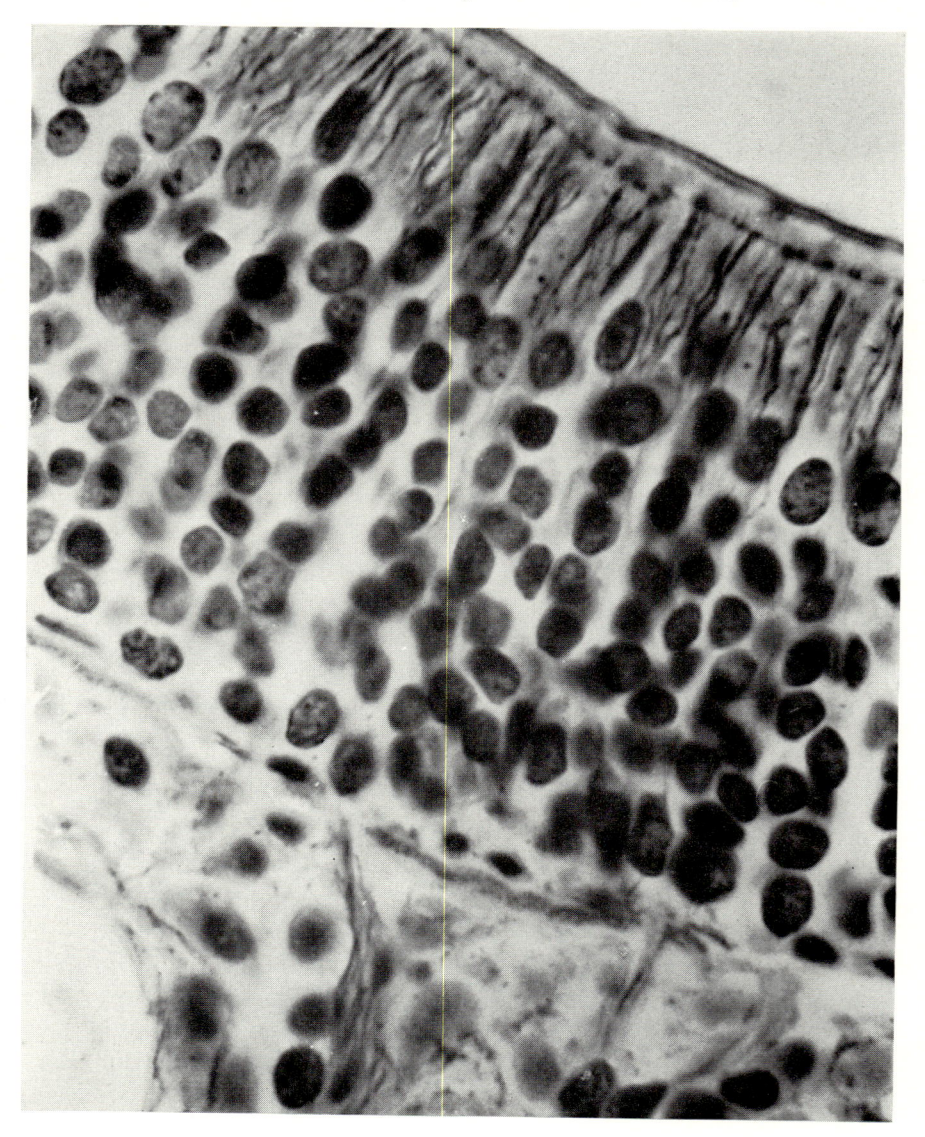

FIG. 1. Mouse olfactory epithelium stained with Bodian. The supporting cells with their large oval-shaped nuclei are prominent in the two most peripheral layers of nuclei (perimeter zone). The nuclei of the receptor cells are smaller, rounder and more dense. They lie in the mid-zone below the supporting cell nuclei. The nuclei of the basal cells lie peripheral to the basement membrane (basal zone) but are often difficult to distinguish from receptor cell nuclei in light micrographs. Note the primary neurons penetrating the base of the epithelium and the terminal knobs of receptor cell dendrites projecting into the mucus sheet. × 1293

However, the mice were killed at intervals of 20 minutes, 8, 20, 44 and 90 days after injection.

The remaining details were as described for the initial series, with the following exceptions. The heads were perfused with 10 per cent formalin, fixed for 3 days in Bouin's solution and then returned to 10 per cent formalin for 9 days before being embedded in paraffin. All sections were cut at 5 μm and all were stained with gallocyanin chromalum before coating with emulsion.

Results

In describing the distribution of labelled cells, it is convenient to consider the olfactory epithelium as having two major zones or subpopulations of cells, although the demarcation is not always sharp (Fig. 1). The *basal zone* consists of the cells forming the base of the epithelium and of the cells immediately adjacent peripherally. This includes most of the true basal cells but could conceivably include a few receptor cells. The *peripheral zone* comprises all cells whose nuclei are peripheral to the basal zone. This last

TABLE I

TOTAL NUMBERS OF LABELLED NUCLEI IN THREE REGIONS OF THE OLFACTORY EPITHELIUM OF MICE AT VARIOUS TIMES AFTER INJECTION OF TRITIATED THYMIDINE

Survival time after injection	Experimental series	Basal	Mid	Perimeter	Total labelled nuclei
20 minutes	2	1481	126	35	1642
1 hour	1	2039	135	70	2244
12 hours	1	1472	57	31	1560
24 hours	1	1548	91	69	1708
4 days	1	1502	70	60	1632
8 days	2	1753	933	95	2781
16 days	1	105	130	20	255
20 days	2	207	580	63	850
30 days	1	44	182	11	237
44 days	2	147	728	72	947
90 days	2	95	614	34	743
Totals		10 393	3646	560	14 599

The radioactive index for the entire epithelium was 0.9 per cent one hour after injection.

zone can be further subdivided into a *perimeter zone* consisting of the most peripheral layer of cells, together with the layer immediately adjacent, distally, and a *mid-zone* lying between the perimeter and basal zones. The nuclei in the mid-zone may well be primarily those of receptor cells with perhaps a few basal and supporting cells, whereas those lying in the perimeter zone may be primarily supporting cells. The use of zones to designate populations avoids implications about cell types, which cannot always be identified with confidence with light microscopy.

Since essentially similar results were obtained in the two series of experiments the data are combined in Table I. This shows that labelled nuclei occurred in all zones of the epithelium at all survival times. However, for the short survival periods the majority of reaction sites were among the nuclei of the basal cells.

As can be seen in Fig. 2, the distribution of radioactive foci was not at random: among the basal cells labelled nuclei were found in nests of about 4 to 13 cells. These clusters were thinly scattered throughout the basal zone. There was no evidence of a correlation between the distribution of nests of basal cells and either the thickness of the epithelium—which ranged from 30 to 130 μm—or the occurrence of labelled nuclei in the associated peripheral zone of the epithelium.

As well as in the epithelium, labelled nuclei were occasionally found at the level of the mucus sheet external to the olfactory epithelium, while scattered radioactive foci were also found throughout the submucosa. In some instances these cells lay immediately below the basal zone.

The number of labelled cells per thousand cells (radioactive index) for the entire epithelium one hour after injection was estimated as nine. However, meaningful counts for specific zones are not easily obtained. The clustering of labelled basal zone cells particularly complicates interpretation, and such estimates were not attempted.

There is no significant change in the relative proportion of labelled nuclei

---

FIG. 2. Olfactory (a-d) and vomeronasal (e) epithelia of adult mice at various periods of survival after injection of tritiated thymidine. Stain: gallocyanin chromalum.

(a) Survival time: 20 minutes. Transverse section through a turbinal showing olfactory epithelium on the outer margin. (Contrast with respiratory epithelium on inner surface). The scattered "nests" of labelled nuclei are concentrated in the basal zone of the epithelium while occasional radioactive foci lie more peripherally in the mid and perimeter zones. × 518

(b) Survival time: 20 minutes. × 1250

(c) Survival time: 8 days. A larger proportion of the labelled nuclei now lie in the mid-zone. × 1060

(d) Survival time: 90 days. Few labelled nuclei remain in the basal zone, the majority being concentrated in the mid-zone. × 14000

(e) Paired vomeronasal organ seen in transverse section. Survival time: 44 days. The neurosensory epithelium is restricted to the medial wall. The uptake of label is restricted to a few cells at the dorsal pole of the right-hand organ (arrow). The low radioactive index is characteristic of this organ and is associated with the absence of basal cells. The dorsal pole is generally the region showing greatest mitotic activity. × 370

(a)

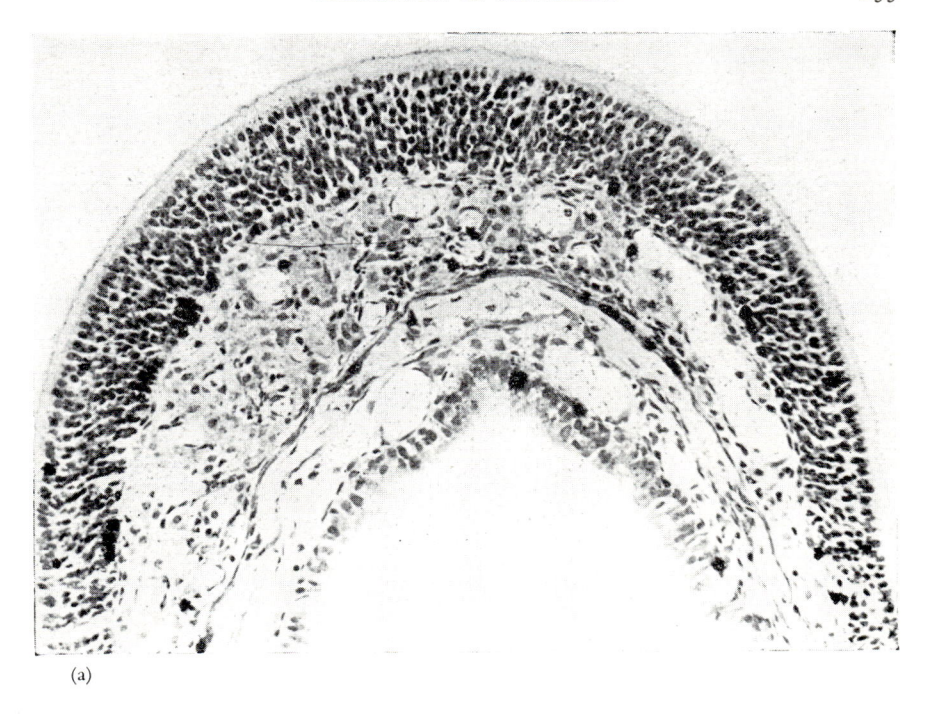

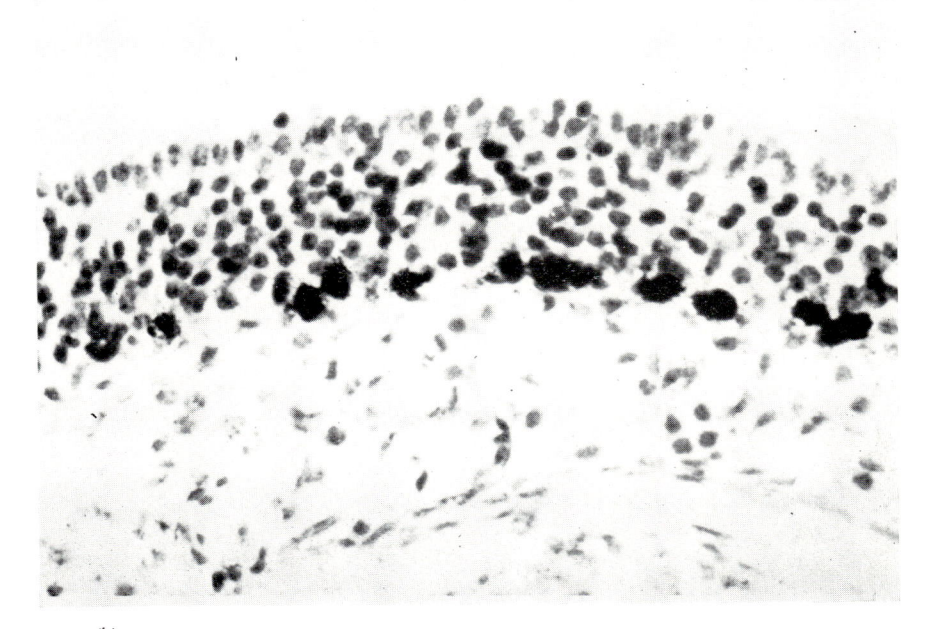

(b)

(c)

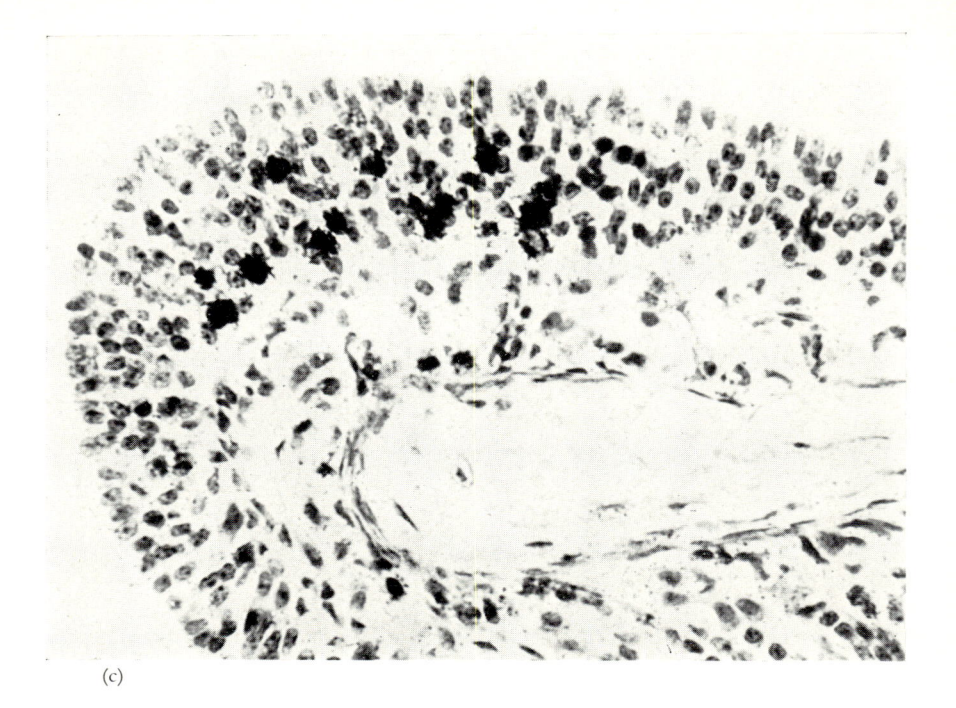

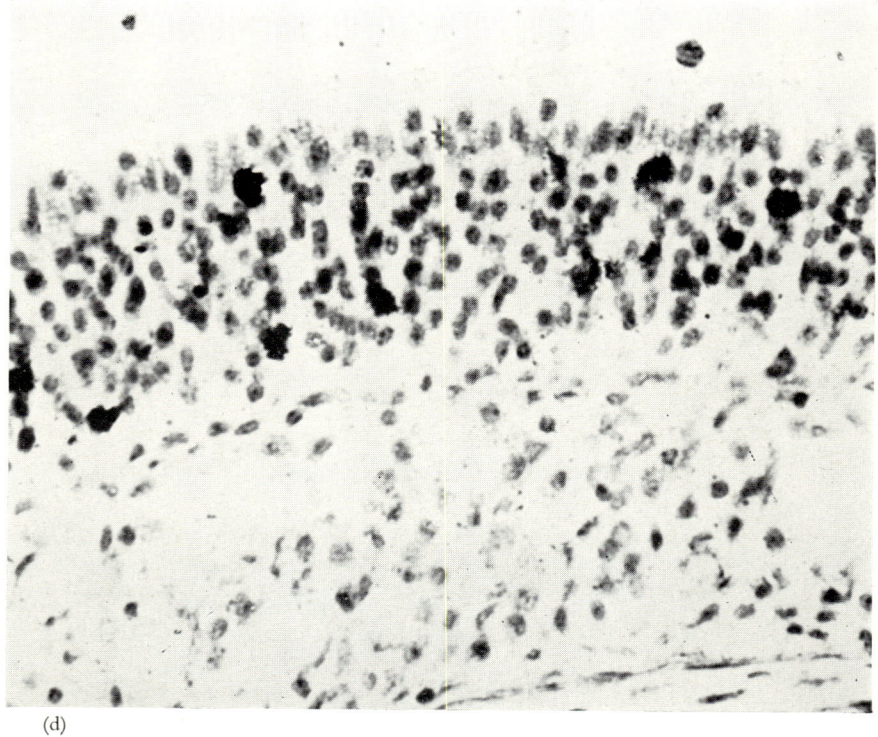

(d)

(e)

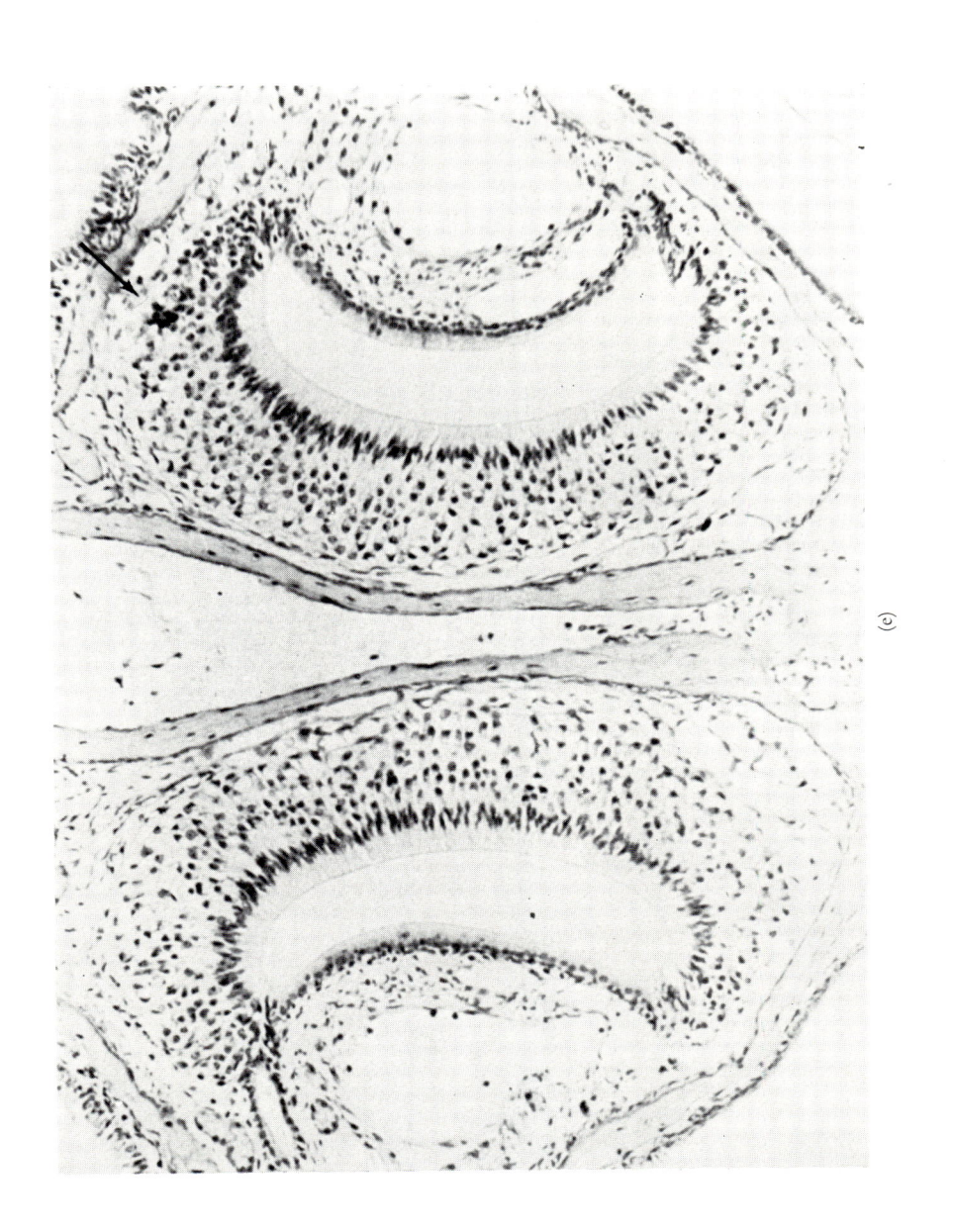

in the different epithelial zones during the first four days after the injection of tritiated thymidine. Thereafter, however, there is a marked and progressive alteration in this distribution which is most rapid between the 4th and 16th days. Thus the number of labelled basal cells diminishes while there is an increase in the relative number of labelled cells in the mid-zone. Only a slight increase occurs in the proportion of radioactive foci in the perimeter zone (Table I). In Fig. 3 these relative distributions are expressed as percentages, the peripheral zone being considered as a single group.

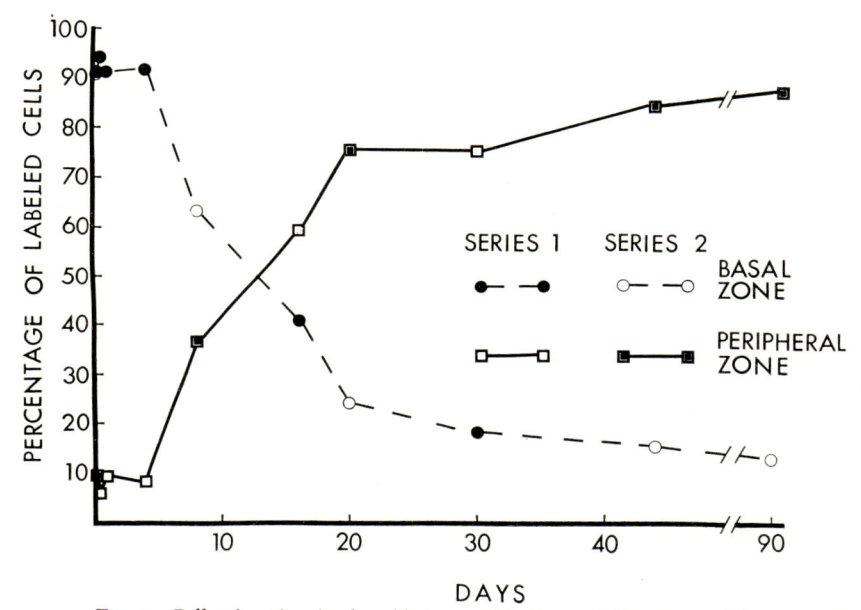

FIG. 3. Cell migration in the olfactory epithelium of the mouse. The relative distribution of labelled nuclei in the two zones of the epithelium at varying intervals after injection of tritiated thymidine is expressed as a percentage of the total number of labelled cells throughout the epithelium.

Although the vomeronasal organ was not systematically studied, some sections were examined since the sensory epithelium of rodents is said to lack basal cells (Bimes and Planel, 1952). The low density of radioactive foci was similar to that seen in the peripheral zone of the olfactory epithelium and there was no evidence of restricted clusters showing high mitotic activity.

Discussion

These findings support the earlier reports of Lams (1940), Bimes and

Planel (1952), Andres (1965) and others, in so far as they show that there are two main populations of dividing cells within the olfactory epithelium, in the basal and peripheral zones. However, there are marked differences in the rates of cell proliferation in these zones. Thus the low radioactive index for the entire epithelium (0·9 per cent) is determined largely by the relatively low initial density of radioactive foci among the peripheral cells. This suggests that, in the absence of input of cells from the basal zone, we can account for no more than an expanding cell population. In such a situation new cells are added at a rate comparable to that at which body weight increases; there is no replacement of cells and migration is rare. An example of a population of this type is the epithelial cells of the lens (see Messier and Leblond, 1965).

In contrast, the dense clusters of radioactive foci among the much smaller population of basal cells, together with their outward migration, is consistent with continuous cell renewal. Such patchy, non-random labelling is apparently not unique, having also been reported for the epithelium lining the jejunum (Hughes *et al.*, 1958). The rate of migration is relatively slow, since no significant changes can be detected within four days after injection. After this period, however, there is a marked movement of cells into the mid-zone and by 30 days the bulk of the migration has been completed leaving a stem population of labelled cells in the basal zone.

Thus the olfactory epithelium seems to be similar to certain other renewing cell populations with restricted proliferation sites in which new cells are also formed predominantly in the basal layer, for example the stratified squamous epithelium of the oesophagus and tongue (Messier and Leblond, 1965). Although the present study provides no evidence to support the claim of Bimes and Planel (1952) that the basal cells replace the cells of Bowman's gland, it is conceivable that the supporting cells undergo continuous renewal. On the other hand, the olfactory epithelium is distinct in that it contains bipolar receptor cells, and it is generally assumed that such cells do not divide. For example, Leblond and Walker (1956) could find no evidence of mitotic activity in the retina. However, Beidler and Smallman (1965) have shown that the cells of the rat taste buds are continuously renewed, and there is evidence of a continuous growth and decay of nerve fibres supplying the sensory field of the pig's snout (Fitzgerald, 1962). But again, there is an important distinction: the olfactory receptor is a primary bipolar neuron, not an accessory cell or a free nerve ending. If it dies, both the cell body and the axon connecting it to the olfactory bulb also die. The axon of a replacement cell must therefore find its way back to the bulb—possibly to a rather specific site. Furthermore, there seems to be little or no reliable evidence that neurons regenerate—at least in mammals—after loss of the cell body. Nevertheless, the possibility exists that continuous renewal of the receptors occurs, as Andres (1965) claims. Further studies, using reliable techniques for labelling and identifying cells, are clearly needed.

But regardless of whether supporting cells or receptor cells or both are renewed, the replacement of cells within this closely interrelated system can hardly take place without considerable disruption of function at those sites where it occurs.

Insofar as the olfactory epithelium is exposed to the moving respiratory air stream it is in contact with the external environment and may undergo continual attrition. Thus some means of renewing its cell populations may be essential. The vomeronasal organ, however, is protected by its secluded location and possibly by having a fluid environment. Consequently replacement of cells may not be so necessary. This could explain the reported absence of the basal cell layer in its epithelium and the low density of radioactive foci observed.

OLFACTORY SENSITIVITY IN MAMMALS

Absolute sensitivity

It is widely held that the olfactory powers of certain mammals are far superior to those of man. Evidence frequently cited is that of Neuhaus (1953, 1955), who found the dog's sensitivity to butyric acid to be 10^5–10^8 times greater than man's. In contrast, Moulton, Ashton and Eayrs (1960) found canine thresholds for this compound to be only 10–100 times lower than human thresholds, while Becker, King and Markee (1962) reported that the dog was about ten times more sensitive than man to oil of cloves.

Until recently, few reliable quantitative data have been available (for species other than man) that would allow a reappraisal of this question of species differences in detection thresholds. In the course of a study involving the telemetry of multiunit activity from the olfactory bulb of rabbits engaged in an odour detection task, we have obtained stimulus–response curves for amyl acetate. Since threshold data for amyl acetate are now available for certain other species, including non-mammalian forms, a more extensive comparison is possible. But before summarizing our findings we shall briefly outline the techniques involved, since they are especially critical in assessing the validity of olfactory threshold data.

We used a two-step air dilution olfactometer allowing delivery of a minimum concentration of 10^{-7} (Fig. 4). A known fraction of the output was monitored by a Varian Aerograph gas chromatograph whose fractionating column was replaced by a direct connexion to the flame ionization

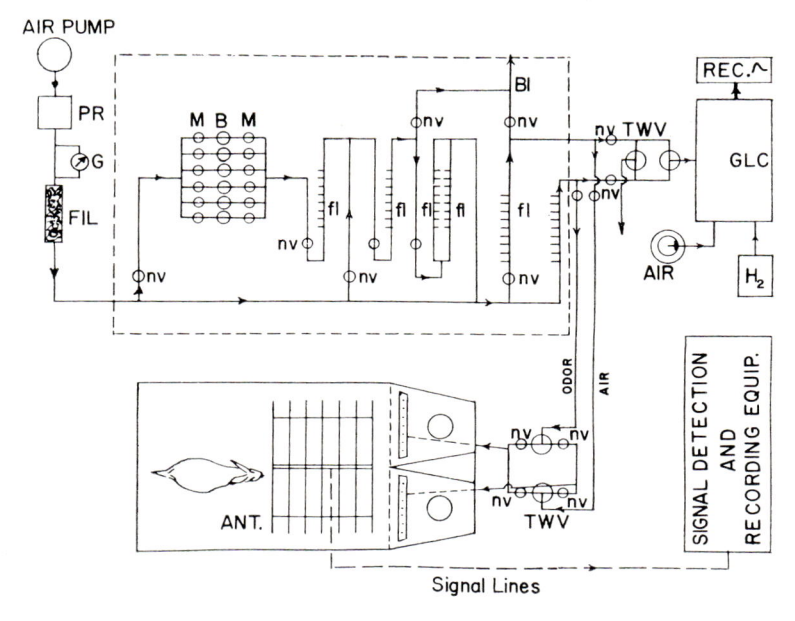

Fig. 4. Olfactometer and behavioural apparatus for determining sensitivity to odours in the rabbit.

Air at controlled pressure is dehumidified and its odorant content attenuated (if not eliminated) by passage through a column of silica gel and activated charcoal (FIL). Manifolds with stopcocks (M) distribute one stream of odorant to a gas dispersion bottle (B) for saturation with the test odorant, while the other stream supplies two dilution steps and an air channel. The olfactometer is enclosed within a chamber (area enclosed by dotted rectangle) whose temperature is maintained at 20 °C. A small fraction of either the air or odour stream enters the gas chromatograph (GLC). The response of the flame ionization detector is signalled by the recorder (REC). Electrophysiological activity from electrodes chronically implanted in the olfactory bulb can be telemetered to the omnidirectional antenna (ANT) on the roof of the test chamber. Bl, bleeder; fl, flowmeter; nv, needle valve; TWV, three-way valve; PR, pressure regulator; G, pressure gauge.

detector. This was installed in parallel with the behavioural test apparatus to provide continuous or periodic sampling of the odorant reaching the animal. For a quantitative determination of concentration the chromatograph was first calibrated at a known concentration. The olfactometer output was then passed to the detector for a known duration and flow rate.

The unknown concentration was estimated by a comparison of the recorded areas. (Discrepancies between concentrations estimated by this method and by calculation from flowmeter readings were relatively small. For example a value of $1 \cdot 28 \times 10^{-2}$ was obtained by the flowmeter and of $2 \cdot 5 \times 10^{-2}$ by gas chromatography for the same sample.)

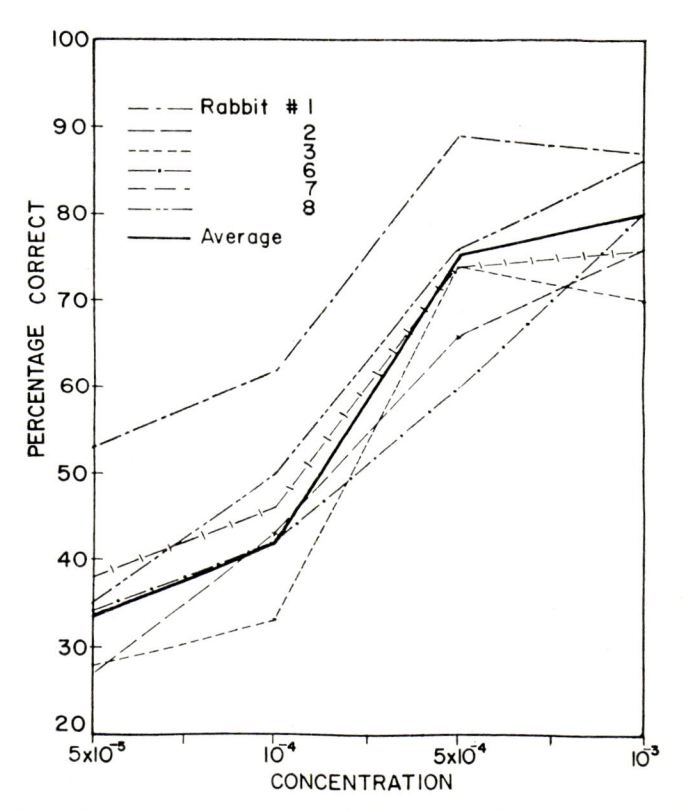

Fig. 5. Detection of n-amyl acetate by the rabbit. Data for six animals were obtained at concentrations of 10^{-5}, 5×10^{-5}, 10^{-4}, 5×10^{-4} and 10^{-3} of vapour saturated at 20 °C. However, no rabbit scored significantly above chance at 10^{-5}. Each point represents an average of 125–400 trials.

The rabbits were tested in a chamber lined with glass, overlain with Teflon sheeting on the floor. Two compartments at one end could be isolated from the main chamber by sliding glass doors and the odorant or filtered air delivered from the olfactometer to the base of each compartment. During trials these inputs were interchanged according to a Gellerman series, so that the odorant and the air were delivered an equal number of times to each compartment. Roof ducts exhausted the air to outside the building.

At the end of the chamber opposite the compartments was a glass treadle. Pressure on this by the rabbit initiated a sequence of events, beginning with the raising of the compartment doors. The rabbit was trained to sample the two compartments, determine which was associated with the odour and signal a choice by interrupting a light beam (directed horizontally across the compartment to a photo-cell at the other side) for 7 seconds. If the choice was correct, 10 ml of water were delivered to a cup in the hinged compartment floor. At the end of 7 seconds this floor moved up to expel the rabbit. If the choice was incorrect the floor swung up and the doors of both compartments were lowered.

Initial training was to concentrations of 10^{-3} or 10^{-2}. When the animal achieved more than 80 per cent correct responses on five successive days, definitive trials began. On any given day 25 trials were presented at each concentration according to a balanced incomplete block design.

During trials, mitral cell activity was simultaneously recorded from electrodes chronically implanted in the olfactory bulb of one rabbit (see Moulton, 1967b, 1968a, b; Goldberg and Moulton, 1969). This was

TABLE II

MINIMUM DETECTABLE CONCENTRATIONS OF n-AMYL ACETATE FOR VARIOUS VERTEBRATE SPECIES

Species	Method	Absolute threshold	Reference
Gopher tortoise	E	10^{-3}	Tucker (1963)
Pigeon	B	10^{-3}	Henton (1969)
Rat	B	$10^{-4.5}$	Moulton (1968a)
Rabbit	B	$10^{-4.5}$	Moulton, Çelebi and Fink, (this paper)
Man	B	10^{-3}	Mullins (1955)

The data for the gopher tortoise (*Gopherus polyphemus*) are responses of the primary olfactory neurons (E); the remaining data were derived behaviourally (B) by techniques which include conditioned suppression (pigeon) and free operant conditioning (rat and rabbit).

transmitted through telemetry units carried on the rabbit's head to an omnidirectional antenna on the roof of the chamber, thus eliminating interference with the rabbit's movements. However, the electrophysiological aspects of these experiments will not be considered here.

The stimulus-response curves derived in this way are shown in Fig. 5. With the exception of data for one rabbit, the curves fall in a rather narrow range. Since no animal scored significantly above chance at 10^{-5}, thresholds must lie in the range $10^{-4.5}$–10^{-5}. Thus $10^{-4.5}$ can be taken as a conservative estimate of the threshold.

In Table II this figure is compared with equivalent data for other species. In addition, D. A. Marshall (personal communication, 1969), using a free operant conditioning procedure, has found that the opossum responds well to amyl acetate in concentrations of 10^{-4}, although this is not a threshold estimate. Furthermore, Tucker (1965) has published figures showing primary nerve responses to 10^{-3} amyl acetate for a number of species. In the sparrow hawk, goose, black vulture and chicken the increase in activity at this concentration is well defined. In the rat, however, there is a barely perceptible change.

Insofar as these data all appear to be derived under well-controlled experimental conditions they are probably comparable. However, electrophysiological techniques may not always extract the most sensitive response to an odour that the system is capable of achieving. For example, the electro-antennogram of the silk moth (*Bombyx mori*) shows thresholds to bombykol, the sex attractant, at 3×10^5 molecules per ml, whereas the behavioural response is $1 \cdot 4 \times 10^4$ molecules per ml (Schneider, 1969).

The striking feature of these various thresholds for amyl acetate is that they all fall within a range of less than two log units of concentration. This is small when taken in the context of the discrepancy of 10^{-5}–10^{-8} between thresholds for butyric acid for man and dog reported by Neuhaus (1953). Indeed, it is sufficiently small for us to question whether it necessarily reflects species differences in the absolute sensitivity of individual receptors. Even intraspecific differences of this order of magnitude have been reported, although for a different odour (Moulton, Ashton and Eayrs, 1960).

Are there, then, other properties of the system which might confer on rats, rabbits and opossums greater sensitivity to amyl acetate than has so far been demonstrated in other species? The size of the organ alone does not offer a convincing explanation, since despite a smaller area of olfactory epithelium, the rat achieves thresholds comparable to the rabbit. However, one variable particularly significant in determining the level of human olfactory performance (Engen, 1960), as well as that of other species, is the amount of training received. A species that relies extensively on odour cues to guide its daily activities may be capable of reaching a higher level of performance than one for whom these cues have restricted and less vital significance. Furthermore, the greater complexity of the mammalian olfactory bulb may be associated with a greater efficiency in detecting weak signals, or signals buried in "noise".

At the threshold levels we have reported, amyl acetate is still present in relatively high concentrations (in the order of 10^{14} molecules per ml). However, because of adsorption of molecules on the walls of the nasal passage and deflection of the main flow of the inspiratory air stream away from the olfactory chamber, the fraction of those molecules entering the nares that finally reaches the olfactory organ is relatively small, and the number of molecules available to each receptor smaller still. Taking these factors into account, Vries and Stuiver (1961) estimated that nine, or less, molecules of butyl mercaptan suffice to stimulate a single olfactory receptor cell in man. *Bombyx mori* shows great sensitivity to the sex attractant, bombykol, as shown by adsorption measures of tritiated bombykol on the antenna (Schneider, Kasang and Kaissling, 1968). From these data it was estimated that at behavioural thresholds each of the receptor hairs on the antenna receives on average a maximum of one molecule during a 2-second stimulus (Schneider, 1969). Thus individual olfactory receptors of diverse species may possess a sensitivity to one or more particular odorants which approaches—and may even reach—the limit theoretically attainable: the detection of one molecule.

We can conclude that, for amyl acetate, any species differences which exist are probably small, with mammals tending to have the lower thresholds. This does not, of course, mean that species differences for certain compounds having unusual biological significance (pheromones) could not be large.

Differential sensitivity

The most probable advantage conferred by an increase in the area of the olfactory epithelium is an enhancement of the organ's discriminative capacity. This includes the ability not only to distinguish between two or more odours compared successively but also to analyse a complex odorous mixture into its constituents. Estimation of the relative proportions of these constituents is a further refinement.

Although man's differential sensitivity to odours has been the subject of several studies, we have little quantitative knowledge about the resolving power of this organ in species where it is most highly developed. A recent study is a step towards meeting this need (Marshall and Moulton, 1969). Five tame opossums were trained to discriminate between two concentrations of one compound and one concentration of another. In one series amyl acetate was paired with other compounds. Animals were consistent in ranking the discriminability of the odorants. As might be anticipated, C_2 and C_7 acetates were the most easily distinguished while C_4 and C_6 were more difficult, with *iso*-amyl acetate lying between these groups. The highest percentages of correct scores (86 per cent and 90 per cent) were attained when DL-menthone and cineole were paired with amyl acetate.

It is interesting that the ability to discriminate among C_4, C_5 and C_6 acetates was more dependent on concentration than that among C_5 and either C_2 or C_7. While there are no data that would allow a comparison with other species, the ease with which the opossum makes such discriminations is striking and in no case did the performance drop below 59 per cent median correct score (in the case of C_5 compared with C_4).

A further aspect of sensitivity to odours that has received little attention is directional sensitivity. The unusual extensions of the nares in some species of bats suggest that further studies might be rewarding (see Moulton, 1967*a*).

SUMMARY

Proliferation and migration of cells in the mouse olfactory epithelium

(1) Studies on the regenerative capacity and on the continuous renewal of cell populations in the olfactory epithelium are briefly reviewed.

(2) Mice were injected with tritiated thymidine and the olfactory epithelium examined for radioactive foci at various periods of survival.

(3) Twenty minutes after injection 90 per cent of the labelled nuclei were found in the basal zone, often scattered in groups of 4–16 cells, while the remainder lay more peripherally. The overall radioactive index was 0·9 per cent.

(4) Between the 4th and 20th day after injection the proportion of labelled nuclei in the basal zone declined to 23 per cent, due to migration of cells peripherally.

(5) We conclude that the products of basal cell division migrate peripherally where they may replace receptor cells or supporting cells or both. The low density of labelled nuclei initially among the receptor and supporting cells suggests that (in the absence of cell input from the basal zone) these cells constitute expanding rather than renewing cell populations.

(6) The vomeronasal organ, which is said to lack basal cells in the rodent, showed a low density of radioactive foci with no concentration in the basal area of the epithelium.

Sensitivity to odours

(7) Six rabbits were trained to interrupt a light beam for seven seconds when they detected amyl acetate delivered from an olfactometer, calibrated by gas chromatography.

(8) Thresholds were in the range $10^{-4.5}$–10^{-5} of vapour saturated at 20 °C.

(9) This figure is compared with those derived for other species, including rat, pigeon and man, and the implications are discussed. These threshold values all lie within two log units of each other.

(10) Differential sensitivity to odours is considered in the context of a recent behavioural study on the opossum.

Acknowledgements

The authors are indebted to Dr J. Altmann for demonstration and advice on the use of radioautographic techniques, and to Drs D. Stevens and D. A. Marshall for valuable discussions on the design of the behavioural apparatus. Preparation of this manuscript and the research described was sponsored by the following agencies: Institute of Neurological Diseases and Stroke, United States Public Health Service, through Grant No. NB-06860-03; the Air Force Office of Scientific Research, Office of Aerospace Research, United States Air Force (Grant No. AFOSR 1056-68); and the United States Veterans Administration.

REFERENCES

ANDRES, K. H. (1965). *Naturwissenschaften*, **17**, 500.
ANDRES, K. H. (1966). *Z. Zellforsch. mikrosk. Anat.*, **69**, 140–154.
BECKER, R. F., KING, J. E., and MARKEE, J. E. (1962). *J. comp. physiol. Psychol.*, **55**, 773–780.
BEIDLER, L. M., and SMALLMAN, R. (1965). *J. Cell Biol.*, **27**, 263–272.
BIMES, C., and PLANEL, H. (1952). *C.r. Ass. Anat.*, **38**, 199–204.
ENGEN, T. (1960). *Percept. Mot. Skills*, **10**, 195–198.
FITZGERALD, M. J. T. (1962). *J. Anat.*, **95**, 495–514.
GOLDBERG, S. J., and MOULTON, D. G. (1969). *Am. Zool.*, **9**, 593. (abst.)
HENTON, W. W. (1969). *J. exp. Analysis Behav.*, **12**, 175–185.
HUGHES, W. L., BOND, V. P., BRECHER, G., CRONKITE, E. P., PAINTER, R. B., QUASTLER, H., and SHERMAN, F. G. (1958). *Proc. natn. Acad. Sci. U.S.A.*, **44**, 476–483.
LAMS, M. H. (1940). *Bull Acad. r. Méd. Belg. VI*, **5**, 110–135.
LEBLOND, C. P., and WALKER, B. E. (1956). *Physiol. Rev.*, **36**, 255–275.
LE GROS CLARK, W. E. (1956). *Yale J. Biol. Med.*, **29**, 83–95.
LE GROS CLARK, W. E. (1957). *Proc. R. Soc. B*, **146**, 299–319.
MARSHALL, D. A., and MOULTON, D. G. (1969). *Fedn Proc. Fedn Am. Socs exp. Biol.*, **28**, 275. (abst.)
MESSIER, B., and LEBLOND, C. P. (1965). *Am. J. Anat.*, **106**, 247–285.
MOULTON, D. G. (1967*a*). *Am. Zool.*, **7**, 421–429.
MOULTON, D. G. (1967*b*). In *Olfaction and Taste II* (Proceedings of the Second International Symposium, Tokyo, 1965), pp. 109–116, ed. Hayashi, T. Oxford: Pergamon Press.
MOULTON, D. G. (1968*a*). *Olfactologia*, **1**, 69–75.
MOULTON, D. G. (1968*b*). In *Theories of Odor and Odor Measurement* (Proc. NATO Summer School, Istanbul, 1966), pp. 483–491, ed. Tanyolaç, N. London: Technivision.
MOULTON, D. G., ASHTON, E. H., and EAYRS, J. T. (1960). *Animal Behav.*, **8**, 117–128.
MOULTON, D. G., and BEIDLER, L. M. (1967). *Physiol. Rev.*, **47**, 1–52.
MULLINS, L. H. (1955). *Ann. N. Y. Acad. Sci.*, **62**, 249–276.
NAGAHARA, Y. (1940). *Jap. J. med. Sci. Trans. Abstr. (Sect. V)*, **6**, 165–199.
NEUHAUS, W. (1953). *Z. vergl. Physiol.*, **35**, 527–552.
NEUHAUS, W. (1955). *Z. vergl. Physiol.*, **37**, 234–252.
SCHNEIDER, D. (1969). *Science*, **163**, 1031–1037.
SCHNEIDER, D., KASANG, G., and KAISSLING, K.-E. (1968). *Naturwissenschaften*, **8**, 395.
SCHULTZ, E. W. (1960). *Am. J. Path.*, **37**, 1–19.

SMITH, C. G. (1938). *Can. med. J.*, **39**, 138–140.
TAKATA, N. (1929). *Arch. Ohr.-, Nas.-u. KehlkHeilk.*, **121**, 31–78.
THORNHILL, R. A. (1967). *J. Cell Sci.*, **2**, 591–602.
TUCKER, D. (1963). In *Olfaction and Taste* (Proceedings of the First International Symposium, Stockholm, 1962), pp. 45–69, ed. Zotterman, Y. Oxford: Pergamon.
TUCKER, D. (1965). *Nature, Lond.*, **207**, 34–36.
VRIES, H. DE, and STUIVER, M. (1961). In *Sensory Communication*, pp. 159–167, ed. Rosenblith, W. A. Cambridge, Mass.: M.I.T. Press.

DISCUSSION

Beets: Dr Moulton, in your Table II (p. 241) you assembled threshold data from a number of sources. I tend to be somewhat suspicious of many of the threshold data published in the literature. You mentioned, for instance, Mullins' work (1955), but this was done before modern purification techniques, such as gas liquid chromatography, were available. Mullins worked with a series of C_4 hydrocarbons (butane, isobutane, butadiene and the butenes), which in pure form are odourless. Since purification without G.L.C. is extremely difficult, I wonder what to think about these threshold figures. What were the other methods used for determining these thresholds?

Moulton: The pigeon data were obtained by the conditioned suppression method; Tucker's data were derived electrophysiologically from primary olfactory neurons. The free operant conditioning method that we used to derive the rat thresholds is essentially the one I described for the rabbit. I agree that Mullins' data are of questionable value and, as I pointed out, I would expect the figure for amyl acetate to lie much closer to that of the rat and rabbit.

Beets: In most of these cases the thresholds measured refer to the concentration in the air current, but not the concentration of the odorant in the air in immediate contact with the receptors?

Moulton: That is so. I am aware of these problems and have stressed them in the past. However, the central question that we are asking is whether we can find species differences in the ability of animals to detect amyl acetate—differences in the order of, say 4–8 log units, such as Neuhaus has found. Clearly we cannot. In this context, then, the data may be useful as a rough guide to the relative sensitivities among different species to concentrations of amyl acetate in the air. If we ask what the equivalent concentrations are at the receptors, we are asking a slightly different question, and must apply corrections of the kind proposed by de Vries and Stuiver (1961).

Amoore: Dante G. Guadagni at the U.S. Department of Agriculture has measured the threshold value of *n*-amyl acetate for man, using his "squeeze-bottle" method. The threshold value was about 5 parts per 10^9 parts of water (Teranishi *et al.*, 1966). The saturation concentration in water is about 0·2 per cent (v/v). These figures yield an absolute threshold of about $10^{-5 \cdot 5}$ in Moulton's units. This would bring man closer to the rat and rabbit (cf. absolute thresholds of $10^{-4 \cdot 5}$, obtained by Dr Moulton).

Davies: What is the final conclusion about the sensitivity of dogs; are they so much more sensitive than other animals or man?

Moulton: There is no good evidence to suggest that the dog's performance

FIG. 1 (Andres). The *upper half* of the illustration shows the possibility of sensory cell propagation and regeneration from blastema cells (BL). Mitoses (M) occur at the level of the sensory cell perikarya (OC). The young olfactory sensory cell sends one ending (D) to the olfactory border (OS). At the apical pole of the ending, division of centrosomes takes place which later on arrange themselves as basal granules of the sensory flagella. Renewal by mitosis diminishes with age and eventually ceases completely in adult animals, but regeneration of sensory cell dendrites is still possible (*lower half*). It is probable that the olfactory vesicles undergo a continuous moulting process which is compared in examples 1–5. Since phases 1–3 predominate at the olfactory border, phases 4 and 5 would be very brief indeed. SC, supporting cells.

is unusual for a "macrosmatic" mammal. It is probably superior to man in differential sensitivity to odours but there are not enough data on absolute sensitivity, over a wide enough range of odorants, to allow a final conclusion. Our estimate that the dog is ten to a hundred times more sensitive than man for butyric acid (Moulton, Ashton and Eayrs, 1960) may not be far out, in view of the data I presented for amyl acetate. But well-controlled experiments, in which the performance of the dog is compared directly with that of fully trained human subjects in the same experimental situation, are particularly needed.

Andres: We have studied the frequency of mitosis in olfactory epithelium of cats and dogs. Mitoses were found in the basal cell layer and in the supporting cells but seldom in the sensory cells. I think now that mitosis of my so-called *Blastemzelle* is common in young animals but very rare in adults (see Fig. 1), but that the peripheral dendrites of the sensory cells moult and regenerate all the time, so that the cells show young, medium and old processes. In some species, such as the cat, this moulting goes on at a very high rate and in others, for example the rat, I don't see any moulting. However, I wonder whether a virus disease could be causing a shortened life cycle of the olfactory vesicle, and therefore the rate of regeneration of the dendrites of the sensory cells has to be increased.

Secondly, I think that the basal cells in the olfactory epithelium are connected by desmosomes to the outer supporting cells, and that the basal cells in which mitosis is frequent provide the new supporting cells, their nuclei then moving into the outer zone. There the supporting cells may divide again, but with a special kind of division like that of glia cells. As in an oligodendrocyte the nucleus of the supporting cell divides to produce another undifferentiated supporting cell within a lateral cytoplasmic protrusion without any change in the supporting cell structure. It is similar to the division of the myelinated Schwann cell where the nucleus migrates to the distal pole of the cell into a protrusion of undifferentiated cytoplasm to produce the daughter cell.

Murray: In continuously growing animals such as the rat and mouse one would expect some degree of continuing growth of olfactory epithelium. This does not mean that bipolar cells are increasing, however. Cell increase and turnover may be confined to sustentacular cells. The gradual decrease of label in the basal cells and increase in the upper layers of nuclei is compatible with replacement from the basal cells of the sustentacular cells, with or without absolute increase. Have you any way of detecting whether your label is over a sustentacular or over a bipolar cell?

Moulton: Since sustentacular cell nuclei tend to lie among the receptor rods forming a clearly defined perimeter zone, it is very probable that the occasional labelled nuclei which we found in this zone at early survival periods are those of sustentacular cells. (But at later survival periods interpretation is complicated by the possibility that there may be shorter-lived labelled receptor cells migrating through this zone.) The nuclei in the mid-zone are primarily those of receptor cells. However, although I know of no evidence for it, we cannot rule out the possibility that occasional sustentacular cells divide in this zone. At later survival periods any labelled nuclei destined to be sustentacular cells might, in any case, pass through this zone. It is perhaps significant that although there is a marked increase with time in the number of nuclei migrating into the mid-zone, there is little, if any, significant increase in the proportion of labelled nuclei found in the perimeter zone. In other words, cells migrating from the basal zone tend to be "trapped" in the mid-zone.

A further criterion for distinguishing one cell type from another is the size and shape of the nucleus. Thus receptor nuclei tend to be small and round, and the sustentacular nuclei, larger and more oval. In most cases the label obliterates the shape and size of the nucleus. However, we have seen small, round labelled nuclei in the mid-zone. Unfortunately, one cannot be certain that this in itself is a sufficient, or entirely reliable criterion.

Since the receptor dendrite and axon are selectively stained by Bodian's silver stain we have attempted to combine this approach with autoradiography. But it is unusual, in the mouse, that a section includes both the dendrite, or axon, and the nucleus of the same receptor cell. Since the labelling is sparse, we would be unlikely to see (and have not seen) the combination of a labelled nucleus which is continuous with a stained dendrite or axon (should this combination occur). However, we are now working on the frog where the simpler structure of the epithelium may favour such a combination.

Ottoson: You mentioned several factors which may influence the absolute sensitivity of the olfactory system. One of these factors is undoubtedly the receptive membrane area. Would it not be useful, in comparing the sensitivities of different animals, to correct for differences in the area of the olfactory membrane?

Moulton: The correction would be useful if a relation between sensitivity and receptive area was known to exist. But I think this is no more than an assumption, although it would be interesting to investigate.

Reese: Another and perhaps equally important measurement is the thickness of the olfactory epithelium. Thickness is largely determined by the number of bipolar receptor cells per unit area and this density could be as important for olfactory sensitivity as the area of the epithelium. How many layers of bipolar cells are there and what is their density in the mouse?

Moulton: In adult mice the epithelium varies between 30 and 130 μm in thickness, representing about 4-15 layers of nuclei of which perhaps two-thirds are receptor cell nuclei.

Døving: The interesting point would be the number of receptors per unit area. Have you any figures for mammals? In a teleost fish (*Lota lota,* L.) it is about 73 per 1000 sq. μm.

Moulton: The density of receptors per square millimetre is estimated to be 120 000 in the rabbit (Allison and Warwick, 1949), 61 800 in the pig (Gasser, 1956) and 42 000-75 000 in the eel, depending on body length (Teichmann, 1954, 1959).

Reese: In the shark the number of receptors per unit area is very low, roughly 10 per 1000 sq. μm.

Beidler: What is the evidence for a relationship between area and threshold? In insects it is clear that you don't need large numbers of active receptors for high sensitivity.

Moulton: There is no good evidence that there is an increase in absolute sensitivity with an increase in area of receptor surface.

Beidler: To put it differently, why does a rabbit have 100 million receptors?

Moulton: I suspect that as the number of receptors increases, so does the ability to resolve a complex odorous mixture into its constituents. The bulb receives a more detailed picture which enhances differential rather than absolute sensitivity. It is true that absolute sensitivity may be marginally enhanced insofar as the net is spread wider, trapping molecules that might otherwise escape detection. But as you point out this is unlikely to be a significant effect in view of the low thresholds reported for some compounds for insects, where the number of receptors is in the order of 10^3–10^4 rather than 10^6–10^7, as in mammals.

REFERENCES

Allison, A. C., and Warwick, R. T. T. (1949). *Brain,* **72,** 186–197.
Gasser, H. S. (1956). *J. gen. physiol.,* **39,** 473–496.
Moulton, D. G., Ashton, E. H., and Eayrs, J. T. (1960). *Animal Behav.,* **8,** 117–128.
Mullins, L. H. (1955). *Ann. N.Y. Acad. Sci.,* **62,** 249–276.
Teichmann, H. (1954). *Z. Morph. Ökol. Tiere,* **43,** 171–212.
Teichmann, H. (1959). *Z. vergl. Physiol.,* **42,** 206–254.
Teranishi, R., Flath, R. A., Guadagni, D. G., Lundin, R. E., Mon, T. R., and Stevens, K. L. (1966). *J. agric. Fd Chem.,* **14,** 253–262.
Vries, H. de, and Stuiver, M. (1961). In *Sensory Communication,* pp. 159–167, ed. Rosenblith, W. A. Cambridge, Mass.: M.I.T. Press.

DETECTION OF HIDDEN OBJECTS BY DOGS

E. H. Ashton and J. T. Eayrs

Department of Anatomy, University of Birmingham

It is widely accepted that the world of the dog differs from that of man in so far as olfaction seems to replace vision as the dominant modality of sensation. Man has made frequent use of the dog's ability to orientate to non-visual cues as a means of supplementing his own sensory capacities. Some of these uses, such as hunting, have relied largely on innately organized patterns of behaviour; others have required recourse to special methods of training, usually based on conditioning techniques. Among these the use of dogs to find human beings and livestock buried under snow, or to detect sites beneath which truffles may profitably be sought, are well known. Less well-known, perhaps, is the employment of the dog, practised on the Somerset coast, to indicate rocks under which fish may be found. It was knowledge of the abilities of the dog under these circumstances that undoubtedly led to attempts by the Armed Forces at the close of the Second World War to use this animal as a biological mine detector in situations where the Polish mine detector, by reason of its inability satisfactorily to discriminate between mines and ground-contaminating material such as shell-fragments, or to detect non-metallic mines at all, was proving inadequate. Thus, in the absence of an alternative device, the only sure method of detection was that of laborious probing of the ground by soldiers with its wastefulness in man-power and its attendant dangers from anti-personnel devices.

Early reports on the use of dogs to detect not only mines but caches of hidden arms were encouraging, though some of the claims advanced for the dog's powers of detection made considerable demands upon credulity. Many of the available reports were anecdotal and tended to emphasize those occasions on which hidden objects had been discovered by dogs to the exclusion of those occasions when such "discoveries" had proved negative. Clearly what was lacking was a scientifically controlled investigation of the problem directed not only to an objective assessment of the extent to which the dog could be relied on as a mine detector but also, by determining how the dog achieved its success, to providing biological information of potential use in the development of a suitable physicochemical analogue. In 1949, Professor Sir Solly Zuckerman was asked to set up a project to this end and it is with some of the findings that emerged from the ensuing study that this communication is concerned. The work was carried out, with the assistance of military personnel, initially at the Royal Army Veterinary Corps Training Centre and Depôt, Melton Mowbray, and later was transferred to the Signals Research and Development Establishment of the then Ministry of Supply at Christchurch.

TRAINING

The basic method of training the dogs used throughout the work to be described followed the pattern laid down in the relevant Army Manual. A hungry dog is led to a number of mines scattered randomly on the ground surface on each of which is placed a small piece of meat. It is taught to sit by the mine before eating the meat. Later the meat is fed from a pouch carried by the handler each time the dog sits at a mine and, during subsequent training, the mine is concealed from view in successive stages, for example partial followed by complete burial, the dog being rewarded on each occasion that it sits at an appropriate site. Such correct and rewarded responses are generally referred to as "points", and incorrect responses as "false points".

PRELIMINARY EXPERIMENTS

Using this method of training, our early findings proved far from encouraging. Only when objects had been buried in a manner involving recent and local disturbance of earth did the dogs achieve significant success; in other circumstances, although certainly the dogs found buried mines, when the number of false points was taken into account, those discovered proved little more numerous than could be accounted for by guesswork. Such findings inevitably laid emphasis on the site of burial rather than the buried object itself as the source of stimulus guiding the dogs' behaviour, and led to a reformulation of the problem. Thus our further work was designed to answer three questions, which will be dealt with in turn, namely:

(a) Can dogs indeed detect objects hidden from view? If so:
(b) What cues are used?
(c) Has this capacity any practical value?

CAN DOGS DETECT HIDDEN OBJECTS?

The first series of experiments to be described involved a stereotyped situation specifically designed to determine whether stimuli arising as a result of the presence of a concealed object are adequate to elicit a response on the part of the dog. A number of "wells" of size 18 × 18 × 9 inches deep, lined with breeze block, was constructed, each "well" being provided with a lid comprising a metal frame and grid of expanded metal and covered either with hessian or with turf (Fig. 1). These wells were arranged in groups of four in a straight line, the dog being taken to each in turn and

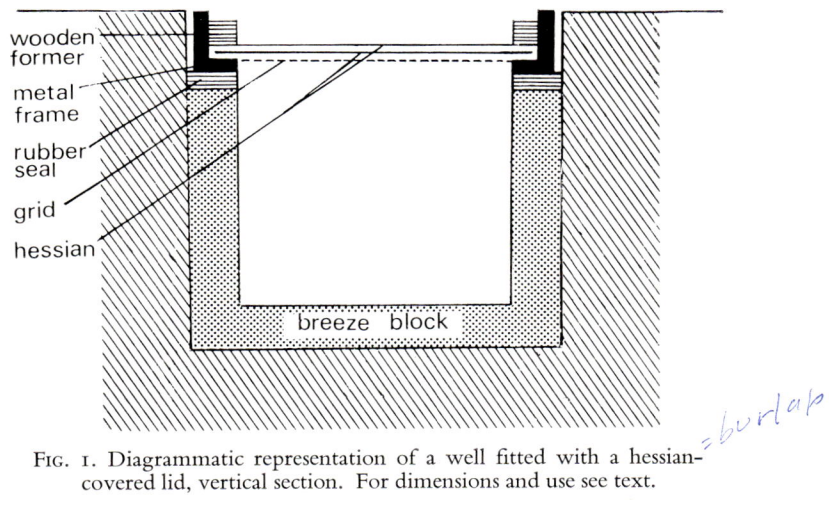

FIG. 1. Diagrammatic representation of a well fitted with a hessian-=burlap covered lid, vertical section. For dimensions and use see text.

required to determine whether or not an object was concealed beneath the lid.

A "block" of trials consisted of placing the objects to be detected in all of the 16 possible ways in which 1-4 items can be placed in four positions. Where different classes of object were to be compared for ease of detectability, these were either tested by means of alternating runs using each type of object in turn or the differing objects were fed into the design to form a mixed pattern of presentation, thus requiring 384 runs (16 × 4!) to compare the relative detectability of the four items. In either case an equal number of empty wells and wells containing objects was presented in each block of trials, thus facilitating statistical validation of the dogs' performance. The results of this experiment (Table I) showed that a wide variety of objects, including anti-tank (A/T) mines, glass casserole dishes and anti-personnel

TABLE I
RELATIVE DETECTABILITY OF LARGE OBJECTS

Type of object	Maximum possible score*	Percentage success (chance level = 50%)	Success with test object relative to filled A/T mine (= 100)
Painted drill A/T mine	640	96·1	98·3
Unpainted drill A/T mine	256	95·3	96·4
Filled plaster cases	640	97·1	100·2
Empty plaster cases	640	95·5	97·4
Glass dishes	640	87·4	96·9
Breeze block	640	69·9	73·2
Filled A/P mine	640	85·4	87·4
Empty A/P mine	640	86·7	88·8

Well technique; turf-covered lids; objects handled by supervisor.
*Combined scores for two dogs both trained to detect filled A/T mines.

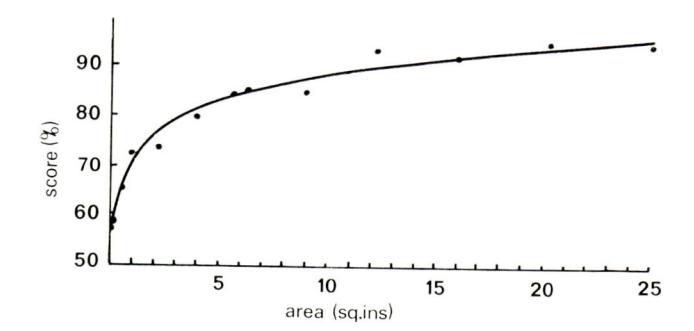

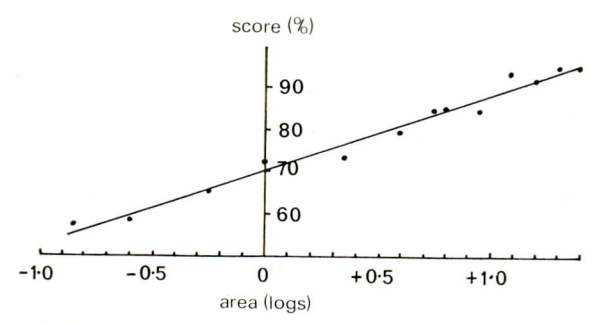

FIG. 2. Relationship between surface area of concealed object and detectability (mean performance of two dogs). *Above:* success score plotted against surface area. *Below:* success score against logarithm of surface area.

(A/P) mines, were readily detected. Only pieces of breeze block proved difficult. Later experiments showed that manipulating the objects by hand influenced ease of detection only in the case of the pieces of breeze block.

It thus emerges that the dog possesses an extraordinary capacity for detecting the presence of objects hidden from view and this finding is reinforced by the additional observation that the detectability of objects is directly related to the logarithm of their surface area (Fig. 2), squares of plywood presenting as small a surface as 0·15 sq. inches being found, though with a low level of success, significantly better than chance. The material of which an object is made also has an influence on its detectability. After careful precautions had been taken to remove all contaminants, wood proved more readily detectable than brass and brass more detectable than glass (Table II).

TABLE II

DETECTABILITY OF OBJECTS MADE OF DIFFERENT MATERIALS

Type of material	Maximum possible score* (=100% success)	Percentage success (chance level = 50%)
Wood (plywood)	640	80·7
Metal (brass)	640	67·6
Glass	640	60·2

Well technique; hessian-covered lids. Three-inch squares of each material were used and handled with tongs. Wooden squares were cleaned with sandpaper, those of metal and glass with ether.
  * Combined scores for two dogs.

Such observations have, in general, been verified in a rather less stereotyped though still artificial environment in which the objects were buried under loosely packed soil. These conditions were similar to those in the wells in so far as the dog was given a clear indication of where its decision was to be made and was presented with an equal number of burial sites containing the objects to be detected and dummy sites. In such circumstances, it became clear that burial sites containing objects such as glass casserole dishes and imitation A/T mines made of plaster could readily be distinguished from the dummy burial sites and that such objects were found nearly as readily as in the wells (Table III).

TABLE III

RELATIVE DETECTABILITY OF LARGE OBJECTS

Type of object	Maximum possible score*	Percentage success (chance level = 50%)
Empty plaster cases	240	85·8
Inverted glass dishes	570	79·9

Objects covered with two inches of loosely packed soil.
  * Combined scores for two dogs both trained to detect filled A/T mines.

HOW DOES THE DOG DETECT THE OBJECT?

The behaviour of the dog during these trials left little room for doubt that the sense of smell played an important, if not exclusive, part in the process of detection. Yet the ability of animals to detect breeze block in wells made of the same material, the smallness of some of the objects detectable and the incidental observation that some animals were able to perform successfully without intensive sniffing suggested that a further investigation of the role of olfaction should be carried out. All investigations along these lines indicated the dominance of the olfactory modality without, however,

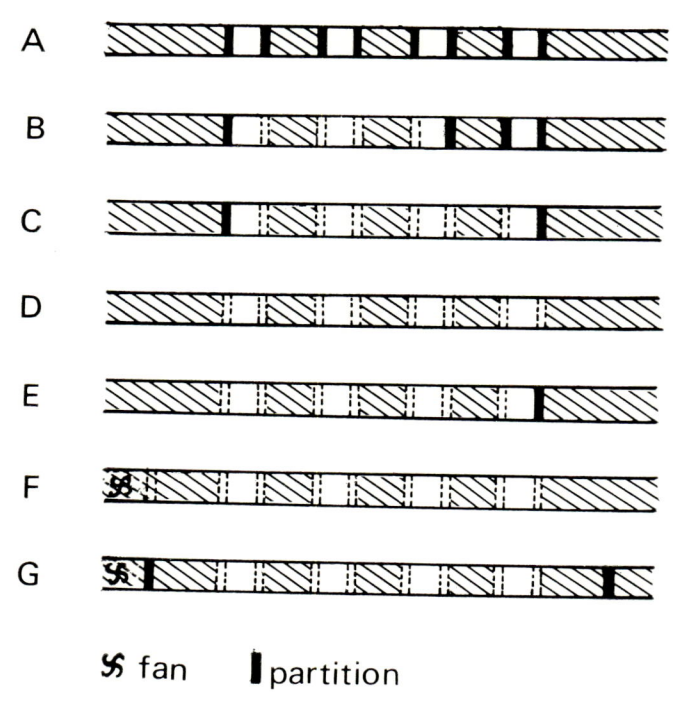

FIG. 3. Schematic arrangement of the partitioning of tunnels to form wells of the type shown in Fig. 1. A: four independent wells; B and C: removal of partitions to test effect of increased air-space and interaction between contained objects; D and E: further increase in air-space by means of open-endedness; F: fan incorporated to set air in motion; G: distraction by fan noise, with partitions to prevent motion of air.

providing an explanation of the anomalies just mentioned. A key experiment in this connexion is illustrated in Fig. 3. Here the standard "well" technique described above was used, with the difference that the wells, instead of being independently dug, were formed by partitioning a continuous tunnel with pieces of breeze block (A), the roof of the tunnel being interrupted by apertures covered, during trials, by hessian-lined lids identical with those used in the earlier study. Removal of selected partitions (B–E) enabled the effects of increasing the volume of still air surrounding the objects placed in the wells to be assessed; the provision of a fan at one end of the tunnel, with or without interrupting partitions (F–G), permitted a comparison of the detectability of the objects surrounded by still or moving air. A/T mines were used and Table IV shows that, whereas

TABLE IV

EFFECT OF CHANGES IN THE VOLUME AND MOTION OF SURROUNDING AIR ON THE DETECTABILITY OF FILLED A/T MINES

Degree of separation	Maximum possible score* (=100% success)	Percentage success (chance level=50%)
Each well separated, air still	3712	84·4
All wells connected, air still	512	59·6
All wells connected, air in motion	896	51·1

Tunnel wells; hessian-covered lids.
* Combined scores for two dogs.

detectability when the tunnel was partitioned to form individual wells was comparable with that previously recorded, this was severely impaired by increasing the air-space and reduced to insignificant levels when the air surrounding the object was set in motion.

Other experiments have supported the view that the stimulus received by the dog was particulate and air-borne. Thus:

(*a*) Dusting one-inch oak cubes, normally readily detected in the wells, with powdered charcoal used as an adsorbent, prevented their detection.

(*b*) Incorporation of an impermeable material or of a water-seal in the hessian covering of the well-lids, or within the soil covering the object, prevented the detection of even large objects.

(*c*) Objects buried under soil were less readily detected when the overlying soil was tightly packed.

(*d*) Investigation of the possible participation of the auditory modality proved negative, since measurement of the change in resonance of a well on the introduction of a sizeable object showed this to be less than could be expected to fall within the dogs' discriminatory capacity.

PRACTICAL VALUE OF DOGS IN MINE DETECTION

At this point we are left with the conclusion that the dog can perform remarkable feats of detection, principally by use of its sense of smell, provided always that it is shown visually where to make its discrimination. We must now enquire whether this capacity is sufficient to enable it so to perform in the open field where no visually oriented choice point is available, an aspect of the investigation which brings us back to those early studies which first led us to doubt the successes claimed by the military. The problem was approached by studying the effects of varying the duration and mode of burial of the test objects, as factors determining detectability. Table V shows that, as a general trend, the level of performance stands in inverse relationship to the duration of burial and, by inference, to the consolidation of the disturbed earth which has taken place since burial. Thus, where no localized disturbance was present, as in 2-year-old minefields which had been laid by removing all the turf and topsoil of the area, placing the mines on the ground and restoring the soil and turf (serial ∞ in Table V), finds were negligible and the number of false points extremely high. In a minefield of the same age, but laid by individual burial of objects,

TABLE V

PERFORMANCE ON OPEN MINEFIELDS OF DIFFERENT AGES

Age of field (months)	Percentage of total objects found	Percentage of total points that were false	Percentage of success attributable to dog + handler	Dummy burial sites Number presented	Dummy burial sites Percentage located
∞	12·0	94·6	4·2	—	—
24	38·1	90·5	28·1	—	—
9	28·0	84·2	16·6	18	11·1
7	33·9	85·1	24·1	20	0·0
5	45·7	80·1	36·8	11	9·1
0·5	85·0	24·1	83·8	44	70·5

Combined scores for two dogs; five types of object.

performance was somewhat better and continued to improve with more recently laid fields. However, only on the fields laid for about two weeks, in which localized disturbance was combined with recent burial, could any confidence be placed in the dogs' performance with regard to the number of mines detected and the fewness of false points.

The limitations of the dog trained by standard methods thus become apparent, but it may properly be asked whether, if differently trained, the dog could make better use of its sensory capacities. As a preliminary to this enquiry, it is important to know the nature of the cues to which the animal successfully responds in such open-field tests; in other words, with what modality does it establish its choice point where islets of disturbed earth are available as a guide?

The problem was first approached by studying the part played by vision and a series of trials was conducted, using recently laid minefields, to determine the effects of blindfolding the dog, its handler or both. Analysis of the results (Table VIA) showed that blindfolding had a slightly adverse effect on performance, most of which was attributable to blindfolding the handler. Detection remained at a high level, however, and was presumably maintained by use of the dog's olfactory modality. The question of whether the dog detected the burial site or the mine was examined by including dummy burial sites in the test-minefields. Such sites were detected slightly less well than sites containing mines (Table VIB), indicating that the mine itself may provide a marginal stimulus. Could then, the dog be trained to respond to this stimulus alone?

Several differing techniques were employed in an attempt to do this. In the most successful of these, mines were buried in beach sand leaving

TABLE VI

OLFACTORY AND VISUAL CUES RECEIVED BY THE DOG AND HANDLER

A

	Total mines presented	Total mines found	Percentage of success attributable to dog + handler
Handler and dog blindfolded	180	130	72·3
Handler alone blindfolded	180	138	77·1
Dog alone blindfolded	180	146	81·8
Neither blindfolded	180	159	85·7

B

	Mines or dummy sites presented	Total found — Mines	Total found — Dummy sites	Percentage of success attributable to dog + handler — Mines	Percentage of success attributable to dog + handler — Dummy sites
Handler and dog blindfolded	360	212	174	56·6	48·7
Handler alone blindfolded	360	233	202	59·8	54·5
Dog alone blindfolded	360	226	168	62·9	50·8

Combined scores for two dogs, both trained by standard military techniques.

prominent sites of local disturbance and dogs trained to detect them in the way described earlier. Once a plateau for detection rate had been reached, the areas of surface disturbance were increased in size from trial to trial by raking the sand until eventually they had coalesced and the whole minefield appeared visually homogeneous. In the absence of any obvious choice point the whole method of assessing performance had to be changed since finds attributable to chance were related to the combined number of points and false points made by the dog rather than to the 50-50 probability of a dog sitting at a given choice point. When due correction had been made for false pointing it turned out that the dogs achieved a success rating for buried A/T mines by use of their own discriminatory abilities of 30-40 per cent and were detecting less than 20 per cent of mines present. Continuous over-training failed to improve this level of performance. Further investigations designed to see whether wooden schu-mines could be better detected than metal A/T mines were not very encouraging; wet schu-mines were found significantly more readily than dry, but no better than A/T mines.

CONCLUSIONS

These studies clearly underline the extreme sensitivity of the dog to olfactory stimuli and show that discriminations based upon the emission of such stimuli can be used to detect the presence of objects buried or otherwise hidden from view. But they also indicate the difficulties in exploiting this sensitivity to practical ends, since stimuli of an intensity which were adequate to engender correct responses at a given choice-point proved inadequate in open-field situations where no such choice point was available. Clearly, and not unexpectedly from the point of view of orientation to olfactory stimuli in the open field, there is a relationship between the strength of stimulus and its capacity to command the attention of the dog, as shown by the relative ease with which the blindfolded animal is able to detect unconsolidated local burial sites. In the absence of such major olfactory cues, however, the levels of the dog's performance shown by these experiments provides little support for some earlier claims that the dog could be of practical value as a detector of concealed or buried objects except in circumstances where the stimulus is large or the zone of detection circumscribed. The experiments tell us little, however, about the behaviour of the dog in relation to olfactory stimuli of a type which might be expected to form part of its natural environment, for the demands made upon it must be regarded as artificial in the extreme. Moreover, the findings cannot be extrapolated into other situations such as tracking, where the olfactory capacities of the dog have, *prima facie*, been successfully exploited and in which the olfactory stimulus might be expected to provide a self-reinforcing continuum rather than an isolated source as the basis for single choice discrimination.

SUMMARY

This communication has attempted to assess the degree of reliability with which dogs can detect the presence of objects hidden from view and to analyse the factors which modify this reliability. Experiments have shown that, in stereotyped situations in which the animal is presented with a clearly defined "choice point", dogs, by use of their sense of smell, are able to detect the presence of concealed objects of widely varying type and size with a high degree of success. Burial of the object under earth provides no complete obstacle to detection provided that, within a narrowly circumscribed area, the dog is required to decide between the presence or absence of a buried article. The olfactory stimulus arising from this object itself, however, has proved insufficient to guide the dog in situations where no such "choice point" is available and the dog's abilities in such circumstances rely on stimuli arising as a result of the act of burial rather than from the buried material.

DISCUSSION

Zotterman: In 1960 Iriuchijima and I recorded from infrared receptors and C fibres in the dog's nose; Iggo and Hensel did the same thing at just about the same time (Hensel, Iggo and Witt, 1960). Just then I received a letter from Alexander von Muralt from Jungfraujoch in which he described experiments on St Bernard dogs, which were finding buried people in the snow thanks to their infrared receptors, and not through their sense of smell, evidently, because they could locate living bodies but not cold corpses.

Ashton: This is fascinating, because it shows once again the falsity of the basic premise upon which our own work was instituted—namely that dogs can *necessarily* locate buried objects. But in our experiments using the stereotyped situation there was overwhelming evidence that the animals were responding primarily to a particulate stimulus.

Zotterman: In your film the dogs did not always sniff at the ground, so they did not seem to be sensing on either infrared radiation or smell; could it be mechanoreceptors in their feet which were the informers, because with recently buried objects the turf may have been much softer? Also, the areas of disturbed earth seemed to be fairly visible.

Lowenstein: The dog that did not apply its nose to the ground looked to me as if it was touching the ground with its paw.

Ashton: On the question of the dog's tactile sense being employed, we looked at this very carefully from high-speed ciné shots and we found that the dog invariably "pointed" (i.e. sat) *before* his paws reached the islet of disturbed ground. Sometimes, in the experiment in the open field, he put his nose right down to the ground and might have had some tactile stimulation through his rhinarium, but, so far as possible, he was trained not to do this.

Concerning the possibility of vision playing a part, we did an extensive experiment in which dogs were worked in these situations sometimes with neither dog nor handler blindfolded, sometimes with the dog blindfolded, sometimes with the handler blindfolded, and sometimes with both. Although the performance was highest when neither was blindfolded it fell only marginally when either or both were blindfolded. But it is biologically interesting that vision was playing a part, even though it was a marginal one.

Wright: Did you try smelling the areas where objects were buried with your own nose? It's a very undignified test but a most illuminating one.

Ashton: Yes, we did actually try this, and obviously human beings can quite readily smell an islet of freshly disturbed earth; on the other hand the people who tried this were quite unable to locate islets that had been made two or three weeks previously, and on which the dogs could still score a measure of success, of the order of 75–80 per cent.

Other of our experiments were relevant to this point. In these we inquired into the dogs' ability to locate mines, parts of whose structure were above ground—the antennae of anti-personnel mines, trip wires and similar devices that human beings often have great difficulty in locating. After specialized training the dogs were effective in that type of situation, and much more so than a man could hope to be.

Zotterman: Have dogs been used for detecting minefields by the Israelis in the desert, for example?

Ashton: I would doubt it, for the reason that when ground disturbance is minimized, as happens very readily in sand, the performance of dogs trained by normal military techniques falls off almost down to chance levels. Only in the one situation of relatively recently laid fields, where there are clearly defined islets of disturbed ground, can the dog trained in this way attain anything like an acceptable level of performance.

Amoore: The dice are really loaded against the dog for detecting a buried object emanating an odour of its own; earth is an incredibly good filter for any odorants added to domestic gas in order to facilitate detection of a leaking main. Usually the trees die long before anyone can smell the leak of gas. In Egypt the police actually employ native *human* trackers who use their sense of smell for tracking thieves across the desert.

Ashton: Tracking is rather different; there is a more or less continuous trail, as distinct from discrete points. We certainly agree with your observation about the capacity of the earth as a filter, but nevertheless it doesn't act as a complete filter, as evidenced by the high levels of success of the animals in the discrimination plots in which there were clearly defined choice points but in which the object was covered by up to 2 or 3 inches of soil.

Zotterman: We recently have had a valuable service by dogs in Sweden where we train them in hunting smugglers of hashish, concealed in motor cars, for example. They are exceedingly successful and it is unquestionably the smell that is the stimulus.

Ashton: The British authorities have, I believe, also been successful in training dogs to locate even quite small quantities of drugs. But the situation is quite different from that confronting a mine-detecting dog.

Duncan: Your interesting experiment was the one where the dog could detect a piece of breeze block inside a breeze-block well. How would you explain that, and were these tests included among those in which activated charcoal removed the effective stimulus?

Ashton: In those early experiments we used fairly large chunks in the tests, comparable in size to anti-tank mines, which in those days were some 9 inches in diameter. The objects were put in by hand, so what the dog was probably picking up was the scent of the handler's hands, as the inside of a well would not be contaminated to nearly such an extent as the object. These were early experiments in which we had not evolved such refinements as cleaning off with ether and handling with tongs, thus narrowing down the type of odour to which the animal was responding. An alternative explanation is that the dogs were detecting a generalized difference between occupied and empty wells.

Duncan: You said that the dogs were *trained* on anti-tank mines, and yet they were giving a positive response to breeze block the first time they were presented with it?

Ashton: Yes, but the anti-tank mines had also been handled.

Hellekant: You used a number of different breeds of dogs; did you find any consistent differences between their abilities?

Ashton: No, there didn't seem to be any clear-cut differences between the breeds but we were never able to accumulate enough information to give a real answer to that point.

REFERENCES

HENSEL, H., IGGO, A., and WITT, I. (1960). *J. Physiol., Lond.*, **153**, 113–126.
IRIUCHIJIMA, J., and ZOTTERMAN, Y. (1960). *Acta physiol. scand.*, **49**, 267–278.

RECENT DEVELOPMENTS IN THE "PENETRATION AND PUNCTURING" THEORY OF ODOUR

J. T. DAVIES

Formerly Beit Memorial Fellow for Medical Research
Now at Department of Chemical Engineering, University of Birmingham

OVER the centuries men have been greatly puzzled by the extraordinary sensitivity of the sensory nerves of living creatures. Indeed, as long as 2000 years ago Lucretius (47 B.C.) tried to unify and explain the concepts of heat, light and odour by an atomic theory. He wrote: "You cannot suppose that atoms of the same shape are entering our nostrils when stinking corpses are roasting as when the stage is freshly sprinkled with saffron of Cilicia and a nearby altar exhales the perfumes of the Orient... You may readily infer that such substances as agreeably titillate the senses are composed of smooth round atoms. Those that seem bitter and harsh are more tightly compacted of hooked particles and accordingly tear their way into our senses and rend our bodies by their inroads." The human sensory system, for example, can respond to quantities of odorous substances (called odorants) as low as 10^{-14} gramme. The latter quantity, although it still contains a very large number of molecules, is far below the level of the best analytical balances. The eye is likewise very sensitive, perceiving as little light as a few quanta (that is, a total energy of about 10^{-18} J or 10^{-11} erg); the ear detects sound waves of amplitudes of the order of atomic dimensions. The astonishing sensitivity of the nose cannot be understood until we know the detailed mechanisms by which the primary events (e.g. the adsorption of the molecules of the odorant) are converted into the electrical impulses which constitute the excitation of the nerve.

For many years it was believed that the chemical senses, smell and taste, were among the more complicated. Certainly their behaviour had long defied numerical analysis, while studies of light and sound have proved more readily amenable to quantitative investigations. A theory of odour should be physiologically reasonable; it should explain the great sensitivity of the human nose (for example, to as little as 10^8 molecules of β-ionone);

and it must account for the tremendous discriminating power of the nose for odorants of slightly different types.

STIMULATION ACCORDING TO THE "PENETRATION AND PUNCTURING" THEORY

During the excitation of the giant axon of the squid, it is known from the studies of Hodgkin and Katz (1949) that the normal excess concentration of potassium ions inside the nerve cell is able to escape through the plasma membrane bounding the nerve cell. Simultaneously the deficiency of sodium ions within the nerve is made up locally by an intake of sodium from the higher concentration in the aqueous phase surrounding the cell. During its resting period the nerve cell, by what is known as "active transport", removes sodium ions from the cell and allows in excess potassium, both these processes occurring against the concentration gradient.

Recent work on olfactory membranes by Takagi, Wyse and Yajima (1966) suggests that here the process of depolarization depends *primarily* on an influx of chloride ions, though potassium ions may also contribute to this process. Studies of a series of anions showed that only F^-, Cl^-, Br^- and HCO_2^- ions can penetrate olfactory membranes: ions such as I^- or ClO_3^-, which are larger (in the unhydrated state), cannot pass across the membrane when it is stimulated by chloroform vapour. In their fully hydrated states the size order of these ions is $HCO_2^- > F^- > ClO_3^- > Cl^- = I^- > Br^-$, so that on the assumption that punctures in the membrane act as sieves, the anions which pass through cannot be fully hydrated. The relative sizes of the naked ions are: $ClO_3^- = I^- > HCO_2^- > Br^- > Cl^- > F^-$, consistent with a hole or sieve hypothesis.

Though it seems perhaps unlikely that such strongly hydrated ions are *completely* desolvated at the receptive membrane before they pass through it, it seems reasonable that the anions should lose *some* of their water of hydration before passing through a hole in the membrane. Indeed, to minimize the energy involved when the ions pass through a predominantly hydrocarbon environment, they may well lose all their *outer* layers of water of hydration, keeping only a strongly held *single layer* of water molecules arranged around each ion. In this way the observed size sequence is explicable, since the naked ion radii are all increased by the same amount, namely the diameter of two water molecules.

The lipid layer in most cell plasma membranes is very thin: even bi-molecular "sandwiches" of the type shown in Fig. 1 would be enough to confer, in normal circumstances, the necessary impermeability to the cell wall and maintain the concentration gradients. Such a layer preserves its continuity by being partly liquid: the hydrocarbon chains of the fatty molecules can move about with a certain degree of freedom, although this will be limited by the presence of rigid molecules such as cholesterol.

Suppose that into this membrane there penetrates a bulky, awkwardly shaped and rather rigid odorant molecule (Fig. 2a). This may either desorb again out of the membrane, or it may diffuse through (Fig. 2b). If the latter occurs, the hole left behind the diffusing odorant molecule may heal only relatively slowly, so that ions can pass through the hole. Once this leakage

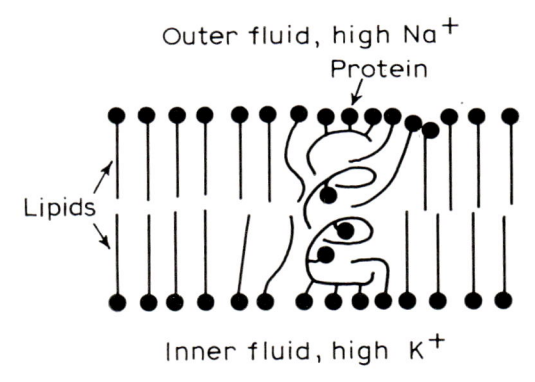

FIG. 1. Mosaic representation of cell membrane. The protein (enzymic) parts of the surface act as sodium pumps, whereas the lipid parts are normally impermeable to ions. Hydrocarbon chains represented by —; polar groups by ●; and the protein backbone by a spiral. The idea of the bimolecular nature of the lipid part of the membrane was due originally to Davson and Danielli (1943) and Danielli (1958).

occurs, it could initiate excitation of the nerve. This puncturing theory (Davies, 1953a,b) is the modern counterpart of the atoms of Lucretius "tearing their way into our senses and rending our bodies by their inroads".

Recent work has shown that the *olfactory nerve impulses* do not begin in the cilia or even at the tips of the olfactory receptors to which they are attached. The neural impulses as such originate where the olfactory receptor narrows to the olfactory fibre. Presumably, just as with the better-understood mechanoreceptors studied by W. R. Loewenstein (1959, 1961), the tips and cilia of the receptor cell become more permeable to ions. This would occur where the odorant molecules penetrate and diffuse: the resulting increased permeability would partially short-circuit the membrane resting potential, setting up a generator current. Such currents do not themselves travel along the nerve fibre, but serve merely to trigger the nerve impulses (which start at the first node of Ranvier in mechanoreceptors, the frequency of these nerve impulses depending on the intensity of the generator current; that is, on the number of channels formed in any given membrane).

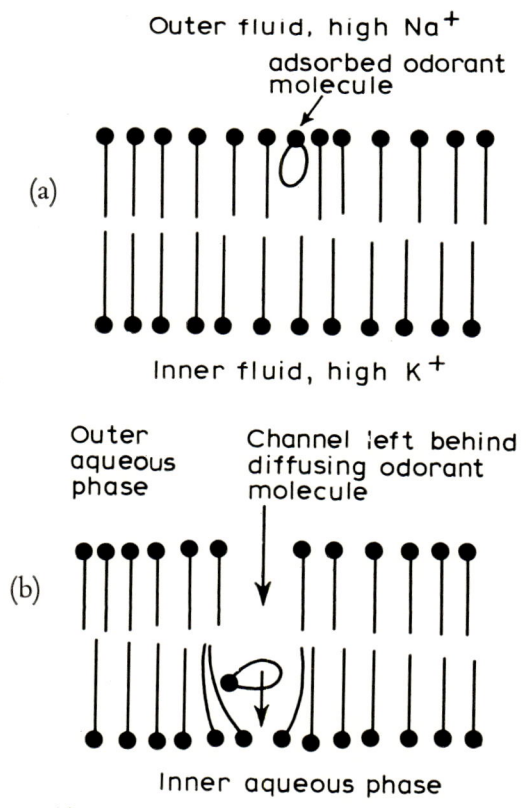

FIG. 2. In (a) an organic odorant molecule is adsorbed, penetrating a lipid region of the cell wall. In (b) the odorant molecule is shown diffusing rapidly through the bimolecular lipid membrane, a sharp hole being left behind it. Through this channel ions may then exchange, initiating the generator current which in turn releases the nervous impulse. (From Davies, 1969.)

If this "puncturing" mechanism of olfactory stimulation is realistic, we should expect a simple relation between the numbers of odorant molecules necessary for threshold stimulation and the number of cells lining the olfactory epithelium in the nose. The latter number is about 2×10^7 in man, a figure in remarkable agreement with the value of 10^8 molecules for threshold stimulation by large molecules such as β-ionone. This suggests that for these, the most effective of odorous substances, perhaps only one molecule of the odorant is sufficient to cause significant ionic leakage through certain very thin membranes.

For less powerful odorants, such as ether or chlorophenol (whose molecules are smaller and less rigid), it appears to be necessary for several odorant molecules to penetrate the cell wall side by side. Such an assumption explains the enormous differences in sensitivity to molecules only slightly different in solubility and chemical properties. For example, if three molecules of odorant have simultaneously to penetrate the same region of cell wall before ions can leak through, the concentration of odorant required must, by probability theory, be greater by several powers of ten than if only one molecule were required to cause leakage. The known million-fold differences in sensitivity of the cell wall to (say) phenol and xylene musk are thus explicable, as is the experimental finding that the olfactory threshold (molecules of odorant per cubic centimetre of air required for the odour to be just detectable) of ethane is about 10^{10} times greater than the threshold for a strong odorant such as β-ionone.

It is not necessary, of course, that *all* the olfactory nerves should be stimulated before an odour is perceived. It may happen (and this seems to be true for moths) that only a few fibres need to be stimulated before the smell is detected (Boeckh, Kaissling and Schneider, 1965), whereas man needs a large fraction of the sensory cells to be affected. Perhaps this apparently greater amplification within the central nervous system of the moth can also explain the high olfactory sensitivity of dogs. The latter, however, have also a relatively great total area of the olfactory epithelium, up to 150 cm^2, compared with only about 5 or 10 cm^2 in man.

On this view that large organic odorant molecules can "puncture" the lipid part of a cell membrane, the mechanism of odour is different from that of taste. Recent work has suggested that adsorption on to specific protein groups in the taste buds is responsible for bitter and sweet tastes, whereas the theory for odour involves a rather non-specific adsorption into the lipid part of the membrane. Perhaps this difference accounts for the known differences between taste and smell: the more lipid-soluble materials have an aroma (or a "flavour", which is lost if we have a heavy cold), while many water-soluble materials of strong taste (mineral acids, salts and sugars, for example) have no appreciable odour. Others, such as ammonia, hydrogen sulphide or sulphur dioxide do have strong odours, presumably originating in the enzymic and proteinaceous parts of the receptor membranes (Fig. 1). Such molecules, particularly sulphides, may well poison the sodium "pumps" of the active transport process, resulting in back-flow of ions (just as there can be a back-flow of water when a centrifugal pump is stopped). One might also remember (in considering the odours of large organic molecules) the Overton–Meyer–Collander theory of penetration rates through the plasma membrane and of narcosis: the effects of the more oil-soluble molecules are related to their oil solubility. Solutions of small polar molecules, such as salts and urea, apparently permeate cell membranes via channels in proteinaceous parts of the surface.

Mathematical model

To place the "puncturing" theory of odour on a quantitative basis, Davies and Taylor (1957) and Davies (1962) assumed that two factors control the extent of dislocation of the nerve membranes. The first is the adsorption coefficient (or adsorption energy) of the odorant molecules in passing from the air to the lipid–water interface constituted by the lipid wall

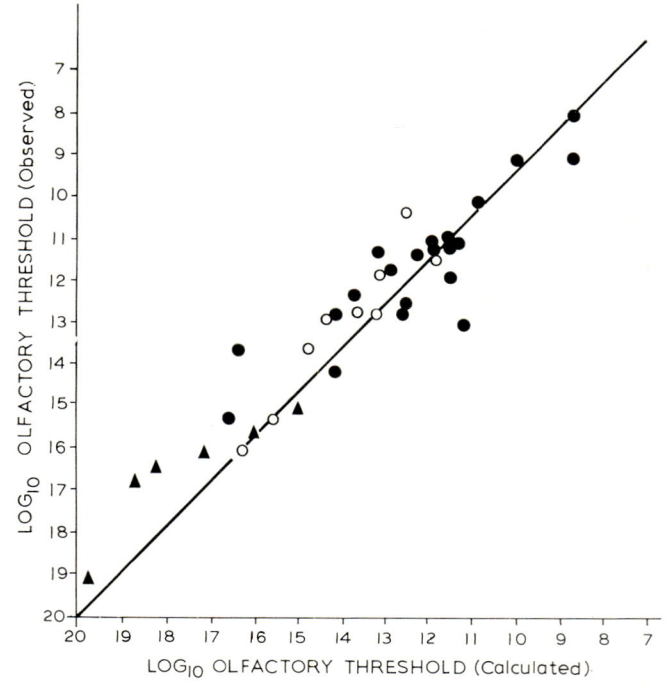

FIG. 3. Measured olfactory thresholds for humans are compared with those calculated from the "penetration and puncturing" theory. Open circles refer to normal alcohols, triangles to *n*-hydrocarbons, and closed circles to other organic compounds of various types. Units are molecules per cm^3 of air. (From Davies and Taylor, 1959.)

of the hairs (which form the sensitive tips of the olfactory nerves) and the thin layer of mucus in which they are bathed. This adsorption energy may be calculated from the free energies of the odorant molecules in first passing from air to water, and then being absorbed from water at the lipid–water interface. The second factor is the effectiveness of each adsorbed molecule in "puncturing" the olfactory nerve membrane at some point. Some large molecules may be effective singly, but with others a number p, which may be up to 5 or 10, must be adsorbed close together on the membrane. From these considerations we (Davies and Taylor, 1957) derived a mathematical equation which correlates olfactory thresholds in terms of the molecular sizes and adsorption energies of the odorant molecules. The appropriate adsorption energies can be approximated from laboratory experiments on model systems, so that olfactory thresholds can be predicted and compared with experimental results (Fig. 3).

This theory makes possible an understanding of the enormous range of olfactory thresholds, with a factor of about 10^{10} between a strong odorant such as β-ionone and a very weak odorant such as ethane. Of this factor, about 10^5 can be accounted for by the difference in adsorption coefficients (i.e. free energies) of the molecules, while the other 10^5 arises from the probability that for the smaller molecules a number p must be adsorbed simultaneously in the same region of the membrane. The range of thresholds is so great because *both* terms vary in the same way: large molecules are both more strongly adsorbed and *also* are each more efficient at puncturing the membrane, so that only a few (p being 1 perhaps) need to be adsorbed (and to diffuse) simultaneously in one region (about 64 Å^2) to cause ionic leakage.

ODOUR TYPE ACCORDING TO THE "PENETRATION AND PUNCTURING" THEORY

The "penetration and puncturing" theory postulates that diffusing molecules of odorant, after being adsorbed, diffuse in the lipid membrane leaving behind them holes through which ions may possibly leak, as in Fig. 2b. But whether such channels in the membrane *will* persist long enough behind the diffusing odorant molecules for this ionic exchange to occur must depend on the ratio of the diffusion rate of the odorant molecule through the membrane to the rate of healing of the lipid cell membranes in different parts of the olfactory epithelium.

In practice, not all types of odorant molecule will be equally effective in leaving a sharply defined hole in a given membrane, and so the quality of the odour will depend on the rate of diffusion of the odorant through the cell membrane relative to the subsequent rate of healing of the appropriate membrane (Davies, 1965). That is, the intensity of the odour must depend on a term such as

$$\left\{1 - \frac{t_{diff}}{t_h}\right\}$$

where t_{diff} is the time for diffusion of the odorant molecule through a certain

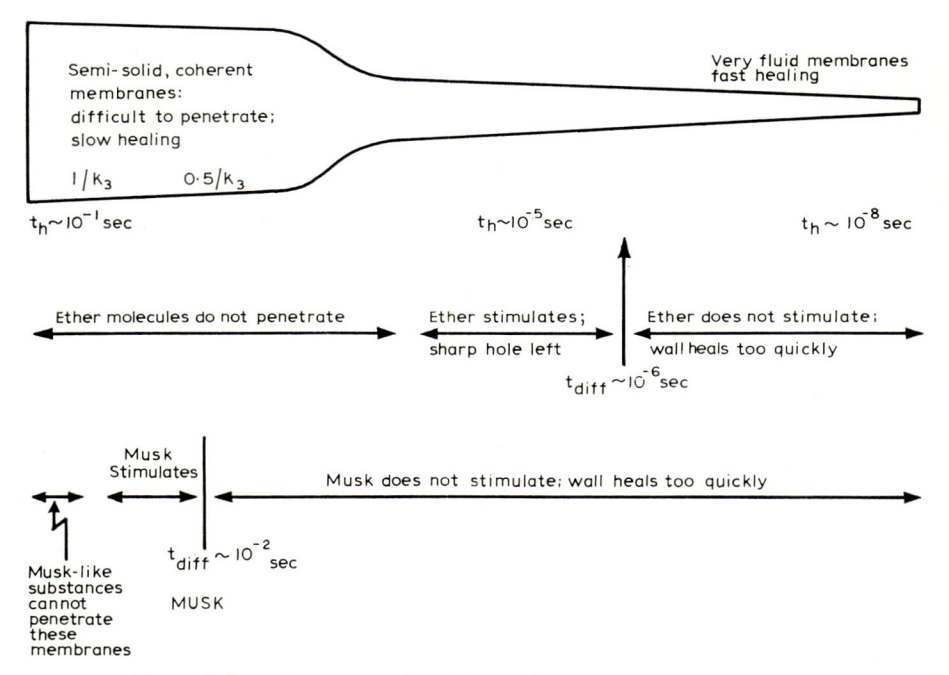

FIG. 4. Schematic representation of range of properties of sensory endings of different olfactory nerve cells. The cell distribution may be such that musk odours (for example) form a fairly well-defined group, with but little overlap with other groups. (From Davies, 1969.)

membrane and t_h is the time of healing of that membrane. Typical values for t_{diff} might be 10^{-6} sec for ether and 10^{-2} to 10^{-1} sec for a macrocyclic musk, but exact values are not required at this point. If for a given odorant and membrane the time for diffusion is small compared with the time for healing of the membrane, a sharp hole will be left after the odorant molecule has diffused through the membrane. Through this hole ions may then leak, so initiating the nervous impulse. In these circumstances the bracketed term above (called the "hole sharpness factor") tends to unity.

But if the odorant molecule is larger, or more strongly adsorbed, then its time for diffusion through the cell membrane is relatively great, and the "hole sharpness factor" is reduced. This factor will reach a limit of zero when the time for diffusion through the membrane is equal to the time for healing of the lipid membrane. The membrane would then heal completely behind the moving odorant molecule; that is, the latter would *not* leave a sharp hole through which ionic exchange might then occur.

On the other hand t_h, the time of healing of the membrane, may have different values for different olfactory cells. Thus, the membranes of certain nerve endings may be relatively coherent and slowly healing, and also very resistant to penetration by the odorant molecules (this is shown schematically to the left of Fig. 4). Other nerve endings may have membranes which are very fluid and easily penetrated, so that any "puncture" heals very rapidly. Such an assumption appears not unreasonable in the light of the known tissue specificity of lipids in mammals: the molecules of the membrane phosphatide have different distributions of their fatty acids (palmitic and oleic) in different tissues of a given animal (Veerkamp, Mulder and van Deenen, 1962). Indeed, it has been suggested in a different context (De Gier and van Deenen, 1961) that there may be a relation between the lipid composition and the permeability properties of red cells, depending on the ratios of sphingomyelin, lecithin and kephalin.

Even single olfactory nerves of the opossum and frog are known (Beidler and Tucker, 1955; Gesteland *et al.*, 1963; Gesteland, Lettvin and Pitts, 1965) to exhibit *some* specificity: a given fibre will not respond so strongly to some particular odour as will another fibre, though the relative responses of the two fibres can be reversed if a different odour is chosen. This is therefore some direct physiological support for a more or less continuous distribution of properties between different olfactory nerves: it appears that no two cells have identical responses, but that every cell responds in one way or another to a number of different odours. There is no *sharp* specificity such as would arise from a chemical or enzymic process, but only the rather blurred specificity and overlap which are characteristic of physical adsorption.

Fig. 4 shows in detail why, on this basis of a distribution of physical properties, an odorant such as a musk will stimulate only the cells with slow-healing membranes. Smaller molecules (such as ether) are too weakly adsorbed (on this theory) to penetrate these musk-sensitive, strongly coherent membranes, just as laboratory experiments (Dean and Fa-Si Li, 1950) have shown that benzene vapour cannot penetrate a monolayer of stearic acid spread on a water surface at a surface pressure of $17 \times 10^{-3} \mathrm{N\,m^{-1}}$, although the more strongly adsorbed molecules of *n*-hexane can penetrate into this stearic acid monolayer. On the other hand musk, for example, would not stimulate appreciably the cells particularly sensitive to small molecules such as ether because the very fluid membranes of such cells will heal much too rapidly behind the relatively slow-moving musk molecules: no sharp hole persists in such membranes behind the diffusing odorant molecule.

The quantitative treatment which follows was developed originally (Davies, 1969) to explain musky odours, but the treatment can be generalized. Let N be the number of stimulated sensitive regions (or non-specific "sites") on a nerve ending, so that on each of these N regions the requisite number p of adsorbed molecules to cause ionic leakage are concentrated. Probability theory then shows that for some constant value of N the mean adsorption x of odorant on to the membrane will also be constant if p is constant. For the various musks, for example, p is about 2, and hence for these it will be true that for a given N, the adsorption x (e.g. in molecules per cm^2 of surface) will also be constant.

We now wish to interpret the adsorption on the membrane in terms of concentrations of odorant in the air, since these are easily studied experimentally. From adsorption theory (Davies and Rideal, 1963; Davies, 1969) for adsorption into a layer of lipids one can write:

$$x = cdK/(1+kc) \qquad (1)$$

where c is the concentration of odorant in the air, K is the adsorption coefficient for the odorant molecule between air and the lipid–water interface, and d is the thickness of that part of the membrane penetrated by the odorant molecules. The constant k depends on molecular cross-sectional area.

We have now related N, through x, to the concentration c of odorant in the air. We next relate N to the response R of the olfactory cells, by assuming that above threshold the response R is proportional to N, so that when we achieve some given (standard) physiological response R, N is constant, and hence x is constant, and c is given by equation (1).

Rather than use K in this equation, it is often convenient to use the related property of the desorption rate z, since this can be very readily measured on model systems in the laboratory (Theimer and Davies, 1967), such as the desorption of odorant molecules from a monolayer-covered water surface into air. The desorption rate z is known from physical chemistry to vary as $1/K$.

Then equation (1) can be rearranged with constant x (i.e. constant N and constant response R) as

$$1/c = k_1/z - k_2 \qquad (2)$$

where k_1 and k_2 are constants relating respectively to the sensitivity of the cells and the cross-sectional areas of the odorant molecules. Under these conditions of constant response, c is the concentration of odorant in the air to evoke some standard psychological response R from a panel of trained judges. An experimental study was undertaken (Theimer and Davies, 1967) with a range of substances having relatively pure musk odours, and the corresponding reciprocal concentrations were conveniently designated "Intensities of Muskiness", abbreviated to I.M. Hence,

$$\text{I.M.} = k_1/z - k_2 \qquad (3)$$

So far this simple treatment has not allowed for the "hole sharpness factor", given above as $(1 - t_{diff}/t_h)$. This factor may also be expressed in terms of the measured *in vitro* desorption rate z if we allow that the diffusion and desorption processes are basically similar, involving similar activation energies: the "hole sharpness factor" then becomes $(1 - k_3/z)$ where k_3 includes the term t_h for the healing of the appropriate sensory membranes. It has been explained above that if this factor is zero, no sharp passage is left behind the diffusing odorant molecule. Accordingly, equation (3) should be corrected by multiplying by this factor:

$$\text{I.M.} = (1 - k_3/z)(k_1/z - k_2) \qquad (4)$$

or

$$\text{I.M.} = (k_1 + k_2 k_3)/z - k_1 k_3/z^2 - k_2 \qquad (5)$$

Comparison with the intensity of muskiness data of Theimer and Davies (1967) shows (Fig. 5) that equation (5) fits the form of their results with the empirical values of 8·1 for $(k_1 + k_2 k_3)$, 3·26 for $k_1 k_3$ and 3·0 for k_2. Hence $k_1 = 6·6$ and $k_3 = 0·49$, these relating to those cells particularly sensitive to musks. We should expect the k_3 value in particular to be characteristic of such cells, and the musk odour will be found only when the experimental desorption rates z are appreciably greater than k_3, of the order $1·5 k_3$ to $2k_3$. The I.M. will be zero, according to equation (4), when $z = k_3$, the corresponding *times* of diffusion being proportional to $1/k_3$ (see Fig. 4). The I.M. will also be zero when $z = k_1/k_2$, these limits of I.M. corresponding numerically to $z = 0·5$ and $z = 2·2$ for musks.

The "strongest" musks (those with the maximum intensity of muskiness) should all have desorption rates of $2k_1 k_3/(k_1 + k_2 k_3)$ if equation (5) is generally valid. Numerically, this ratio is about 0·8, so that the strongest musks should *all* have desorption rates of about this figure.

It is important to note that strongly adsorbed musky molecules (z low or K high) will have lower olfactory *thresholds*, but under perfumery conditions (far above the thresholds), the *strongest* musks will be those with z around 0·8.

The data in Fig. 5 are from known musks of many different chemical types, including macrocylic musks, indane musks, tetralin musks, aromatic carbonyl musks and isochroman musks. For nitromusks, the rather ready solubility in water precludes direct measurements of desorption rates, but the energies of desorption, as found indirectly (Davies and Taylor, 1957), are indeed very close to those for macrocyclic musks. Fig. 5, like the above equations, represents a necessary but not a sufficient condition for muskiness. A known musk will give data round the correlation, but some substances

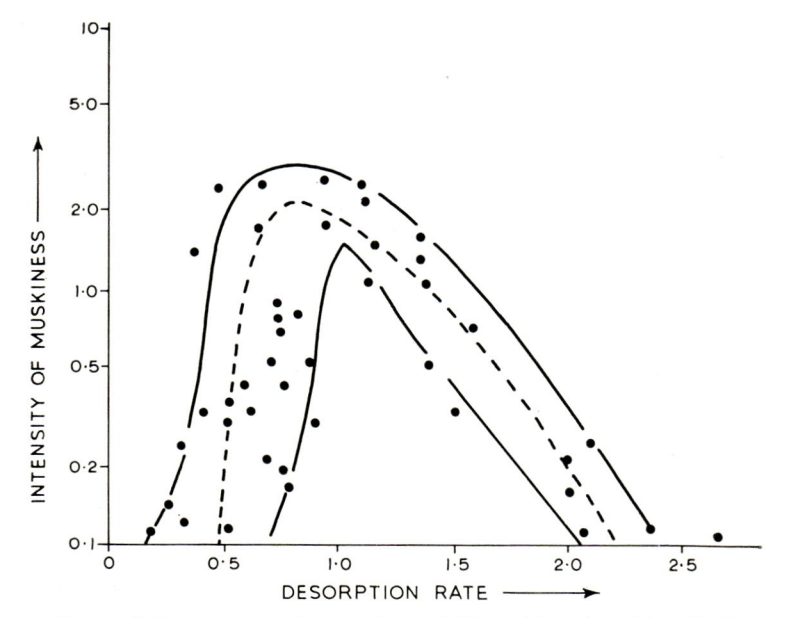

FIG. 5. Points represent the experimental "intensities of muskiness" of known musk compounds (of various chemical types) plotted against the measured desorption rates from the film–water interface into air (Theimer and Davies, 1967). The broken line represents equation (5), the constants being chosen to give the best fit to the experimental points.

may give points on the correlation yet be odourless or have a non-musk odour (e.g. a "burnt" odour).

For the existing distribution of olfactory cells, the limits to any given odour quality will be imposed firstly by the ability of the odorant molecules to penetrate certain nerve cell walls, and secondly by the ratio of the rate of healing of the membrane to that of diffusion of the odorant molecules. Desorption rates, giving a measure of each of these factors, would therefore be expected to be a significant *first factor* in correlating odour quality. But it is necessary to take another factor into consideration: this *second factor* is the geometry of the odorant molecule—its size and the balance between its non-polar and polar portions. The size can best be expressed as A, the cross-sectional area of the oriented molecule (Davies, 1965); while the ratio of hydrocarbon to polar material in the molecule is given (Theimer and Davies, 1967) by the ratio of the length (L) of the oriented molecule to its breadth (B) measured at the polar end. For example, for a molecule to be a musk, besides the desorption requirement, it is necessary that A lies in the range 40–57 $Å^2$ and that L/B lies in the range 2·8–3·3 inclusive. For the moth sex attractants, whose specificity is very high (Boeckh, Kaissling and Schneider, 1965), clearly the geometry factor is again most important. But for the smaller molecules, particularly those which form less well-defined classes than musks, one would expect L/B to be much less significant.

The geometrical factor for large odorant molecules can perhaps be interpreted in physical terms in the following way. The odorant molecule (say a musk) penetrates into the lipid layer of the cell wall. This is normally liquid, impermeable to ions, and self-healing against disturbances because of the rapid free movements of the hydrocarbon chains of the molecules of the phospholipids, acids and esters in it. However, when a rigid, bulky odorant molecule has penetrated into the film, several neighbouring fatty chains (often unsaturated in olfactory tissue) are constrained to stand up straight, forming a local micro-crystalline region. Thus, by its size and its inertia, and partly by its shape, a large odorant molecule reduces the fluidity of the surrounding membrane into which it has penetrated. This explains how the relatively long times of membrane healing (approx. 10^{-2} to 10^{-1} seconds) can be achieved: the adsorbed musk molecules reduce the fluidity of the surrounding membrane before diffusing through, thus enabling a sharp hole to be left.

Some support for this interpretation comes from the *in vitro* experiments of Adam and Jessop (1928). They showed that large, rigid molecules such as cholesterol can, by their size and inertia, obstruct the movements of several neighbouring "liquid" hydrocarbon chains in monomolecular layers of certain fatty acids, nonylphenol or monomyristin, spread on the surface of water. For example, as many as 16 molecules in a spread film of nonylphenol are considerably immobilized by a single molecule of cholesterol, and with a ratio of one molecule of cholesterol to four or fewer molecules of nonylphenol, the mixed film shows marked elastic afterworking; that is, the re-establishment of equilibrium after a sudden compression or expansion takes up to several minutes. Further studies showed that specific complexes are not responsible: it is the size, rigidity and inertia of the cholesterol molecules which cause the effect; molecules with a single straight chain, even if this is as long as 22 carbon atoms, are ineffective (Adam, 1944). In a similar way, large rigid odorant molecules of the appropriate shape might reduce the fluidity of the cell membrane before diffusing through, while other odorant molecules, if they are flexible, may even act as plasticizers and increase the fluidity of the membranes.

If, however, as a first approximation, we neglect the shape factor L/B,

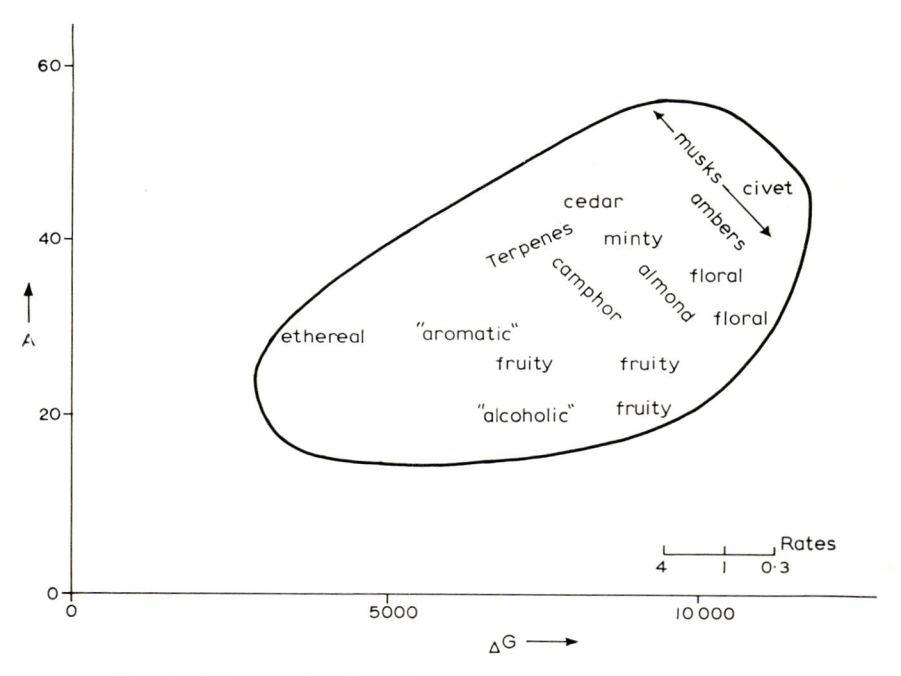

FIG. 6. Odour quality "map"—a correlation of odour quality with measured cross-sectional areas A (in Å^2) of odorant molecules and with free energies (ΔG, expressed in calories per mole) of desorption of the odorant from the lipid–water interface into air. These energies are related to K by the relation $\Delta G = RT \ln K$; and to desorption rates (z) by
$$\ln z = \text{constant} - \Delta G/RT.$$
The various areas indicated should denote uniquely the quality of the odour in physical terms. (From Davies, 1965, 1969.)

the quality of any odorant should depend on z and on the molecular cross-sectional area A; in equation (5) k_2 is a function of the cross-sectional area of the oriented odorant molecules, as is also k_3. Over the proposed range of olfactory cells, one might therefore expect a correlation such as that of Fig. 6, the various areas of which should denote uniquely the quality of the odour in physicochemical terms. The desorption rate z limits the regions in general as designated by equation (5). This suggests (because of the proposed distribution of olfactory cells) that the different odour qualities shade into one another, rather than there being any sharply demarked "primaries". But, as shown in Fig. 4, the cell distribution in the olfactory epithelium may be such that some odour qualities, such as musky, constitute relatively well-defined groups compared with (say) floral. Beets (1966, personal communication) has pointed out how the odours of most macrocyclic molecules shade gradually into one another, from camphoraceous (C_8, C_9, C_{10}) through terpenic, cedary, musky (C_{15}, C_{16}, C_{17}) to civetty (C_{18}, C_{19}).

Another piece of evidence in support of the general merging of odours into one another, as shown in Fig. 6, comes from the similarity studies of Amoore and Venstrom (1966). They reported (from panel tests) appreciable psychological similarities (~2 units) between minty and camphoraceous odorants; between floral and musky; and also between floral and minty. On their scale, completely similar odours score about 6·5 units of similarity.

Yet further evidence comes from electrophysiological studies of similarity (Døving, 1966): there is a significant rank correlation coefficient of 0·684 between his results (Døving, 1966, private communication) and those of a plot (Davies, 1965) which is an earlier version of Fig. 6.

However, among the various musks, Døving (1966) found a relatively close similarity in electrophysiological response: materials studied included an indane musk, several oxa-octahydroanthracene musks, a tetralin musk and a nitro-acetophenone musk. The latter showed rather less similarity to a macrocyclic lactone musk, however, and androstan-3-ol was less similar than most others within the group. But compared with other types of odour, Døving found a "rather strong relationship between musky odours": they constituted a "rather homologous group with significantly high chi-square values between all pairs". Gesteland, Lettvin and Pitts (1965) likewise report that in the epithelium of the frog those receptors particularly sensitive to musk do not respond much to other odorants. On the other hand the other receptors show extensive overlap in their response to different odorants, rather like poorly constructed optical filters, again supporting a physical process of adsorption, rather than a specific or enzymic process, for olfaction.

SUMMARY

The "penetration and puncturing" theory of olfaction not only explains the enormous range of known olfactory thresholds but can (with certain assumptions) explain how ionic leakage leads to generator currents which initiate the impulses in the olfactory nerves.

The primary process envisaged is one of adsorption and diffusion of the odorant molecules in lipid membranes: if a sharp hole persists behind the diffusing odorant molecule, ions will leak across the membrane.

Given a range of physical properties in the lipid membranes of different receptor cells, the tremendous discriminatory power of the human nose can be explained in physicochemical terms.

This paper summarizes an attempt to achieve this, in terms of the interaction of odorant molecules of different sizes and shapes with the postulated lipid membranes. The quality of the odour as perceived depends on at least three odorant parameters—desorption rate, molecular size and molecular geometry.

REFERENCES

ADAM, N. K. (1944). *The Physics and Chemistry of Surfaces*, pp. 70–71. London: Oxford University Press.
ADAM, N. K., and JESSOP, G. (1928). *Proc. R. Soc. A*, **120**, 473.
AMOORE, J. E., and VENSTROM, D. (1966). *J. Fd Sci.*, **31**, 118.
BEIDLER, L. M., and TUCKER, D. (1955). *Science*, **122**, 76.
BOECKH, J., KAISSLING, K. E., and SCHNEIDER, D. (1965). *Cold Spring Harb. Symp. quant. Biol.*, **30**, 263.
DANIELLI, J. F. (1958). In *Surface Chemistry and Cell Membranes*, pp. 243, 254, 258, ed. Danielli, J. F., Pankhurst, K. G. A., and Riddiford, A. C. London and New York: Pergamon Press.
DAVIES, J. T. (1953a). *Inds Parfum. Cosmét.*, **8**, 74.
DAVIES, J. T. (1953b). *Int. Perfumer*, **3**, 17.
DAVIES, J. T. (1962). *Symp. Soc. exp. Biol.*, **16**, 170.
DAVIES, J. T. (1965). *J. theor. Biol.*, **8**, 1.
DAVIES, J. T. (1969). *J. Colloid Interface Sci.*, **29**, 296.
DAVIES, J. T., and RIDEAL, SIR ERIC (ed.) (1963). *Interfacial Phenomena*, pp. 295–298, 303–306. New York: Academic Press.
DAVIES, J. T., and TAYLOR, F. H. (1957). *Proc. II Int. Congr. Surf. Activ.*, vol. 4, pp. 329–340. London: Butterworths.
DAVIES, J. T., and TAYLOR, F. H. (1959). *Biol. Bull. mar. biol. Lab., Woods Hole*, **117**, 222.
DAVSON, H., and DANIELLI, J. F. (1943). *The Permeability of Natural Membranes*. London: Cambridge University Press.
DEAN, R. B., and FA-SI LI (1950). *J. Am. chem. Soc.*, **72**, 3979.
DE GIER, J., and DEENEN, L. L. M. VAN (1961). *Biochim. biophys. Acta*, **49**, 286.
DØVING, K. B. (1966). *Acta physiol. scand.*, **68**, 404.
GESTELAND, R. C., LETTVIN, J. Y., and PITTS, W. H. (1965). *J. Physiol., Lond.*, **181**, 525.
GESTELAND, R. C., LETTVIN, J. Y., PITTS, W. H., and ROJAS, A. (1963). In *Olfaction and Taste* (Proceedings of the First International Symposium, Stockholm, 1962), pp. 19–34, ed. Zotterman, Y. Oxford: Pergamon Press.
HODGKIN, A. L., and KATZ, B. (1949). *J. Physiol., Lond.*, **108**, 37.
LOEWENSTEIN, W. R. (1959). *Ann. N.Y. Acad. Sci.*, **81**, 367.
LOEWENSTEIN, W. R. (1961). *Ann. N.Y. Acad. Sci.*, **94**, 510.

Lucretius, T. C. (47 b.c.). *On the Nature of the Universe*. English translation by Latham, R. (1951). London: Penguin Books.
Takagi, S. F., Wyse, G. A., and Yajima, T. (1966). *J. gen. Physiol.*, **50**, 473.
Theimer, E. T., and Davies, J. T. (1967). *J. Agric. Fd Chem.*, **15**, 6–14.
Veerkamp, J. H., Mulder, I., and Deenen, L. L. M. van (1962). *Biochim. biophys. Acta*, **57**, 299.

DISCUSSION

Wright: Professor Davies has put his finger on the key problem, which is the mechanism by which the permeability of the cell wall is modified by the stimulus molecule. I find it difficult to accept the idea of the molecule penetrating into the cell and presumably being metabolized away, because on that basis one stimulus molecule could stimulate one receptor cell once and that would be all. It is very, very difficult to accept the quite amazing sensitivity of the sense of smell if one molecule can only stimulate one receptor cell once.

I suggest that it may not be necessary for the stimulus molecule to penetrate into the membrane in order to open a hole. All it needs to do is to spread the molecules out enough to allow sodium ions to penetrate (and it is significant that in nerve excitation the sodium and potassium ions are involved and other ionic species which are known to be present do not leak through). The opening of the hole by the right amount is a critical event, therefore, and we have constructed a mechanical model of how this could be done. The cell membrane is known to be the site of a double layer of molecules, protein or lipid, oriented at right angles to the surface. To simulate this we floated a number of small bar magnets on the surface of some water; under mutual repulsion they distribute themselves uniformly. A molecule with a dipole moment is represented by another bar magnet. When it is some distance from the surface it has no effect on the distribution of the dipoles in the surface, but when it is brought near the surface, but not into actual contact with it, the interaction between the dipole moment of the stimulus molecule and the dipole layer in the surface opens a substantial hole which then heals as the magnet is moved away. This mechanism has some advantages in that the molecule is not used up in the process of opening the hole.

Davies: A small molecule like alcohol obviously can penetrate through membranes and when it gets right inside the cell it may complex with proteins and other substances and will not come out again immediately.

With large odorant molecules one will, on the penetration theory, eventually get fatigue (as one does to musk), and then it will take a few minutes of breathing clean air to get rid of the large odorant molecules which are still adsorbed into the membrane.

The idea that depolarization depends mainly on the passage of sodium and potassium ions has been disputed by Takagi, Wyse and Yajima (1966) who have studied the depolarization of olfactory membranes (see p. 266 of my paper). I think their finding has to be looked at by other investigators, to see what happens in general when the olfactory membrane is stimulated.

Wright: We need to know more about the membrane itself. That is the obvious requirement at this stage.

Døving: I have a concrete proposal concerning the lipid composition of the membrane of the olfactory epithelium. A student at our institute, Hanna Mustaparta, recently worked on the distribution of different types of receptors in the olfactory epithelium of the frog, and found that there is an uneven distribution of receptor types. Taking advantage of the fact that you can blot the epithelium by filter paper and tear the olfactory cilia away, one could make a lipid analysis of different parts of the epithelium and see if the composition of lipids or other chemical compounds varies.

Davies: That would be most interesting to do.

Mac Leod: When I was beginning the work on the glomerulus (Leveteau and Mac Leod, 1969) referred to earlier by Dr Døving, I wanted to find very different odours to test. Professor Davies published his diagram at that moment (similar to Fig. 6 of his present paper, p. 278). I was not convinced by his hypothesis, but it offered the possibility of choosing molecules which would be very different from each other in their physicochemical properties. The computation of paired correlations between 12 chemicals as given by the response profiles of 130 glomeruli in the rabbit shows a high degree of correlation between some pairs of chemicals whose representative points are not very close together in your diagram. Neither are these chemicals in the same qualitative class. Conversely, other pairs, which according to your diagram seemed strongly correlated, were completely independent for the glomeruli. Of course this is only one factor which cannot rule out your idea, but it is against it.

I should like eventually to do some experiments *in vitro* which would prove that adsorption of molecules by a lipid layer would change its ionic permeability. Would it be possible to devise *in vitro* studies to show that?

Davies: I am sure it would be, using lipid bilayer membranes formed between two aqueous phases. One would use a Teflon pot with a little hole in it. Dr F. H. Taylor and I (Davies and Taylor, 1954) carried out some experiments with odorants added to erythrocytes in solution.

We did find a weakening of the membrane as reflected in an acceleration of haemolysis by saponin and by other substances. The lower the olfactory threshold (i.e. the stronger the odorant), the more the acceleration of haemolysis, and so presumably the more the weakening of the membrane. Of course the erythrocyte has a much tougher membrane than has the olfactory receptor cell.

Dr Mac Leod apparently finds a strong correlation between the odours of amyl acetate and decanol, for example. Does this agree with psychological correlations?

Mac Leod: There are no similarities to the observer. Olfactory similarities revealed by glomerular analysis appear to be relevant to similar physicochemical properties rather than to psychological proximities. The trouble is that these physicochemical properties are still unknown.

Davies: My chart (Fig. 6, p. 278) agrees reasonably well with the psychological measurements: this ties up with some results of Amoore and Venstrom (1966), as I have mentioned in my paper. Further, Døving (private communication, 1966) found a significant rank correlation coefficient of 0·684 between his electrophysiological similarity results and my physicochemical similarity chart.

Døving: If we accept that the data on the 12 odours which Dr Mac Leod worked on have a good solution in four dimensions, and you have, let us say, 25 possible physicochemical parameters, all dealing with these 12 chemicals, we would have to combine these 25 parameters four by four and that makes $\binom{25}{4}$ or 12 650 combinations.

Davies: It will certainly be a very long time before we can predict that some given new molecular structure—something nobody has ever synthesized before—will have an odour exactly of a certain type and intensity. We can at present only hope to find the three or four most important physical variables. This I have tried to do in my paper.

Adey: If I may return to the problem of membrane structure, I find myself in some respects in a triangular situation between Dr Davies and Dr Wright. Just as they indicate that any model of the membrane should be both necessary and adequate, it is worthwhile looking at some of the evidence about the sort of membranes we are really dealing with in the central nervous system and how inadequate it is to think of them as simply lipid double layers. But let us recall first that the receptor sensitivity in neural structures is extraordinarily high and that for example the retina responds to a single photon, so that we should not expect the incidence of light energy on the retina in an amount adequate for perception to be associated with poking large holes in the membrane. One quite basic question that we must consider is whether or not it is necessary for any penetration to occur through the membrane.

Here the immunologists may have a lesson for us. It now seems clear that the plasma cell, for example, makes antibody in response to an antigen that is fixed on to the membrane but never actually penetrates it. There is transcription from this molecule bound to the outside of the membrane which is interpreted by ribosomes that are on the inside, and the antibodies are prepared; they pass back across the membrane, and the ribosome never "sees" the antigen to which it responds (see Nossal, Williams and Austin, 1967). This is dramatic enough to make us wonder about the need in neural structures for a direct penetration.

When it comes to the membrane impedance to ions, which are much smaller than the large antigenic molecules, some of W. R. Loewenstein's work on cultured cells is revealing (Kanno and Loewenstein, 1966). He showed that the presence or absence of calcium in the medium can critically determine the size of the molecule that can pass through the cell membrane, and if excess calcium is added to cultured cells that are isolated from one another they will bind together and simultaneously the membrane impedance increases. On the other hand if one adds chelating agents like EDTA or EGTA, membrane impedance drops and at the same time very large molecules, as high as 300 000 in molecular weight, can pass across the membrane. So that it is not necessary to say that the size of the molecule alone is the determinant of whether or not it can cross the membrane. We are dealing with a structure that is plastic and responsive in a very curious and complex way to large molecules.

On the question of the membrane structure itself, so far in this symposium we have apparently chosen to ignore much work on central nervous tissues that suggests a quite different type of organization for neural membranes from a simple lipid bilayer. I refer to the work of Schmitt and Davison (1965) and others who suggest that there are as many as five layers: (1) An outer "fuzz" of mucopolysaccharide and mucoprotein. This outer layer can be very much influenced by the presence of calcium; it consists of highly hydrated networks whose volume can be changed up to 100 000 times by minor changes in calcium concentration. Within that is (2) a protein layer, which is closely interfacing with (3) and (4), a lipid bilayer. Inside that is (5) more protein. It seems reasonable to suggest that a succession of changes occur through these layers and that the so-called "fuzz" on the surface is not in adventitious relationship to the cell and has much to do with the excitatory process at its most fundamental point, because a protein conformational change is probably the first step in the excitatory process, rather than something that affects the lipids and requires a major energy exchange. It is conceivable that protein conformational changes can occur in response to very low levels of energy shift.

Evidence about the vital role of protein has come from work by Huneeus-Cox (1964) and Huneeus-Cox and Smith (1965) in Chile on the squid giant axon. They prepared antibodies to axon material which will selectively block either the action potential or the membrane potential. The addition of sulphur-containing amino acids like methionine enhances their effects, suggesting that the excitatory process depends upon a protein change. Search for the so-called electrogenic proteins has not yet been successful, however.

When one comes to the question of the way in which ions move across the membrane, according to Eigen's current work potassium and sodium ions are not moving simply as "little hydrated balls of fluff through holes that may be sharp or rough". They are carried by peptide carriers that grip them much as in a clam shell, and their fit to the "clam shell" appears to be extremely specific (Eigen and de Maeyer, 1969). So I would be dubious about any postulate of the easy flow of ions in a relatively non-specific way through a hole that had sharp or jagged edges. The question of the presence of surface substances on central neurons has had an interesting history. As recently as 7 or 8 years ago it was postulated that there was no extracellular space in brain tissue. Then there was a very large extracellular space, about 20 to 25 per cent of the total volume; then it shrank to 10 or 15 per cent. Then it was suggested to have some organization, and now by a series of elegant studies by electron microscopy by Pease (1966), then by Rambourg and Leblond (1967) and by Bondareff (1967), there is shown to be a well organized mucopolysaccharide–mucoprotein material which fills a great deal of the extracellular space and appears to have specialized appearances around the somata of nerve cells. The electron microscopy of the olfactory surface and mucus layer may equally reveal interesting structure, and we should be more concerned about what occurs during the binding of the odorant molecule, and how it is that even at a distance of 20 or 30 μm from a dendrite it may still be able to exercise an excitatory function.

Martin: Perhaps because I have done no work in this subject and have a theory of my own (see p. 306), I would like to make an attack on the puncture theory! It seems to me that almost all of its basic assumptions are unreasonable. My picture of a cell membrane is essentially that it is liquid and is capable of healing very rapidly. The idea of a hole being made through this, when surface tension would close the hole up again rapidly, seems to me to be totally unreasonable.

What one expects of a liquid film when something enters it is that the film will close up behind the molecule as it goes through. If by chance a hole is formed, provided it is below a certain size surface tension will immediately close it up again. Again a liquid surface shows practically no specificity with regard to different isomers of roughly similar structure, but optical isomers smell entirely different in many cases. There are cases—the pheromones of insects—where the receptor is responsive to practically one single molecule. And there is a vast range of molecules of all different chemical types which one can smell. None of these things seem to be explainable in terms of the conventional picture of the cell membrane. Further, if you are going to involve ions as the stimulating factor, you have an ion pump which is continuously working and is capable of rapidly changing the concentration, and you would therefore have to inject a great deal of any ion in order to make any difference to it. The puncture theory does not explain sensitivity of the order of one or two molecules per cell, which seems to be the type of thing which happens.

Davies: Obviously the membrane does heal, because in the ordinary passage of an impulse down a nerve there is local ionic leakage and fairly free ion movement during local depolarization, as Hodgkin and Katz (1949) showed. Then where the next region becomes depolarized the lipid membrane breaks down and the impulse in this way passes along the nerve. I am only really adapting that idea to the sensory nerve tips, which I assume are more sensitive and fragile to stimulation by the odorant molecules.

Martin: But with a pump which is capable of dealing with enormous fluxes of these ions, a small puncture is not going to make any difference. Thousands of molecules would be needed to excite the cell. An amplifying mechanism is needed, the possible nature of which I hope to suggest after Dr Amoore's paper (see p. 306).

Davies: Amplification arises from the way that the generator current, due to very small ionic leaks on the receptor, is integrated to trigger the nervous impulse at the first node of Ranvier along the nerve, as in the mechanoreceptor (see references to W. R. Loewenstein's work in my paper).

With regard to insects, because of their extreme sensitivity to the molecular form of the odorant the cell wall of the receptor must be rather solid and must contain a lot of (say) cholesterol or some rather solid lipid. It may be a pseudocrystalline or liquid crystalline structure.

The analogy with vision mentioned by Dr Adey is apt: I agree that the effectiveness of one quantum is a most remarkable phenomenon. I think it is agreed that the stimulus comes from the rhodopsin molecule adsorbing light and changing from the 11-*cis* retinene to the *trans* isomer, creating a dislocation in the rod membrane. It is known from experiments on monomolecular films of *trans* long-chain acids that isomerization to the *cis* form causes a dislocation of the structure. (This subject is reviewed by us in Davies and Rideal, 1963.) The effect of calcium ions is what one would expect on this sort of film. There are acid groups and phosphatide groups; the calcium ions bind the negatively charged head-groups together to give a rigid film.

Adey: But it is rather critical where this calcium is, where the protein is and what their precise physical relationship is to the lipid layer (Adey *et al.*, 1969; Wang and Adey, 1969). In other words, if you talk about a hierarchy of processes which start with very low quantum levels of any of these substances and finish up with processes that involve ATP and the burning of carbohydrate energy, you have shifted energy levels through about four gears in order to achieve this last phase that is associated with the initiation of the impulse. I don't think we should exclude or pass over any of these major steps in the hierarchy.

Davies: I agree that they should all be looked at. But I don't think this necessarily disproves what I have been saying.

You mentioned the thickness of the mucus layer on the olfactory epithelium. This is 20 μm thick in man, and Dr Reese tells me that the tips of the cilia may be only a small distance below the surface, about 1 or 0·5 μm, so that this is the distance that molecules have to travel through the mucus to reach the tips of the cilia. This corresponds to a time of diffusion of perhaps 1 or 2 milliseconds. Perhaps another 2 or 3 milliseconds is required for the electrical impulse to go down the cilium. The total process so far thus takes less than 5 milliseconds, compared with an overall time latency for olfaction of around 200 milliseconds.

Adey: The question of the movement of large molecules through a macromolecular net has been considered by Laurent (1966) in Uppsala, and he points out that the size of the molecule is not necessarily the determinant of the rate at which it moves. It is the charge distribution on that molecule in relation to the charge network.

Davies: There isn't a large dipole charge on many of these molecules, such as cyclohexane. They are practically neutral, and they are probably small by comparison with the molecular network structure in the mucus. This is something that one could and should determine experimentally.

Duncan: Comparisons with the photostimulation of the rhodopsin molecule are not really satisfactory because rhodopsin in the retinal rod is arranged on the transverse membranes across the outer segment, and the change in permeability probably does not take place across these membranes. My second point is that I think we all agree that a permeability change takes place; it is the site of this change in olfactory receptors that is uncertain. Perhaps one could test whether it takes place across the free border of the cells (that is, at apical cilia) by using ^{24}Na.

Lowenstein: But electron microscopists have yet to agree upon the fact or the direction of continuity between these stacked membranes and the actual membrane of the rod. That is one point. And its only fair to say that the hole-punching theory is one of *two* rival hypotheses of photostimulation; the exposure of an enzymic site is the second hypothesis (Wald, Brown and Gibbons, 1962).

Zotterman: Steric hindrance occurs in the 11-*cis* isomer of retinene which makes it twisted. Energy is rapidly liberated when light enters. And that is the only elementary process of sensory stimulation where we know the primary process, through the work of George Wald and Ruth Hubbard (Wald, 1968). We don't know anything about the primary reaction in any other sense organ.

Lowenstein: What we wish to know is the nature of the immediate consequence of that primary event.

Zotterman: This must be an effect on the protein, as Dr Adey implied. Then you come to the general idea of excitation, approaching David Nachmansohn's view that it starts with a protein receptor molecule reacting with acetylcholine.

Beidler: We don't yet know much about membranes, as has been said. Most information has come from myelin, erythrocytes and so on. If someone can convince us, and I hope that Dr Ottoson will be able to do so, that the cilia are of great importance, we should investigate these biochemically and see what kinds of proteins are there.

Davies: There are certainly active ionic pumps (perhaps peptides) that cross the membrane, but there can also be a strong propagated "leakage" which is the mechanism of a nervous impulse; there also is some intimate relation between the lipids and the proteins which needs to be investigated.

Adey: An interesting model of the excitatory cell membrane was devised by an engineer, Wei (1966), in which he proposed a membrane amplifier, discussed a moment ago. By analogy with a transistor device he suggested that the base current, which of course could be considered the controlling current in the model, would be a longitudinal current in the membrane, not a transverse one. A good many of us have racked our brains to think how one might look for just such a longitudinal current which was exercising a controlling function in the face of a very much larger transverse current, when one tries to separate this component.

Wright: It remains important to distinguish between the two cases: (1) the stimulus molecule is captured and somehow used up in the process of stimulating the nerve; or (2) the action is a physical one of "touch and go". The possibility has been mentioned that the stimulus molecule may be penetrating the tip of one of the olfactory cilia, which are both small and full of fluid. If the firing molecule does penetrate into the interior of the cilium, then the process of metabolizing it away poses a major problem. So I feel that a mechanism based on one molecule, one cell, once, is a very difficult one to accept.

Zotterman: Some information can be obtained by recording the receptor potential, as was first done by Adrian. He suggested that the generator potential would not follow the all-or-none principle.

Davies: On Dr Wright's point, the penetrating molecule might pass through the membrane into the cell or it might return into the mucus. If it passes through, it will tend to remain fairly near the inside of the membrane. If one is breathing clean air, even large molecules such as musk will tend to return to the atmosphere after 2 or 3 minutes, when one has taken many breaths. Some molecules, however, may have diffused by then well into the interior of the cell and became complexed with proteins. But with smaller molecules like benzene, fatigue is not so important, because they can diffuse *rapidly* away from the vicinity of the membrane.

Martin: If the molecule comes in and does work in making a hole, this is presumably because it has got more work out of its interaction with part of this membrane. And therefore I cannot see any reason why it doesn't stay there and continue to block up the hole.

Davies: It is like water evaporating, perhaps. There are hydrogen bonds with the other water molecules but the water will still evaporate; even at equilibrium there is a continual, dynamic interchange of water molecules between the vapour and the surface layer of the liquid.

Martin: When water evaporates the surface closes up behind it. You will only have time, therefore, for a few sodium ions or other ions to go through the surface before this happens, and this won't be enough to excite the membrane.

Davies: Once the ions start to pass the membrane, enough will move through in the short time elapsing before the whole membrane spreads back, possibly under the influence of the structural proteins. I think there would be sufficient time to get appreciable ionic transfer, just as happens with mechanoreceptors (see Loewenstein's work, quoted in my paper). The mechanoreceptors are easier to study and so are better understood.

Martin: One doesn't know what happens with the mechanoreceptors!

Beets: Professor Davies has attempted to explain fatigue as a process taking place at the epithelium. I do not think this is necessary, since fatigue seems to consist of two components and the major one, characterized by fast recovery, operates in the bulb (Adrian, 1950). The existence of a minor component characterized by slow recovery is suggested by experiments reported by Stuiver (1958) and by Professor Ottoson (1956).

Davies: Is it not true that the larger the molecule (e.g. a musk), the more the fatigue? This would certainly suggest that the sensory receptors are controlling in fatigue.

Beets: Some large molecules cause hardly any fatigue and some do.

Moulton: I believe we are talking about "adaptation" and not "fatigue"; they are different phenomena. At the receptor level, adaptation is seldom, if ever, evident, although I have seen a sustained depression in the excitability of the primary olfactory neurons of the rabbit after stimulation with butyric acid.

Zotterman: I showed adaptation in taste and that goes for the perceptual response too, with the same time-relations. In taste receptors it is not fatigue, but adaptation. If you raise the stimulus intensity you get a new response, as Adrian and I found.

Duncan: I did want to ask Professor Davies how, on the basis of his hypothesis, he would explain adaptation, in either a mechanoreceptor or an olfactory receptor, where you are getting continuous mechanical stimulus or a continuous stream of olfactory molecules. I would have thought that on his hypothesis you would get continual puncturing in a simple mechanical way and therefore you wouldn't get adaptation at the receptor.

Davies: There must be some balance between the amount of ions that leak out and the rate at which the system will recover to its resting state—the latter depending on how fast the sodium pumps work: if the fibre has carried several impulses, it then has to wait for the sodium pumps to restore the ionic levels to the resting state.

REFERENCES

Adey, W. R., Bystrom, B. G., Costin, A., Kado, R. T., and Tarby, T. J. (1969). *Expl Neurol.*, **23**, 29–50.
Adrian, E. D. (1950). *Electroenceph. clin. Neurophysiol.*, **2**, 377–388.
Amoore, J. E., and Venstrom, D. (1966). *J. Fd Sci.*, **31**, 118.
Bondareff, W. (1967). *Anat. Rec.*, **157**, 527–536.

Davies, J. T., and Rideal, Sir Eric (ed.) (1963). *Interfacial Phenomena*, pp. 295–298, 303–306. New York: Academic Press.
Davies, J. T., and Taylor, F. H. (1954). *Nature, Lond.*, **174,** 693.
Eigen, M., and Maeyer, L. C. M. de (1969). *Neurosci Res. Program Bull.*, in press.
Hodgkin, A. L., and Katz, B. (1949). *J. Physiol., Lond.*, **108,** 37.
Huneeus-Cox, F. (1964). *Science*, **143,** 1036–1037.
Huneeus-Cox, F., and Smith, B. H. (1965). *Biol. Bull. mar. Biol. Lab., Woods Hole*, **129,** 408.
Kanno, Y., and Loewenstein, W. R. (1966). *Nature, Lond.*, **212,** 629–630.
Laurent, T. C. (1966). *Fedn Proc. Fedn Am. Socs exp. Biol.*, **25,** 1128–1134.
Leveteau, J., and Mac Leod, P. (1969). *J. Physiol., Paris*, **61,** 5–16.
Nossal, G. J. V., Williams, G. M., and Austin, C. M. (1967). *Aust. J. exp. Biol. med. Sci.*, **45,** 581–594.
Ottoson, D. (1956). *Acta physiol. scand.*, **35,** suppl. 122, 1–83.
Pease, D. C. (1966). *J. Ultrastruct. Res.*, **15,** 555–583.
Rambourg, A., and Leblond, C. P. (1967). *J. Cell Biol.*, **32,** 27–53.
Schmitt, F. O., and Davison, P. F. (1965). *Neurosci. Res. Program Bull.*, **3,** pt. 6, 1–87.
Stuiver, M. (1958). Thesis, Groningen University.
Takagi, S. F., Wyse, G. A., and Yajima, T. (1966). *J. gen. Physiol.*, **50,** 473.
Wald, G. (1968). In *Les Prix Nobel en 1967*. Nobel Foundation. Amsterdam: Elsevier.
Wald, G., Brown, P. K., and Gibbons, I. R. (1962). *Symp. Soc. exp. Biol.*, **16,** 32–57.
Wang, H. H., and Adey, W. R. (1969). *Expl Neurol.*, **25,** 70–84.
Wei, L. Y. (1966). *I.E.E.E. Spectrum*, **3,** pt. 9, 123–127.

COMPUTER CORRELATION OF MOLECULAR SHAPE WITH ODOUR: A MODEL FOR STRUCTURE-ACTIVITY RELATIONSHIPS

John E. Amoore

*Western Regional Research Laboratory, Agricultural Research Service,
U.S. Department of Agriculture, Albany, California*

The sense of smell is characterized by the interaction of a *chemical* compound, the odorant, with a *biological* system, the organism. Hence it would seem entirely logical to apply the concepts and methods of *biochemistry* to its investigation. Only recently, however, have biochemical approaches to the subject been tried. The picture that is emerging is still very incomplete, but it does strongly suggest the central role of specific receptor proteins during the initial events of the olfactory process.

Classical observations on the odours of compounds should have alerted research workers long ago to the presumptive involvement of proteins. Two of the most remarkable facts about smell are its sensitivity and its specificity. Odorous substances can often be detected in extreme dilution, with thresholds as low as 10^{-14} moles of odorant per litre of air being quoted by Laffort (1963). The active principle of green bell peppers, 2-methoxy-3-isobutylpyrazine, has a mean detection threshold in air of 2×10^{-14} moles/litre; this was the lowest threshold encountered in the extensive experience of the authors (Buttery *et al.*, 1969). Such sensitivity is reminiscent of the activities of vitamins, hormones and antibiotics, which are generally agreed to exert their effects through their extremely high affinities for particular proteins and enzymes.

The study of relationships between chemical constitution and odour, especially in the search for synthetic perfumery ingredients, has revealed many examples of slight isomeric modifications in molecular structure causing gross changes of odour quality or intensity (Beets, 1957). Even optical antipodes can differ in odour, according to cogent evidence recently adduced by Friedman (1970) for *l*- and *d*-carvone (spearmint and caraway respectively). Such delicate selectivity in the response of the organism to subtle chemical differences is the hall-mark of a protein.

Here it seems appropriate to recall some almost-forgotten remarks of two great luminaries of nineteenth-century biological chemistry. In discussing Piutti's report that D-asparagine tastes sweet but the L-isomer insipid, Louis Pasteur (1886) commented: "Le corps actif dissymetrique qui interviendrait dans l'impression nerveuse, traduite par une saveur sucrée dans un cas et presque insipide dans l'autre, ne serait autre chose, suivant moi, que la matière nerveuse elle-même, matière dissymetrique comme toutes les substances primordiales de la vie: albumine, fibrine, gélatine, etc." Note that the three examples he cited, as analogous to the nervous material, are all proteins; the first hint of the receptor protein idea. In seeking an explanation for the observation that invertase will split only α-D-glucosides and emulsin attacks only β-D-glucosides, Emil Fischer (1894) felt impelled to conclude: "Um ein Bild zu gebrauchen, will ich sagen, dass Enzym und Glucosid wie Schloss und Schlüssel zu einander passen müssen, um eine chemische Wirkung auf einander ausüben zu können." Here was born the receptor-site theory of biochemistry.

These men had more important matters on their minds than taste and smell, and in fact their concepts were applied over the years with great success in enzymology, immunology and drug action. Left behind in the general advance were the chemical senses, which fell victim to all sorts of weird theories. Now, better late than never, in the last few years the wheel has come full circle, and the concepts of main-line biochemistry are once more being brought to bear on taste and smell.

RECEPTOR PROTEINS

In 1966 Dastoli and Price announced the isolation, from cow tongue, of a purified protein which combines selectively with sugars. The protein exhibits towards each sugar an affinity which is proportional to the relative sweetness of the sugar. There are difficulties with the reproducibility of the preparation and the assay, but such problems are common in biochemistry. Nevertheless, the work has been confirmed and extended by Hiji, Kobayashi and Sato (1968). Further work yielded, from pig tongue, a different protein which appears to form complexes with bitter-tasting substances (Dastoli, Lopiekes and Doig, 1968). Hence the concept is emerging that there may be selective receptor proteins for each of the classical primary tastes: sweet, bitter, salt and sour.

Parallel work on the sense of smell has been begun by Ash (1968). He obtained, from the rabbit olfactory epithelium, proteinaceous extracts which exhibited spectrophotometric changes on addition of particular odorants, especially linaloöl and linalyl isobutyrate. An analogous experiment has been done by Riddiford (1970) on the antennae of certain male Saturniid moths. A protein-containing extract was obtained and submitted to electrophoresis, revealing a band which would on occasion carry with it radioactively labelled female sex pheromone.

In principle, a pheromone is simply an externalized hormone; and specific proteinaceous receptors for internal circulating hormones have been isolated—for example, an oestradiol-binding protein from rat uterus (Toft, Shyamala and Gorski, 1967). Imperfect as some of these experiments may be, they do support the idea of specific receptor proteins for each of the various classes of taste and odour, or types of pheromone.

PRIMARY ODOURS

The classification of qualitative responses in the chemical senses is a task varying considerably in difficulty from one sense to another. The four primary tastes have been reasonably obvious since antiquity, mainly because numerous single common substances exist which clearly possess one or other taste. Overlap between two tastes is relatively uncommon, because the corresponding chemical properties tend to be mutually exclusive. Certain pheromones have been identified systematically by extracting the active substance from the gland which produces it, chemically identifying the compound, and recording the characteristic biological response produced in the receptive organism. Again overlap is rare, presumably because it would be biologically undesirable and has been eliminated by natural selection.

With the sense of smell the problem is very much more complicated. An enormous number of chemicals have a smell, and virtually no intuitive classification is possible. Behavioural responses are vague, such as acceptance, rejection, or indifference. Overlapping between the component modalities of smell is probably the rule rather than the exception. Of the many experimental methods that have been proposed for breaking this impasse, only one is showing signs of success. That is the study of specific anosmia, the inborn condition of being unable to detect one odour or class of odours, while all other odours are perceived normally. The phenomenon was described half a century ago by Blakeslee (1918) and proposed as a method for classifying odours by Guillot (1948). Since then the idea had lain dormant again until systematically developed by the present author (Amoore, 1967).

An innate chemical defect in a living organism is usually traceable to a defective protein, often but not necessarily an enzyme. Looking back on the problem of specific anosmia, from the vantage point of our present knowledge and beliefs, it is natural to form the hypothesis that each primary odour may be detected by a selective receptor protein on the olfactory nerve endings. Specific anosmia would then be due to the failure to inherit the appropriate receptor protein for a given primary odour.

This is strictly hind-sight, but the situation with the chemical senses may be exactly analogous to the studies of biochemical defects which contributed so greatly to the knowledge of intermediary metabolism. The inborn errors of human metabolism collected by Garrod (1923) and the radiation-induced metabolic mutants of *Neurospora* exploited by Beadle (1945) are the foundations of biochemical genetics, on which so much of our knowledge of modern biochemistry is laid.

Specific anosmia or "odour-blindness" would appear to be analogous to the well-known taste-blindness to phenylthiocarbamide (PTC). The non-tasting condition, which affects about 30 per cent of the population, is recognized to be an inherited recessive character (Blakeslee, 1932). The evidence for the possible inheritance of specific anosmia is much weaker and still controversial (Amoore, 1970). The balance of opinion would again suggest the expected recessive inheritance, but more experiments are needed.

The superficial resemblance between specific anosmia and other biochemical defects is still tentative. However, it does conform well with current biochemical thought. Furthermore, these "Nature's experiments" afford a unique opportunity to "cut the Gordian bands, too intricate to unloose" by other experimental methods. Here, if the phrase will be pardoned, is a unique opportunity to solve the "olfactory code."

SPECIFIC ANOSMIA TO ISOVALERIC ACID

The only specific anosmia so far examined in detail is that towards the lower fatty acids (Amoore, 1967; Amoore, Venstrom and Davis, 1968). About 2 per cent of the population are unable to smell these acids, unless the concentration is raised to such an extent as to be quite repugnant to normal subjects. The odour detection threshold of each person was measured by means of a dilution series, with a sample-sorting test applied at each level, so as to establish statistically the subject's ability to detect a given concentration of the acid.

The most pronounced deficiency was observed with isovaleric acid. In Fig. 1 are summarized the thresholds for 443 observers. Most of the people fall into a normal or Gaussian distribution of sensitivities, but eight persons have markedly low sensitivities (high threshold concentrations). Two of the specific anosmics have a quite extraordinary defect, being unable to smell the solution until it is raised to a concentration about 30000 times the average threshold level. This is a very interesting observation in its own right because it could be interpreted as meaning that the missing receptor protein can be very highly selective; that is, in its absence none of the other receptor proteins are able to detect this odorant except at greatly increased concentration. However, other explanations are possible. For instance,

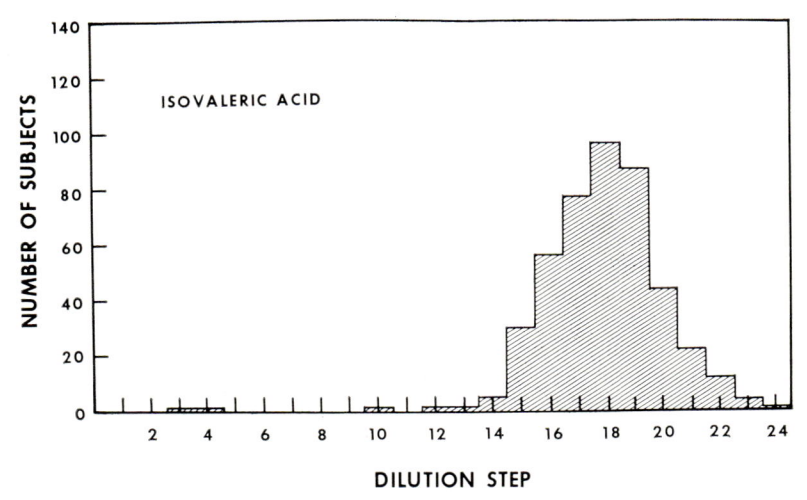

FIG. 1. The odour detection thresholds of 443 persons towards purified isovaleric acid. Each dilution step is one-half the concentration of the preceding step, starting from step "0" for the saturated aqueous solution, 5·12 per cent, v/v. (From Amoore, 1968b, by courtesy of the American Gas Association.)

it is noteworthy that intermediate degrees of deficiency exist. These could imply different deleterious mutations in the fatty-acid sensitive receptor protein. Alternatively it could mean that several different receptor proteins coexist with somewhat overlapping specificities, and only when all of the proteins capable of registering a fatty acid are absent will a very high degree of specific anosmia be observed. Genetic studies should be able to sort out the possibilities.

The chemical range and extent of this specific anosmia was explored by testing the deficient individuals with various chemical relatives of isovaleric acid. The results are shown in Fig. 2. The deficiency is quite selective, and is observed in high degree only with fatty acids of intermediate chain length, especially 4 to 7 carbon atoms (butyric to oenanthic acids). This indicates that the molecular size of the odorant is important. Furthermore, only carboxylic acids exhibit the defect. The corresponding aldehyde, alcohol or ester of comparable molecular size was smelled easily by the specific anosmics. Hence the receptor system is selective towards a particular functional group of the molecule, —COOH. Two of the branched-chain acids, isobutyric and isovaleric, were notably more difficult to smell,

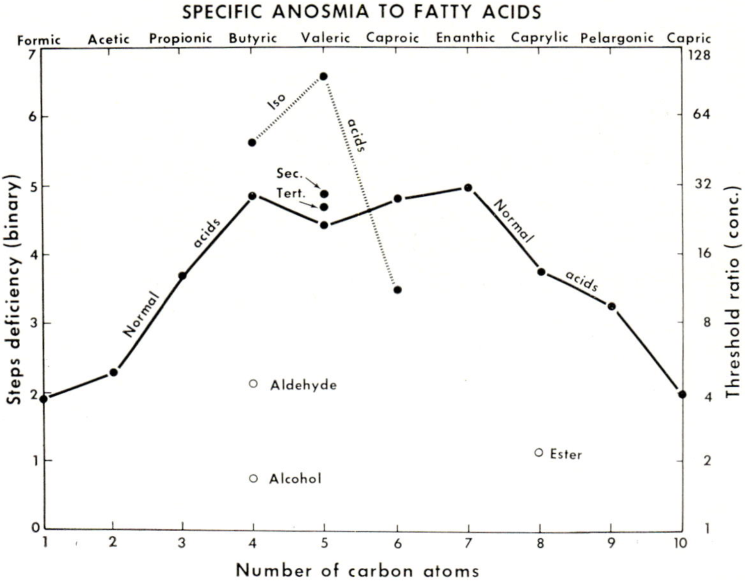

FIG. 2. Influence of variations in molecular structure on the degree of specific anosmia. The right-hand ordinate gives the ratio of the detection threshold concentrations between ten specific anosmics and 97 normal observers tested as controls. (From Amoore, 1968a, by courtesy of Necmi Tanyolaç.)

and isocaproic easier, than the corresponding straight-chain acids. This indicates that even when functional group and molecular size are correct, the detailed shape of the molecule has some influence on the degree of specific anosmia observed.

Applying the concept of Guillot (1948), I feel justified in equating the chemical range of a particular type of specific anosmia to the range of the corresponding primary odour in a normal person's sense of smell. This primary odour is realistically named "sweaty." Just how many other primary odours there are is still an open question. I have carried Guillot's survey much further, and reported 62 different chemicals towards which some person or other encounters a personal smelling defect (Amoore, 1969). Some of these 62 obviously belong to the same class, but it looks as though a comparatively large number of primary odours will emerge in the final analysis, somewhere between 20 and 30. Although this method of identifying and mapping primary odours is successful, it is exceedingly slow. Unless short-cuts or team-work can be applied, it will require many years of detailed research to work out the whole of the olfactory code.

THE CONFORMATION OF A PROTEIN RECEPTOR SITE

A central concept in biochemistry is the receptor site—that specific location on the surface of a protein where it combines reversibly with its reactant, usually a solute of low molecular weight. The receptor site is invoked repeatedly in virtually any discussion of enzymes acting on substrates, antibodies precipitating antigens, drugs approaching acceptors, or hormones exerting their effects. Nevertheless only very recently has any receptor site been laid open to exact atomic delineation. This has now been done by X-ray crystallography for the digestive enzyme carboxypeptidase A, which attacks polypeptide chains from the free carboxyl end, removing one amino acid residue at a time. The results could be extremely illuminating for many students of the chemical senses, who may not be aware of the full import of the receptor site concept, and its probable significance in understanding the olfactory process.

Lipscomb and co-workers (Reeke et al., 1967) have worked out the full three-dimensional structure of the backbone chain of carboxypeptidase, with the results shown in Fig. 3. The polypeptide chain is represented by a folded ribbon. The structure possesses a definite cavity, into which will fit the molecular model of a synthetic substrate, glycyl-L-tyrosine. The receptor site is physically so proportioned that it will accept only L-amino acid residues, and it has a side-pocket that lends a preferential specificity for amino-acids having an aromatic side-chain (tyrosine, tryptophane, etc.)

Certain functional groups in the receptor site are exactly adapted to bind their opposite numbers in the substrate. The zinc atom holds the penultimate carbonyl group, the glutamate side-chain may hold the imino, and an arginine side-chain binds the terminal carboxyl. When this last binding occurs, it pulls the arginine group inwards 0·2 nm (2 Å), and thereby triggers a remarkable conformational change in the protein. A tyrosine side-chain of the enzyme, that had been lying well clear of the receptor site, suddenly swings over through a 1·4 nm (14 Å) arc and closes the receptor site like a lid. The phenolic —OH provides the final catalytic impetus to splitting the peptide bond of the substrate.

I do not mean to imply that the olfactory process includes a catalytic phase. I only wish to use this as an illustration of how well-known principles in biochemistry can readily explain the established facts of olfactory discrimination. An olfactory receptor site with comparable properties could easily distinguish D- and L-antimers, let alone α- and β-epimers or

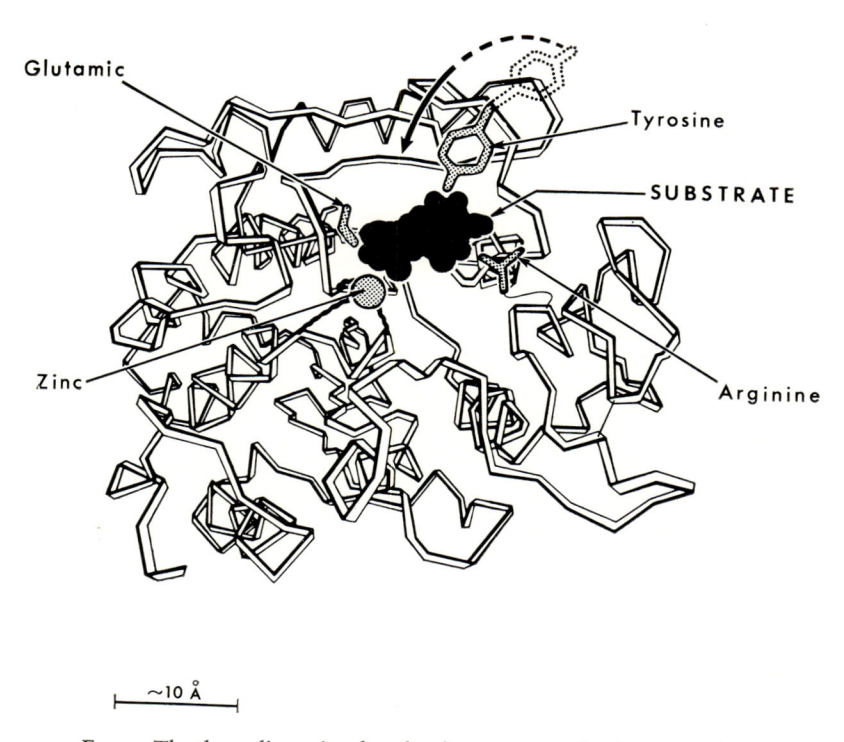

FIG. 3. The three-dimensional molecular structure of carboxypeptidase A, showing the interaction with a model substrate, glycyl-L-tyrosine. (Modified after Steitz, 1968.)

cis- and trans-isomers which may differ in odour. It can explain the possibility of specific functional group recognition, such as the —COOH of the sweaty primary odour. The finding of chemically unrelated substances having similar odours, for example among the camphors, could be a consequence of many different combinations of elements being able to assume a similar external form and size, such as a sphere 0·7 nm (7 Å) in diameter. A movable lid on the receptor site could explain the discrimination between hemi-spherical and full-spherical molecules with the same curvature.

Here we have a challenging idea of how the olfactory and gustatory receptor proteins may achieve their stereochemical selectivity. What has long been a mystery in the context of odour science can now be seen to be rather commonplace in biochemistry. How the reception of an odorant molecule is transduced into a propagated nervous impulse is still, of course, entirely unknown, but it probably has something to do with conformational changes in proteins. An "opening" of the opsin structure at the moment of visual excitation has been demonstrated (Wald, 1968). Recently Koshland (1968) has speculated on the possible involvement of a cooperative effect among protein sub-units as an amplification stage in the olfactory process.

COMPUTER ASSESSMENT OF MOLECULAR SHAPE

It is a rare privilege to be able to "see into" a protein's receptor site, and only possible so far with long-established enzymes that can be prepared in quantity and in suitable crystalline form for X-ray analysis. The infant subject of chemosensory receptor proteins has only just been discovered, and a long wait seems inevitable before anybody will be able to view an olfactory receptor site. Meanwhile we shall have to be content with studying the structure of molecules possessing a given odour, and attempting to elucidate what stereochemical features they have in common. The still hypothetical olfactory receptor site can then be imagined to have properties complementary to the odorous molecule.

For this phase of the work I have used space-filling scale molecular models. The molecule is put into its most likely configuration and arranged in a systematic orientation. Then it is photographed in silhouette, from three directions mutually at right-angles (Amoore, 1964). The silhouette photographs are examined in a unique pattern-recognition machine developed by Palmieri and Wanke (1968) at the University of Genoa. The instrument (acronym "PAPA") consists of a television-type camera linked directly to a special computer. It scans the molecular shapes by means of a reproducible collection of random lines (Fig. 4).

From the number of intersections made by each of the lines with different molecular silhouettes, the similarities between those silhouettes can be quantitatively calculated and printed out by the machine in a few seconds. The molecule in Fig. 4 is isovaleric acid itself, with which the other 14 lower fatty acids were compared (Amoore, Palmieri and Wanke, 1967). It transpired that there is a good correlation between the similarities of molecular shape, and the degree of specific anosmia observed (Fig. 5). The correlation coefficient r was 0·80 for this set of data. This means that as much as 64 per cent of the observed variation in the sweaty odour character

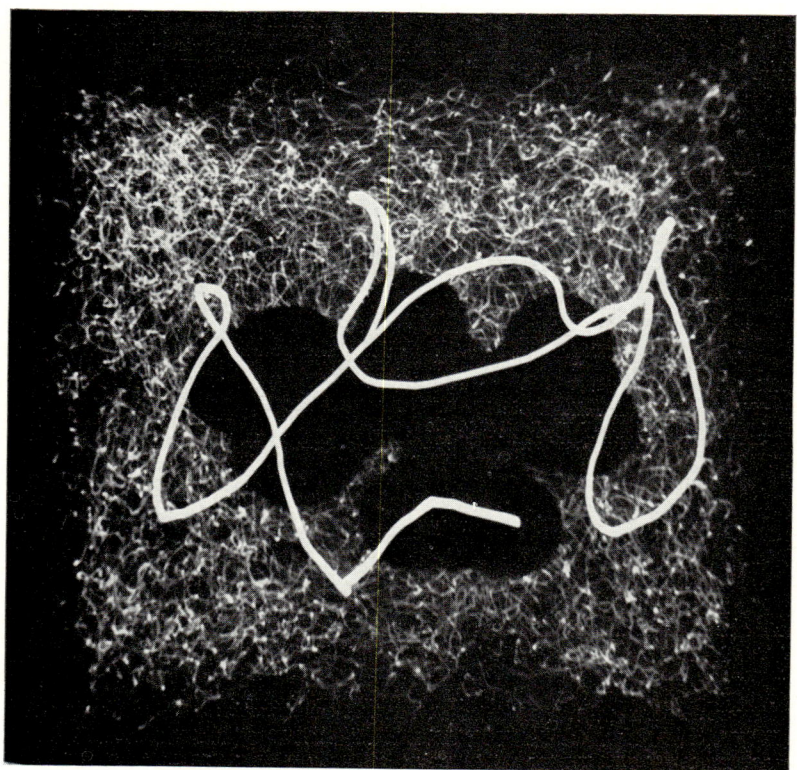

FIG. 4. A silhouette photograph of the molecular model of isovaleric acid is scanned by means of 4096 random lines in the PAPA pattern recognition machine. One of the lines is shown at high intensity to illustrate the scanning principle. (Photo by courtesy of G. Palmieri and E. Wanke, University of Genoa.)

of these compounds can be explained simply by scanning three photographs of a conventional molecular model.

The PAPA computer has recently also been successfully applied to a problem concerned with insect pheromones (Amoore *et al.*, 1969). This was a complicated three-way collaboration between entomologist, biochemist and computer specialist, widely scattered geographically as well as scientifically. Blum at the University of Georgia had gathered a large number of data on the abilities of other compounds to release an alarm reaction in the ant *Iridomyrmex pruinosus*, whose natural alarm pheromone is 2-heptanone.

Without revealing the biological activities observed, he gave lists of the chemicals to me, for preparation of molecular silhouette photographs. The pictures were sent to the Italians for scanning by their PAPA machine. A total of 49 different ketones were tested, and the correlation coefficient between molecular shape and biological activity was 0·57. The procedure

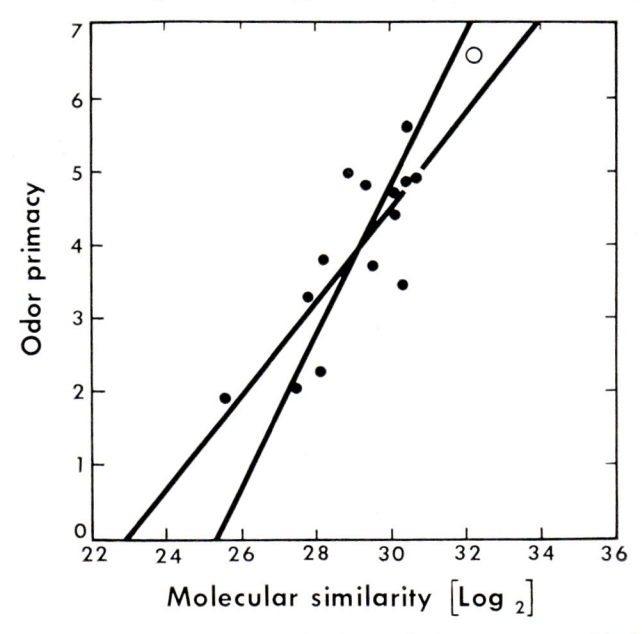

FIG. 5. Correlation between molecular similarity, measured with the PAPA machine, and odour primacy, which is the degree of specific anosmia observed. Each point represents a lower fatty acid; the open circle indicates the "sweaty" standard odorant, isovaleric acid itself. The two regression lines are based on the assumption that either x or y is the independent variable. (From Amoore, Palmieri and Wanke, 1967, by courtesy of Macmillan Journals.)

even worked with an assortment of 35 non-ketones, including alcohols, aldehydes, ethers, esters, amines and halogenated hydrocarbons; the correlation coefficient $r = 0.81$.

FUTURE POSSIBILITIES FOR CORRELATION OF STRUCTURE AND ACTIVITY

The results obtained with the PAPA machine prove beyond doubt that there is a strong correlation between molecular shape and odour quality. This is despite the fact that rather unsophisticated criteria of molecular parameters are being employed. It is a very crude model to consider atoms as rigid billiard-balls, and to pay practically no attention to the distribution of electrons over the surface of the molecule. Somewhat improved results were obtained in the study of ant alarm pheromones by using as a datum point, not the centre of gravity of the molecule, but the centre of its chief functional group of atoms. This permitted a small input of electrical affinity information to be supplied to the computer.

We have not yet applied the full capabilities of the PAPA machine to this problem. So far the machine has been used in a subservient role, simply to carry out very rapidly morphological comparisons that could be done equally well, though 100 times slower, by hand. Actually the PAPA machine is designed to abstract commonalities among related patterns, and even to devise its own systems of criteria for discriminating types of pattern too subtle for visual classification, such as weather maps.

Nevertheless what we have accomplished for the sense of smell is probably more advanced than any stereochemical survey that has been conducted in other areas of chemical constitution/biological activity relationships. It is somewhat paradoxical that the chemical senses, so long neglected as a subject for molecular biological analysis, should now be paving the way for a computerized approach to the much longer established biochemical sciences of immunology and pharmacology.

Of course, a much more modern concept of molecular conformation and electron distribution should be incorporated into our analyses. Happily this is becoming available in the studies of Clementi (1968). It is already possible to make a comprehensive calculation of the electronic molecular orbitals in a compound with up to 20 atoms, by conducting slightly approximated Schrödinger-type calculations with a very large computer. This gives not only the best conformation of the molecule, but also the electronic distribution and chemical affinity throughout the whole molecular volume. At present the computing costs are prohibitively high for re-surveying the several hundred compounds that have been put through the PAPA analysis. However, large-scale computing costs have a history of decreasing, and short-cuts may be discovered which could still yield results sufficiently rigorous for biological applications.

As a parting thought, we might speculate on the possible benefits to be derived from a comprehensive analysis of all known chemical compounds as possible candidates for every worthwhile biological activity. Very many groups of chemicals are known which display some useful application, whether it be as a perfume or flavour, an insect lure, an antibiotic, a hormone or a hallucinogen. If all this information were fed into a gigantic computer which combined Clementi's synthesis of molecular deportment with Palmieri's analysis of recurring patterns, then the machine would ascertain exactly what molecular properties are essential to each biological activity.

When subsequently the molecular specifications of all the million or more known chemical compounds were to be run through the machine, it is very likely that the cost of the project would be recovered many times over through the flagging of existing compounds for which a totally unsuspected but valuable biological activity could be revealed.

SUMMARY

The paper reviews a continuing investigation of the principles underlying olfactory classification and discrimination. The theme of this presentation is the growing evidence that the initial detectors of sapid or odorous molecules are specialized proteins, having specific affinities for particular classes of compounds.

The detailed study of the specific anosmias is expected to lead eventually to a classification of the human sense of smell into a limited number of primary odours. Within each class of odour, the basic stereochemical requirements for discrimination can be assessed by submitting silhouette photographs of molecular models to a pattern-recognition machine.

It is anticipated that still more powerful computer methods could be applied advantageously in the theoretical screening of large numbers of known molecular structures for valuable, but previously unrecognized, biological activities.

REFERENCES

AMOORE, J. E. (1964). *Ann. N.Y. Acad. Sci.*, **116**, 457–476.
AMOORE, J. E. (1967). *Nature, Lond.*, **214**, 1095–1098.
AMOORE, J. E. (1968a). In *Theories of Odor and Odor Measurement* (Proc. NATO Summer School, Istanbul, 1966), pp. 71–85, ed. Tanyolaç, N. Istanbul: Robert College.
AMOORE, J. E. (1968b). *Proc. Operating Sect., Am. Gas Ass., Distrib. Conf.*, 242–247.
AMOORE, J. E. (1969). In *Olfaction and Taste III* (Proceedings of the Third International Symposium, New York, 1968), ed. Pfaffmann, C. New York: Rockefeller University Press.
AMOORE, J. E. (1970). In *Handbook of Sensory Physiology, vol. IV: Chemical Senses*, ed. Beidler, L. M. Berlin: Springer-Verlag. In press.
AMOORE, J. E., PALMIERI, G., and WANKE, E. (1967). *Nature, Lond.*, **216**, 1084–1087.
AMOORE, J. E., PALMIERI, G.. WANKE, E., and BLUM, M. S. (1969). *Science*, **165**, 1266–1269.
AMOORE, J. E., VENSTROM, D., and DAVIS, A. R. (1968). *Percept. Mot. Skills*, **26**, 143–164.
ASH, K. O. (1968). *Science*, **162**, 452–454.

BEADLE, G. W. (1945). *Chem. Rev.*, **37**, 15–96.
BEETS, M. G. J. (1957). In *Molecular Structure and Organoleptic Quality*, pp. 54–90. London: Society of Chemical Industry.
BLAKESLEE, A. F. (1918). *Science*, **48**, 298–299.
BLAKESLEE, A. F. (1932). *Proc. natn. Acad. Sci. U.S.A.*, **18**, 120–130.
BUTTERY, R. G., SEIFERT, R. M., GUADAGNI, D. G., and LING, L. C. (1969). *J. agric. Fd Chem.*, **17**, 1322–1327.
CLEMENTI, E. (1968). *Chem. Rev.*, **68**, 341–373.
DASTOLI, F. R., LOPIEKES, D. V., and DOIG, A. R. (1968). *Nature, Lond.*, **218**, 884–885.
DASTOLI, F. R., and PRICE, S. (1966). *Science*, **154**, 905–907.
FISCHER, E. (1894). *Ber. dt. chem. Ges.*, **27**, 2985–2993.
FRIEDMAN, L. (1970). In preparation.
GARROD, A. E. (1923). *Inborn Errors of Metabolism*, 2nd. edn. London: Oxford University Press.
GUILLOT, M. (1948). *C. r. hebd. Séanc. Acad. Sci., Paris*, **226**, 1307–1309.
HIJI, Y., KOBAYASHI, N., and SATO, M. (1968). *Kumamoto med. J.*, **21**, 137–139.
KOSHLAND, D. E. (1968). *Chem. Engng News*, **46** (41), 38–39.
LAFFORT, P. (1963). *Archs Sci. physiol.*, **17**, 75–105.
PALMIERI, G., and WANKE, E. (1968). *Kybernetik*, **4** (3), 69–80.
PASTEUR, L. (1886). *C. r. hebd. Séanc. Acad. Sci., Paris*, **103**, 138.
REEKE, G. N., HARTSUCK, J. A., LUDWIG, M. L., QUIOCHO, F. A., STEITZ, T. A., and LIPSCOMB, W. N. (1967). *Proc. natn. Acad. Sci. U.S.A.*, **58**, 2220–2226.
RIDDIFORD, L. M. (1970). *J. Insect Physiol.*, in press.
STEITZ, T. A. (1968). *New Scient.*, **38**, 568–570.
TOFT, D., SHYAMALA, G., and GORSKI, J. (1967). *Proc. natn. Acad. Sci. U.S.A.*, **57**, 1740–1743.
WALD, G. (1968). *Nature, Lond.*, **219**, 800–807.

DISCUSSION

Martin: Quite independently 10 or 12 years ago I formed a hypothesis which overlaps with Dr Amoore's. I vented this for the first time in 1968 at a conference in Göteborg (Martin, 1968). The basic facts that I wanted to explain were the extraordinary sensitivity of the nose as the sensor and the extraordinarily wide range of specificity that it has. Because of the high sensitivity it seems to me that one must necessarily postulate an amplifying mechanism in order to explain the excitation of olfactory cells by just a few molecules. An amplifying mechanism with which we are all familiar, and which is the obvious one to invoke, is that of enzymes, so I made my basic postulate the idea of an enzyme with two clefts, one of which will synthesize, or break down something to liberate, a substance which is probably a common excitant of cells, but this enzyme will only be active when an odorant molecule is drawn into the other cleft. The difference between the fundamental or primary smells would lie in this latter part of the enzyme, which would be specific for an odorant; but all odorants would cause the production of the same substance which would excite the cell. Dr Amoore has already covered the way in which you can expect the shape and functional groups of molecules to be drawn into a particular site in a protein, which of course fits in very well with the known properties and specificities of enzymes.

There are one or two other points. Smell differs from all the other senses in the way one becomes tired to it. If you have a sugar in your mouth you taste sweetness continuously; if the light is on you know that it is on. Similarly with sound: if a clock is ticking which you don't normally notice, if you listen for it, you can hear it. But if you sit in a room with a constant concentration of any chemical in the air you soon become tired to it and are unable to smell it. A fairly simple explanation of this could be given if there were a constant production of a precursor substance which accumulated in a pool from which the enzyme drew to convert it into the excitatory substance. If an odorant substance were absorbed in the enzyme the rate of production of the excitatory substance would increase, causing a sensation of smell, and the pool would be depleted. If the concentration of odorant remained the same the depletion of the pool would continue until the rate of conversion to excitatory substance equalled the rate of production. There would then be no sensation. Any rise or fall of the concentration of the odorant would give rise to a positive or negative smell. This kind of negative smell can be observed as the peculiar smell of fresh air on leaving a room in which one has become accustomed to a constant level of, for instance, chloroform vapour.

This hypothesis has the advantage that it can be tested by looking for substances in the olfactory tissue which will combine with the odorant. The pheromones of insects would appear to be the substances to use, since the specificity and sensitivity are both high. Thus by chemical fractionation of the antennae of male *Bombyx mori* a substance would be sought that combined with very low concentrations of radioactive bombykol. The high sensitivity implies that a very stable association of pheromone and enzyme will be formed. Enzyme activity would be sought in any such complex that was found.

One other corollary of the theory is that normally all the types of enzyme for primary sensations would be expected to be more or less activated by substances present in the blood. Thus there would be a steady level of excitation, and this would be changed by exposure to a volatile odorant which could cause either stimulation or inhibition.

Amoore: The experiment you suggest about pheromones has already been partly accomplished by Dr Lynn M. Riddiford at Harvard University. She has extracted a protein from the antennae of male Saturniid moths

308   DISCUSSION

which on two occasions was found to migrate electrophoretically with radioactively labelled sex pheromone from the female (see my paper, p. 295).

Wright: We have been studying the possibility of modifying the insect's response to a pheromone by preadapting it to various odour chemicals.

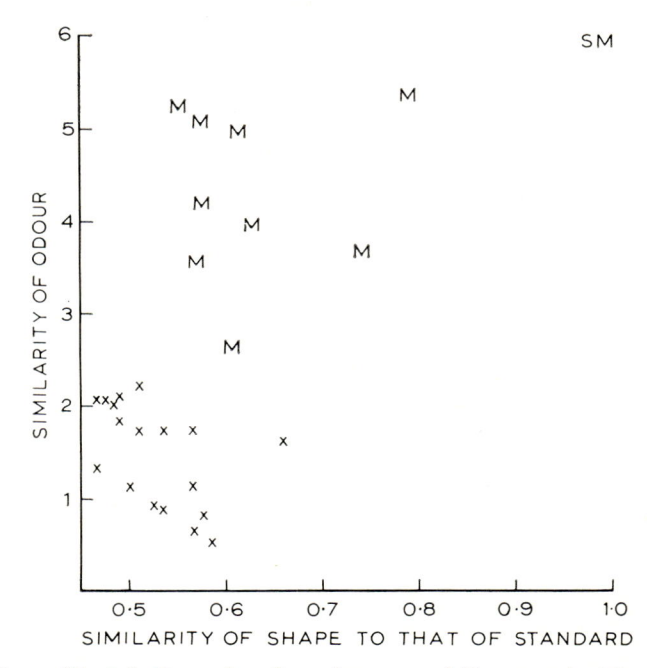

Fig. 1 (Davies). Data taken from Amoore and Venstrom (1967) are plotted to show whether odour quality correlates with a geometrical factor alone, namely the overall molecular shape relative to the shape of a standard substance. Here this standard is a musk, 15-hydroxypentadecanoic acid lactone (point SM). Other substances generally regarded as having musky odours are indicated by M; and non-musks (with little similarity to musks) are indicated by X. It is clear that though there *is* an overall correlation of odour quality with molecular shape, *within the group of musky compounds the correlation is much weaker*. Further, there is considerable overlap of the shape similarity factor between musks and non-musks in the region around 0·57.

We use the Mediterranean flour moth (*Anagasta kühniella*, Zell.). The female emits the pheromone which causes the male to become excited within a second or two, and then the response falls off more slowly, over a few minutes. Pre-exposure of the insect to linaloöl makes the response continue without fatigue. How would that fact fit in with your theory?

Martin: If the first chemical inhibits the enzyme the pheromone response might last longer. I cannot think of a good explanation.

Davies: Dr Amoore's approach using the PAPA type of machine is an extremely valuable one, but I think the results can be interpreted in slightly different ways. I am not sure that the mathematical correlations can be taken entirely at the values that Dr Amoore takes. In Fig. 1 I have replotted some results of Dr Amoore (Amoore and Venstrom, 1967) on similarity of odour against the similarity of shape, referred to a standard musk (SM). The question is whether it is fair to make the correlation with a line through all the points (including the non-musks) or whether you should leave the non-musks out. Using only the known musks (marked as M) one finds a much poorer correlation between the similarities of odour and of shape. So in interpreting these correlations one has to be careful how many non-musks are included. One could conceivably include, for example, substances like hydrogen sulphide and sulphur dioxide and get a much better correlation by doing this.

My other point, on Dr Martin's theory, is that these musks are in general of different chemical types, and whether this would fit in with an enzyme model, I rather doubt: one can have different functional groups such as acetyl, NO_2 or —O— in musks and the odour is more or less the same.

Amoore: When the substance is already a musk you are aiming for a much higher degree of refinement because you are looking for correlations within an odour group. I am trying to split up the whole gamut of odours and get some rough classes. What I hope for is more refinement of the methods and then eventually we shall be able to make quantitative correlations within an odour type. We are approaching this with the fatty acids having the sweat-like odour; the correlation has been working well there.

Beets: The collection of musks used by you has been originated in the course of synthetic work based on the information available regarding the relation between musk odour and molecular morphology. Intensities are rather unpredictable and may vary between odourless and extremely intense. The collection of strong, weak and non-musks which Dr Davies has shown is the result of this type of synthetic research.

Amoore: You are talking of intensities or thresholds, and I am only trying to deal with subjective estimates of quality so far. The same thing applies to the fatty acids; I am not trying to predict the absolute thresholds of the odorants, but rather the difference in threshold between two groups of observers—normal and specifically anosmic. This provides a measure of odour quality.

Martin: In reply to Professor Davies, I would have expected that the musks which differed most in shape would be different in intensity of smell. Secondly, a number of substituted peptides with unnatural groups like fluoride or bromide can nevertheless be hydrolysed by enzymes at rates exceeding those of the natural substrates.

Ottoson: This idea of an enzyme playing a key role in the sensory process might be tested by applying a solution of an enzyme inhibitor to the surface of the mucosa and recording from a bulbar unit or directly from the receptor layer.

Zotterman: Dr Martin's suggestion of the excitatory mechanism of olfaction recalls a study of children with a hereditary condition, one group of whom had no taste buds and a second group had taste buds but no taste responses. The latter group could be enabled to respond by treatment of the tongue with eserine (physostigmine).

Duncan: It is not quite as simple as this. Dr Martin is suggesting, I think, that the odorous molecule produces an allosteric effect in a membrane enzyme when it combines with its receptor site. It seems probable that a similar situation exists when a drug molecule combines with its receptor site. Biochemical pharmacologists are, of course, studying the drug–receptor combination intensively, but even here, where the preparation (e.g. the electroplaque) may be studied biochemically, it is very difficult to demonstrate an allosteric effect conclusively. Ideally, the enzyme should remain *in situ* on the membrane during the experiment, and Dr K. Bowler and I (1970) believe that erythrocyte ghosts may be a suitable preparation. However, in olfaction, we are dealing with a situation that is much more difficult to investigate; the receptor membrane is very much smaller and less accessible for biochemical study. To take up Professor Zotterman's point concerning anticholinesterase drugs and acetylcholine; the difficulty is that Dr Martin's postulated messenger substance is intracellular.

Martin: No; there is no theoretical limitation on where the enzyme should be. It could be intracellular, in the membrane, or even extracellular.

Duncan: There could well be at least two different enzyme steps involved. Let us consider gustatory receptors for a moment. The sequence of events could be (*a*) an allosteric interaction at the receptor membrane, causing the release of (*b*) an intracellular chemical messenger which produces (*c*) a localized permeability change, recorded as the receptor potential (Kimura and Beidler, 1961) which is responsible for (*d*) excitation of the afferent neuron. The synaptic transmission involved may or may not be cholinergic, although I think that the balance of evidence does not favour the view that it is cholinergic (Duncan, 1964*a*). Thus, you may show that gustatory or olfactory sensitivity is impaired if you use certain enzyme inhibitors, but it would be difficult to decide where the enzyme in question fits in such a chain of events.

You can certainly inhibit chemoreceptors (e.g. taste buds, carotid body chemoreceptors) with agents that block the tricarboxylic cycle; examples are arsenite or malonate (Heymans, 1955; Duncan, 1964b). Conversely, dinitrophenol, cyanide and ATP (Dontas, 1955; Joels and Neil, 1963) will *stimulate* carotid body chemoreceptors. It is my belief (Duncan, 1967) that an ATP–ATPase interaction is involved in the generation of the receptor potential.

I think that the sequence of events that I have outlined is (at least theoretically) satisfactory for taste buds. It seems unlikely to me that the combination of the sapid molecule with its receptor site produces a change in cation permeability across the free border of the cell, where the cell membrane will be in contact with the varying and possibly deleterious solutions on the tongue's surface. I agree, therefore, that there is some form of signalling system between the receptor sites and the lateral borders of gustatory cells. However, this signalling system might not be necessary in olfactory cells; combination of the odorous molecule with its receptor may conceivably produce a change in cation permeability directly. The olfactory cilia, where we seem to agree the receptor sites are located, lie in the mucus layer, which might constitute a reasonably constant environment. On this hypothesis there would be a localized depolarization which would spread along the cilia to the body of the cell.

Beidler: The latency for taste receptors is about 12–30 milliseconds but with most inhibitors there is no effect for 5 minutes or so. There is the problem of *where* you are hitting the receptor, of the time-relationships of inhibitor action, and of how specific the inhibitor is. It is not simple.

Wright: The crucial question is still whether olfactory stimulation is one molecule, one receptor, once, or whether one molecule can stimulate a great number of receptor cells. If it is the second, the gain doesn't have to be built in by any enzymic process and the whole emphasis shifts to a different kind of mechanism. The great need is for some experimental test to be devised that will distinguish between these two possibilities.

Davies: Is there something we could deduce from the time-scale of 12–30 milliseconds? Are there any enzymes known to react within this sort of time?

Beidler: We found that inhibitors don't work as fast as that on the taste response to applied stimuli, so the question is, where *does* the inhibitor work? For example if I put cyanide on the tongue, the receptors respond within 30 milliseconds. If you leave the cyanide on for 5 minutes, you have something quite different and the taste responses to stimuli are affected. The question arises, what did cyanide really interfere with? Was it a specific "taste" protein, or some step further along, or was it an effect on the general metabolism?

Zotterman: I can underline that, because I worked on the frog's tongue with an anti-thiamine substance; it completely abolishes the response to water, but it takes a few minutes to develop (von Muralt and Zotterman, 1952).

REFERENCES

Amoore, J. E., and Venstrom, D. (1967). In *Olfaction and Taste II* (Proceedings of the Second International Symposium, Tokyo, 1965), pp. 3–17, ed. Hayashi, T. Oxford: Pergamon Press.
Bowler, K., and Duncan, C. J. (1970). *Biochim. biophys. Acta*, **196**, 116–119.
Dontas, A. S. (1955). *J. Pharmac. exp. Ther.*, **115**, 46–54.
Duncan, C. J. (1964a). *Nature, Lond.*, **203**, 875–876.
Duncan, C. J. (1964b). *Naturwissenschaften*, **51**, 172–173.
Duncan, C. J. (1967). *The Molecular Properties and Evolution of Excitable Cells.* Oxford: Pergamon Press.
Heymans, C. (1955). *Pharmac. Rev.*, **7**, 119–142.
Joels, N., and Neil, E. (1963). *Br. med. Bull.*, **19**, 21–24.
Kimura, K., and Beidler, L. M. (1961). *J. cell. comp. Physiol.*, **58**, 131–139.
Martin, A. J. P. (1968). *International Symposium on Sensory Evaluation of Food*, Kungälv, Sweden, 1968. (Preprinted abstracts only.)
Muralt, A. von, and Zotterman, Y. (1952). *Helv. Physiol. Pharmac.*, **10**, 279–284.

ODOUR SIMILARITY BETWEEN STRUCTURALLY UNRELATED ODORANTS

M. G. J. Beets and E. T. Theimer

International Flavors and Fragrances (Europe), Hilversum, Holland; International Flavors and Fragrances (R & D), Union Beach, New Jersey

The process of olfaction is initiated by interaction of a population of odorant molecules with a population of molecular receptor sites on the membranes of the receptor cells.

Our present knowledge of the interaction process is limited but we know that it is based on a physical contact and we are fairly certain that it does not involve an irreversible structural change of the odorant molecule. In short, it is a special case of an absorption–desorption process and for this reason the resulting interaction pattern can be visualized as a complex assembly of transitory interaction complexes between single odorant molecules and individual sites, each of which may generate an elementary bit of information.

An abundance of evidence is available indicating that the characteristics of olfactory response are determined largely by the morphology of the odorant molecule as it is presented to the site. The significance of molecular shape, the wide variety of odorant structures and the analogy with other biological processes strongly suggest that the efficiency of olfactory interaction on the molecular level depends on the ability of the receptor site to accommodate, or to adapt its morphology to, the steric requirements of the arriving odorant molecule.

The total pattern of molecular signals or elementary bits of information generated on the olfactory membrane, the primary information pattern, is subjected to a number of processing steps in the receptor cells as well as in the glomeruli and the mitral cells of the olfactory bulb. The final product, the olfactory code message, is delivered to the brain for translation into patterns of association and recognition and, eventually, into a verbal expression of the odour sensation.

It is attractive to assume that the various processing steps lead to elimination of non-essential complexity and to retention, in some processed form, of all essential informational elements which are present in the primary pattern.

Olfactory interaction is, like any other absorption process, a statistical phenomenon. Consequently, even if the population of odorant molecules is structurally homogeneous, the primary information pattern must be highly complex since a population of identical molecules is presented to the sites in a variety of orientations and conformations which are distributed randomly over the various site types represented in the membrane.

Some types of odorant molecules, particularly those with a rigid structure and a sterically accessible polar group, can be expected, by analogy with a conventional absorption process at an interface, to be presented to the receptor membrane as a relatively homogeneous collection in which a single orientation predominates. The ensuing interaction in such cases will produce a primary information pattern consisting of one predominating type of information accompanied by other types occurring with lower frequencies. Since the type of informational structure is probably retained throughout the various processing steps we may assume that the olfactory code message and the resulting odour sensation also consist of a predominating component and a number of secondary components.

Odorants of this type with similar molecular shapes can be expected to produce information patterns and consequently odour sensations in which the same component predominates in different degrees. Such odorants belong to the same primary group, a concept defined by Amoore in 1952, and these can be said to have odours which are fundamentally similar. They are comparable to similar colours, in the spectrum of which the same frequency predominates. The analogy can be extended to similar colours with unrelated spectral composition. We have indications that such cases of pseudo-similarity occur also in olfaction.

Several techniques such as cross-adaptation (Le Magnen, 1942/1943, 1948; Cheesman *et al.*, 1953, 1956), electrophysiology (Døving, 1966, 1968) and psychological methods (Døving and Lange, 1967) are available for the detection of fundamental odour similarity.

Amoore (1967, 1968) and Amoore, Venstrom and Davis (1968) showed that specific anosmia can also be used for this purpose. Although specific anosmia may have a purely psychological basis, a more attractive explanation is that one or a few types of receptor site are lacking or not functioning properly in the anosmic membrane which, consequently, is unable to generate the predominating component in the information pattern of a primary odorant. Possible as this may be for a single site type, it is extremely improbable that the site types responsible for the gamut of secondary components are lacking also. For this reason we must assume that in cases where the anosmic person observes no odour at all, the frequency of occurrence of the secondary components in the information pattern is below the threshold required for the formation of an odour sensation.

On the other hand, anosmic persons with good olfactory acuity are able to observe the secondary notes. They describe the odour sensation, which usually has a much lower intensity level, in terms of key-words which are totally unrelated to those used for the predominating component. Guillot proposed the word parosmia for such cases. We may assume that specific anosmia and parosmia are both cases of the inability of the receptor sites to generate the predominating component in the olfactory information pattern.

On the basis of this reasoning, we may attempt to predict the types of reactions obtained from a panel evaluating a series of odorants belonging to the same primary class.

(1) Normal observers will describe the predominating component for all odorants in terms of the same or closely related key-words. Some of them will also be able to detect the secondary notes.

(2) A second group of panel members will be either anosmic or parosmic for all odorants. A few may show total anosmia for the whole collection but most of them will observe for all or some odorants a much weaker sensation which is described in terms of key-words which are unrelated to those used by the normal observers for the predominating component.

(3) Key-words used to describe secondary notes may vary considerably between odorants, either in nature or in frequency of occurrence. No significant difference can be expected between the collections of these key-words used by normal and parosmic observers.

I

II

III

Since these characteristics are probably indicative of fundamental odour similarity we have applied this method to a series of three ketones with structures I–III.

Of these, the steroid ketones I, Δ^{16}-androstenone-3 and II, $\Delta^{4,16}$-androstadienone-3 have been described by Prelog and co-workers (1945) as having the same characteristic urine odour. The monocyclic ketone III was synthesized by Beets and Van Essen (1969) and was found to have an extremely intense odour, very similar to that of the steroids I and II. The

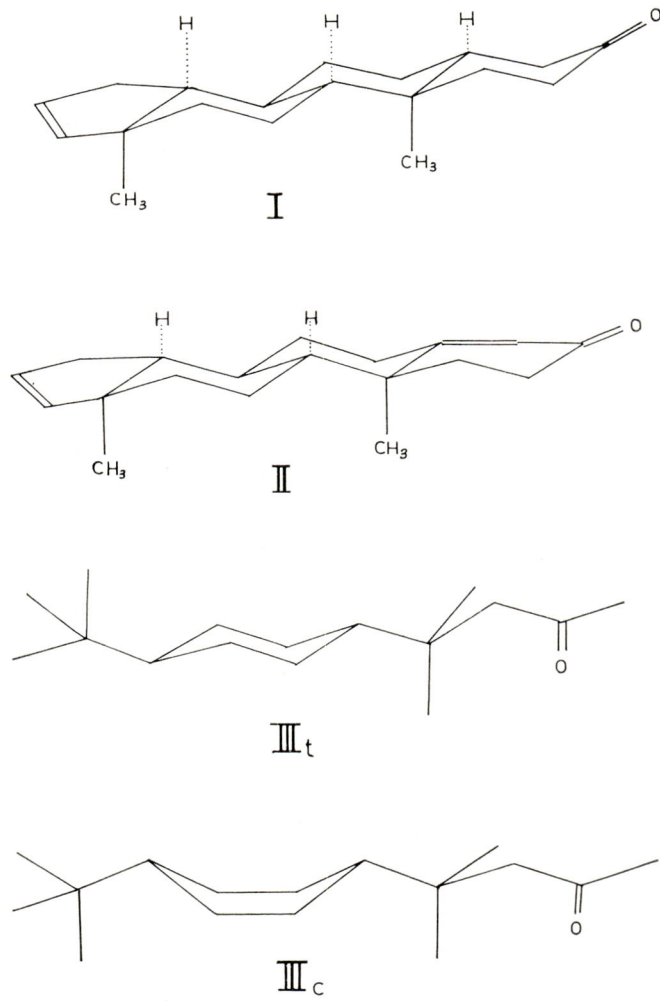

FIG. 1. Configurations of steroids I and II and of the *cis* (IIIc) and *trans* (IIIt) forms of the monocyclic ketone III.

question which interested us is whether the olfactory information pattern for the monocyclic ketone III is dominated by the same component as the patterns of I and II or, in other words, whether III belongs to the same primary class as I and II.

The configuration of the monocyclic ketone III used for this study has not been proved unambiguously. However, both possible stereomers, the *cis* (IIIc) and the *trans* form (IIIt), have very similar stretched shapes which are not too different from those of the steroid ketones I and II (Fig. 1).

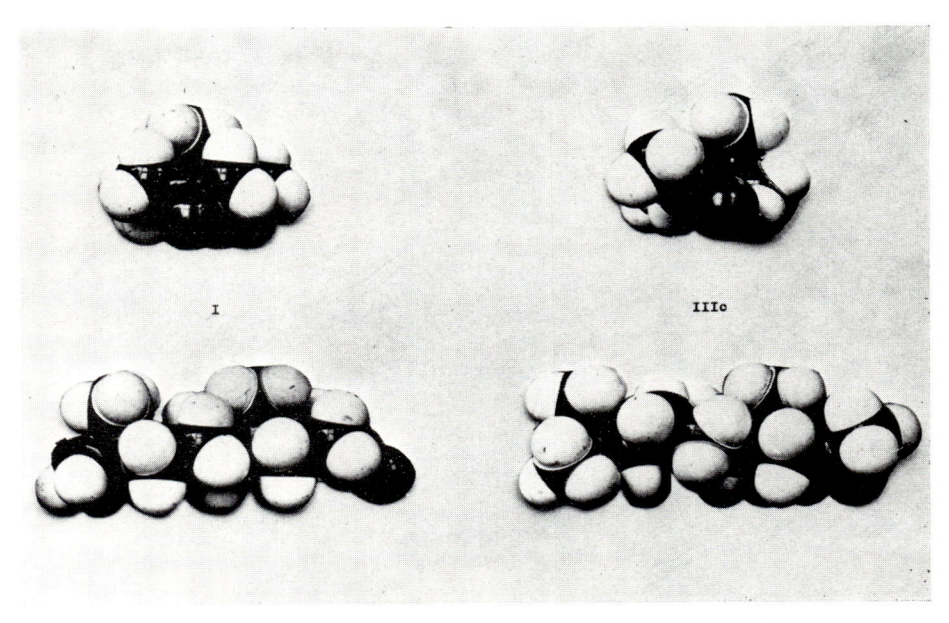

FIG. 2. Molecular models of steroid ketone I and *cis* configuration of the monocyclic ketone IIIc, seen in directions along (*upper*) and perpendicular to (*lower*) the dipole axis.

The molecular models of the steroid ketone I and of the *cis* configuration of the monocyclic ketone IIIc, shown (Fig. 2) in directions along and perpendicular to the dipole axis, demonstrate that the length of the molecules, as well as their profiles viewed from the latter direction, are rather similar. The profiles in a third direction, perpendicular to the ones shown in the picture, indicate that the steroid profile is somewhat bulkier than that of the monocyclic ketone. We may conclude that the profile similarity between the steroid ketones and the monocyclic ketone is sufficiently close to render fundamental odour similarity possible.

The three ketones were evaluated by a panel consisting of 100 members, 65 men and 35 women. A smelling strip which had been dipped in a dilute alcoholic solution of one of the ketones was presented, after evaporation of the solvent, to each panel member with the request to estimate the odour intensity and to describe the odour quality in terms of key-words. Evaluations of the several ketones by each panel member took place at large intervals. A small number of observers evaluated only the ketones I and III. These evaluations were included in view of the extremely close structural and odour similarities of the two steroids.

The results are summarized in Table I. Of the normal observers, 21 were

TABLE I
DISTRIBUTION OF RESPONSES TO THREE KETONES

Type	Observers	Men	Women
Normal	51	31	20
Anosmics	6	3	3
Parosmics	29	21	8
Inconsistent	14	10	4
Total	100	65	35

able to describe secondary notes.

The predominating odour component was characterized in practically all cases by the five key-words shown in Table II or by semantic equivalents. Many observers mentioned more than one descriptor to describe the odour of a substance. Table II also shows the frequency of occurrence of each key-word.

TABLE II
KEY-WORDS USED TO CHARACTERIZE ODOURS AND THEIR DISTRIBUTION

Key-words	Monocyclic ketone Total	Women	Men	Androstenone Total	Women	Men	Androstadienone Total	Women	Men
Urine	28	9	19	27	10	17	26	9	17
Perspiration	15	8	7	16	8	8	16	8	8
Animal	7	2	5	11	2	9	9	2	7
Amine	7	3	3	4	1	3	3	1	2
Sex	5	1	4	3	—	3	3	—	3

These frequencies show the same trends for the three ketones. The ratios between the frequencies found for the descriptors "urine", "perspiration" and for the total of the remaining key-words are roughly the same for the three odorants. However, male observers seem to have a significant preference for the key-word "urine" over "perspiration" which is lacking in the evaluations by the female group.

Only six panel members found all three ketones odourless. In the much larger group of parosmics, four showed total anosmia for androstenone and one for both steroids.

Fig. 3 is a graphic display of the major key-words used by normal observers and parosmics to describe the secondary notes. Although the frequency of occurrence for these descriptors varies widely between odorants, the descriptor patterns for the three ketones consist of the same key-words.

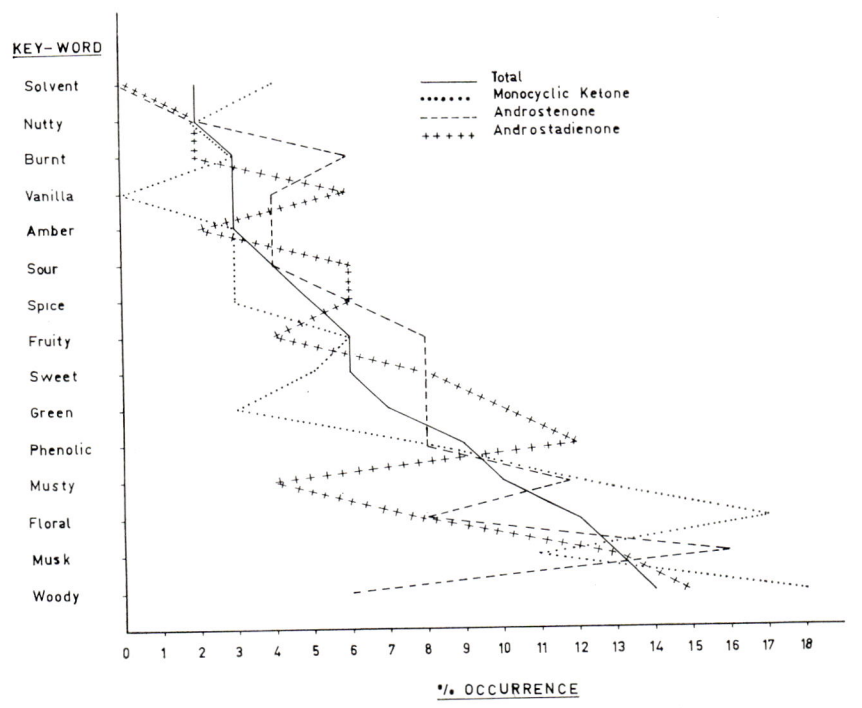

FIG. 3. Graphic display of key-words used to describe the secondary notes by normal observers and parosmics.

The latter certainly do not all represent primary types but the recurrence of the same descriptors for different odorants could mean that the number of receptor site types cannot be very large, a conclusion which is plausible for biochemical reasons also.

Fourteen per cent of the panel members did not belong consistently to any of the above-mentioned categories and presented descriptors characteristic of the predominating component only for one or two of the ketones. Key-words mentioned by this group for the secondary notes practically all belong to the collection shown in Fig. 3.

The reproducibility of these inconsistent evaluations has not yet been tested. However, since the evaluations obtained from 86 per cent of the panel members are consistent for the three ketones and agree with the classification predicted on theoretical grounds, the conclusion seems to be justified that the three ketones belong to the same primary group, or, in other words, that the intensive "urine" note in the odours of the three ketones is derived from the same component in the primary information patterns generated by their interaction with the olfactory membrane.

The panel test described here offers a convenient means to obtain a preliminary conclusion. Confirmation should be sought by application of the more elaborate methods mentioned earlier in this paper.

The striking and probably fundamental odour similarity between the steroid ketones and the monocyclic ketone suggests two lines of research.

In the first place, the role of profile similarity is not limited to olfaction and it is possible that various pharmacological characteristics of steroids can be reproduced more or less effectively by compounds of much simpler structure, such as the monocyclic ketone and its structural variants.

In the second place it may be interesting to investigate whether the incidence of anosmia for the "urine-perspiration" note is correlated with specific medical or psychological characteristics. Since anosmia is probably caused by structural variations in the proteinaceous olfactory membrane, it may have a genetic source. This, as well as the obvious link between the urine-note, steroid structure and human metabolism, suggests that anosmia for this note could have some diagnostic value.

SUMMARY

A monocyclic ketone is described of which the intense odour is closely related to the urine odour of certain steroid ketones reported in the literature.

A preliminary test with a large panel shows that the descriptor patterns as well as the cases of specific anosmia found for the monocyclic ketone and two steroid ketones coincide to a large extent. The stereochemistry and molecular shape of the three ketones, as well as the theoretical aspects of these phenomena, are discussed.

Acknowledgement

We wish to thank Syntex Research (Palo Alto) and Schering Corp. (Bloomfield, N. J.) for samples of steroid ketones and Miss L. Trip (Hilversum) and Miss M. R. McDaniel (Union Beach, N.J.) for their able assistance.

REFERENCES

AMOORE, J. E. (1952). *Perfum. Essent. Oil Rec.*, **43**, 321–330.
AMOORE, J. E. (1967). *Nature, Lond.*, **214**, 1095–1098.
AMOORE, J. E. (1968). In *Theories of Odor and Odor Measurement* (Proc. NATO Summer School, Istanbul, 1966), pp. 71–85, ed. Tanyolaç, N. Istanbul: Robert College.
AMOORE, J. E., VENSTROM, D., and DAVIS, A. R. (1968). *Percept. Mot. Skills*, **26**, 143–164.
BEETS, M. G. J., and VAN ESSEN, H. (1969). U.S. patent application 849.192.
CHEESMAN, G. H., and MAYNE, S. (1953). *Q. Jl exp. Psychol.*, **5**, 22–30.
CHEESMAN, G. H., and TOWNSEND, M. J. (1956). *Q. Jl exp. Psychol.*, **8**, 8–14.
DØVING, K. B. (1966). *Acta physiol. scand.*, **68**, 404–418.
DØVING, K. B. (1968). In *Theories of Odor and Odor Measurement* (Proc. NATO Summer School, Istanbul, 1966), pp. 493–508, ed. Tanyolaç, N. Istanbul: Robert College.
DØVING, K. B., and LANGE, A. L. (1967). *Scand. J. Psychol.*, **8**, 47–51.
LE MAGNEN, J. (1942/1943). *Année psychol.*, **43/44**, 249–264.
LE MAGNEN, J. (1948). *C.r. hebd. Séanc. Acad. Sci., Paris*, **226**, 753–754.
PRELOG, V., RUZICKA, L., MEISTER, P., and WIELAND, P. (1945). *Helv. chim. Acta*, **28**, 618–627.

DISCUSSION

Lowenstein: Were the descriptive terms suggested to your subjects in any way?

Beets: No, not at all. We leave them to say what they want.

Lowenstein: But you accept semantic substitutes, of course?

Beets: Yes, in order to streamline the collection of key-words. For instance, cedarwood and sandalwood are classified as woody.

Lowenstein: I noticed the sophisticated term "amine" in Table II. This term would only be used by a scientist.

Beets: Yes; less sophisticated people would say "ammonia" in this case.

Davies: The photographs of the molecular models of your steroid ketone I and the monocyclic ketone confirm that the length of the molecule is important in determining its odour, particularly in its ratio to the breath of the polar end. I would agree that, as for musks, these dimensions are important, and that the third dimension is not so important.

Beets: I can only say that the total profile, oriented with respect to the polar axis of the molecule, seems to be important. All dimensions, including the length, play their part in this.

Wright: I have recently completed a re-examination of the far infrared spectra of 47 compounds which have been certified, mainly by Dr Beets, as having a musk-like smell. The spectrum is in the range of wavelengths from 500 to about 100 cm^{-1}. I plotted the number of times a peak occurred in a moving band 7 cm wide. One sees concentrations of peaks which are statistically significant near 160, 255, 350 and 405 cm^{-1}. In the preliminary statistical evaluation, 38 out of 47 musks (81 per cent) showed a peak at

160 cm^{-1}, whereas out of 109 control substances only 35 per cent showed this peak. The ratio is approximately two. The same was found for the other three bands. One comes to the conclusion that the musk sensation is definitely associated with an infrared spectrum pattern depending upon the simultaneous presence of these four frequencies, bearing in mind that the spectra are by no means a perfect criterion, but the best available at present, of how the molecule is vibrating. The statistical significance is very high. One could suggest that some of the results described by Dr Beets and Dr Amoore may be due to the fact that some people lack the receptor for one particular frequency, let us say the one at 160 cm^{-1}. The pattern of the remaining three would be perceived as a different sensation. It would be very interesting to see whether the compounds which are rated by the parosmics as having this other type of odour conform to such a residual pattern. This would be a good test of the whole concept and would also provide a fairly simple numerical explanation of the kind of results that Dr Beets has described.

Amoore: On the relation between the urine and the musk descriptions of odours, I also have smelled these same steroid ketones (Δ^{16}-androstenone-3 and $\Delta^{4, 16}$-androstadienone-3). To me they are merely a rather cheap musk; I don't seem to experience this urine smell, so I am perhaps one of the parosmics. But I have found that about 4 or 6 per cent of people fail to perceive true musks such as pentadecalactone; and the one such person that I tested with these two steroids couldn't smell them at all. So I wonder if there is a close relationship between the urine odour and the musk odour.

Beets: The only thing I know on this is that among the secondary components of the information pattern of the odour sensation of these steroids, musk is represented, because the key-word "musk" was obtained fairly frequently from the parosmic observers. So it is clear that in at least some of these compounds the urine component and the musk component both occur. I don't think they are the same thing, mainly because you observed the steroids as a rather weak musk of low quality.

Amoore: They are quite weak for me. I am not a musk anosmic; I may be a urine anosmic.

Beets: That may be right. I would classify you as a parosmic for this class of odorants; that is, as an anosmic for the urine component. Normal observers are struck by the unique intensity of the latter. I have never experienced an odour of comparable strength.

Amoore: That must mean that two primaries at least are involved here, a musk primary and a urine primary, which I do include in my extended list of possible primary odours.

Moulton: We have recently made a preliminary investigation of the sensitivity of rats to the monocyclic ketone which Dr Beets studied. Eight rats were trained to detect both the cyclopentanone and metaxylene in an odour test apparatus until they achieved a stable performance. The rats were then transferred to the ketone. Two of them, however, were never able to achieve scores above chance. When these experiments were completed, seven of the rats (including the two which could not be trained on the odour of the ketone) were examined electrophysiologically. Single bulbar units which responded to amyl acetate were identified. The percentage of these units which also responded to the ketone ranged from 11 per cent in one animal to 50 per cent in another. Similar results were obtained with Δ^{16}-androstenone-3, both in this group and in a further group of 11 rats. All rats thus possessed units sensitive to amyl acetate and androstenone, although the number of units responding to amyl acetate was generally much higher (D. G. Moulton and N. Ai, unpublished observations).

Beets: Rats may be parosmic to these odorants, which means that you can expect only weak responses to the secondary components of the odour.

SPECIFIC PHYSICOCHEMICAL MECHANISMS OF OLFACTORY STIMULATION

R. H. WRIGHT[1] AND R. E. BURGESS[2]

[1]*British Columbia Research Council;* [2]*Department of Physics, University of British Columbia, Vancouver, Canada*

THERE is at present no adequate theory of the mechanism of olfactory stimulation. Proposed correlations between the intensity of the sensation and such factors as the volatility or solubility of the odorous material cannot really be termed theories, nor can the term be applied to correlations between the quality of the sensation and the configurational or vibrational properties of the molecules.

The essential prerequisite to an adequate theory is a secure correlation between some specific property or properties of the odorous molecule (the *osmic stimulus*) and the specific nature of the evoked sensation (the *olfactory response*). Without such a correlation we can do no more than speculate; given such a correlation (which must necessarily be quite empirical at the outset) the details of the process of stimulation can be explored.

The most fruitful point of entry into the problem is likely to be through a consideration of odour specificity. As Hainer, Emslie and Jacobson (1954) pointed out, the ability of the nose to discriminate clearly and quickly between closely similar odours implies an information-handling capacity of a high order and an informational content of more than 20 "bits" in any given odour sensation. This consideration is perfectly general and is sufficient to rule out many of the proposed correlations, such as those seeking to equate osmic specificity with molecular size or shape or the presence of particular structural elements. Moreover, it is now clear that a given odorous molecule normally presents a plurality of mutually independent stimulus qualities or primary osmic stimuli (Wright, 1968).

The existence of such discrete osmic characters was implicit in Dyson's (1937) suggestion that they are associated with the various vibratory motions of the molecule. This idea was made more specific by Wright (1954) when he gave reasons for supposing that the relevant vibrations are of low frequency and likely to have a "whole molecule" character.

From an examination of the common features in the far infrared spectra of families of compounds having generally similar odours, statistically significant correlations have now been established between certain combinations or patterns of vibratory frequency and the odours of musk, of bitter almond, and of cumin (Wright, 1967; Wright and Robson, 1969). Such correlations provide the necessary point of entry into a consideration of the process by which a specific osmic stimulus can generate an olfactory response of equivalent specificity.

The purpose of this paper is to trace the course of an odorous molecule as it moves into contact with the receptor organ and to examine the possible ways in which its vibrational character might enable it to exercise a selective stimulation and thereby impress its own specificity upon the organism.

TRANSPORT OF MOLECULES

It is generally agreed that substances are smelled in a state of molecular dispersion, either as gas molecules dispersed in air or dissolved in a watery, mucous or lipid medium. This means that the osmic character must be looked for at the molecular level. It follows that it is mass movement of the air that conveys the stimulus molecules into the nose, and that diffusion can play no significant part until the molecule reaches the immediate vicinity of the receptor surface. The diffusivity of a typical odorous molecule in air is about 0.1 cm^2 s^{-1}, making it a very slow process over any but the shortest distances. Electric fields in the vicinity of the nostrils cannot appreciably affect the transport of molecules which will be almost wholly un-ionized.

When the molecule reaches the interface between air and the watery film which may be presumed to cover the receptor cells, we are first concerned with the probability of adsorption on to the surface and then with the probability of an onward passage into the water. Given sufficient time, an equilibrium may be reached between the concentrations of the odorous molecules in the gaseous and liquid phases (Davies and Taylor, 1957; Laffort, 1963). However, it is by no means certain that equilibrium will indeed be attained in the time available. When the matter is considered kinetically rather than statically the time factor becomes significant in more than one way.

To cite a single example, when a molecule such as that of a carboxylic acid enters the aqueous phase it may be a candidate for electrolytic dissociation. This would not happen instantaneously but after a time, t_i, which depends upon the collision frequency and the activation energy. For a molecule of 0.2 nm (2 Å) radius in water at body temperature and assuming an activation energy of 0·4 eV (Gurney, 1953), t_i is about 10^{-6} s. If the molecule can pass from the air–mucus interface to the mucus–receptor surface in a time shorter than this it will probably arrive in the un-ionized state. The diffusivity, D, in water of a molecule of the size considered is of the order of 10^{-5} cm^2 s^{-1}, from which it appears that the transit time through a film 1 nm (10 Å) thick would be about 5×10^{-10} s. With a film ten times thicker the time would be increased 100-fold, which would still allow the majority of the molecules to arrive un-ionized. The actual thickness of the film has not been recorded, but it is likely to be fairly thin. The ultimate sensory elements are thread-like cilia less than 0·2 μm thick and as much as 200 μm long (Bloom, 1954; Gasser, 1956). The significant distance as it relates to diffusion is that from the air–mucus interface to the nearest parts of these cilia, the extremities of which may actually lie in the surface and be held there by interfacial tension.

Comparable in importance with the translational diffusion of the molecule is the rotational diffusion, for unless the molecule becomes firmly adsorbed on to the receptor the aspect it presents to the surface will vary as it rotates.

For a spherical molecule of radius 0·2 nm (2 Å) the mean time for rotation through an angle of one radian is 3×10^{-11} s (Debye, 1929). This is a minimum estimate because the viscosity of the mucus may be appreciably greater than water, and molecules are generally not spherical. Moreover, the forces between a polar molecule and a polar solvent complicate the viscous drag of the liquid. Finally, the rotation of the molecule may be composed of large angular jumps rather than a quasi-continuous diffusive rotation (Frenkel, 1954; Ivanov, 1964). The matter is important because it may set a lower limit to the vibrational frequencies likely to be osmically significant.

Using the value for the rotation time given above, and assuming that the molecule must execute at least 100 vibrational oscillations while turning through one radian if its vibrations are to interact significantly with a receptor, we have for the period of one oscillation 3×10^{-13} s, and for its frequency, 110 cm^{-1}. Experimentally it has not been possible to identify osmically significant frequencies lower than about 100 cm^{-1} (Wright, 1964).

ELECTRICAL DIPOLE OF THE MOLECULE

It has been conjectured (Müller, 1936) that the electrical dipole moment of a molecule could be relevant to its odour and Müller remarked that substances with zero dipole moment usually tend to have rather weak odours. Evidently the dipole moment cannot by itself confer the observed degree of odour specificity but it is not therefore to be wholly disregarded. As an un-ionized but polar molecule approaches the electrical double layer that exists in the nerve cell membrane, the first forces of interaction will be the electrostatic ones which can exert an orienting effect if the field is non-uniform and may become strong when the separation is small.

Monosubstituted benzene derivatives of the type C_6H_5X may provide an example of this. When X is —CN, —NO$_2$, —CHO, —COCH$_3$, —NCS or —I, the odour is strong, but when X is —CH$_3$, —OCH$_3$ or —NH$_2$, it is somewhat weaker and has a quite different quality. In the first group the substituent is at the negative end of the dipole and would probably be oriented in the opposite way to the members of the second group where the substituent is at the positive end of the dipole. This may be the physical basis of the role of the "functional group" in Beets' "profile—functional group" theory of the molecular basis of olfaction (Beets, 1964).

A problem mentioned but not resolved by Müller was that substances like benzene, naphthalene or *p*-dichlorbenzene, which have zero dipole moment, nevertheless have distinctive odours but often with high olfactory thresholds. In many cases such molecules have a non-zero quadrupole moment, and furthermore in a watery medium an induced dipole moment may be produced by the influence of the polar solvent (Smith, 1955).

The situation in polar solvents cannot yet be analysed precisely because both the dipole moment and the polarizability of the solvent molecules must be taken into account and the available models are still not satisfactory. What is clear is that a molecule in a polar solvent such as water has an apparent static dipole moment that may be greatly different from its value in the vapour phase. There will be corresponding effects of still undetermined magnitude on the dynamic properties of the dipole, including its vibrational frequency, damping and interaction with other dipoles.

MOLECULAR VIBRATION

A molecule regarded as an elastic structure can vibrate in ways and with frequencies that depend upon the masses and interconnexions of the atoms. Two experimental techniques are available for determining the characteristic vibration frequencies: Raman spectroscopy which depends upon the vibrational modulation of the polarizability, and infrared absorption which relates to the modulation of the dipole moment. For a given molecule, some lines are common to both spectra while others appear in one but not the other. Therefore it would be desirable to use both techniques in investigating the possible relevance of vibration to odour, but various purely practical difficulties stand in the way, especially if vibrations with frequencies as low as 100 cm^{-1} are to be studied.

Before 1962, when far infrared spectrophotometers first appeared on the market, Raman spectroscopy was the only available technique and it had only limited applicability because of the destructive photolytic effect on organic substances of the intense and rather short-wave radiation normally used to excite the spectra. Far infrared spectroscopy has therefore been the method of choice in recent years despite such disadvantages as the impossibility of recording spectra in aqueous solution. This is a serious defect because, as explained above, the vibrational frequencies in a polar solvent may be different from the values found in such non-polar solvents as benzene or heptane which are normally used in infrared spectrophotometry. This may explain why the vibrational correlations so far observed are not as precise and invariable as was originally expected. Very recently, Raman spectrographs using intense laser radiation of longer wavelength (632·8 nm, 6328 Å) have become available and it is to be hoped that both techniques can be applied in future.

If a single vibrational mode of a molecule is represented in terms of a harmonic oscillator, its energy (including the zero-point energy) in any assigned state is well-defined and constant in time, but the displacement together with such dependent properties as the dipole moment or polarizability are constantly varying and can only be defined statistically. The vibrations of real polyatomic molecules are not perfectly harmonic and the anharmonicities permit the appearance of overtone and combination bands in the spectra, arising out of the linear combination of fundamental bands. The spectral intensities of these are weaker than those of the fundamentals. In seeking correlations between the vibrational frequencies and the odorous character of a molecule it is entirely possible that overtone and combination bands which are weak or even absent from the spectra may nevertheless be significant because of the different nature of the interaction.

INTERACTING DIPOLES

If two vibrating dipoles are electrically coupled the interaction modifies the vibrations of both. The dipoles are assumed to be separated by a distance smaller than the wavelengths of interest which, for frequencies of 100 to 500 cm^{-1}, are 2×10^4 to 10^5 nm, so that the coupling is without retardation and is inversely proportional to the cube of the separation. A classical analysis of their interaction suggests that the vibrational energies and frequencies are modified because of the coupling (Hirschfelder, Curtiss and Bird, 1954). The effect is most pronounced close to resonance and for a given magnitude of effect the necessary coupling is smaller the closer molecular vibrations are to resonance. By a change of notation the same considerations are applicable to two molecules which have no dipole moment and interact by virtue of their polarizability. Interactions via the polarizabilities are generally weaker than the interactions via the dipole moments but are not negligible in either case.

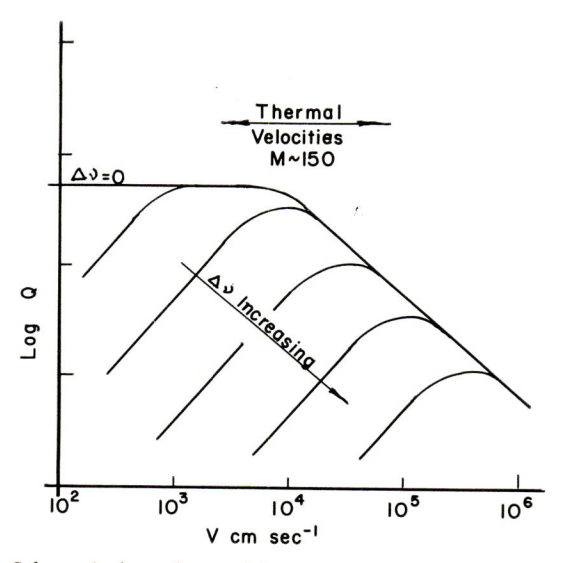

FIG. 1. Schematic dependence of the "cross-section of vibrational transfer" Q between two molecules with vibrational frequency difference $\Delta \nu$ and relative velocity V.

In view of this conclusion it is immediately important to ask how resonance of the vibrational frequencies of two molecules could result in a stimulation of the receptor.

In the case considered above, the two dipoles were assumed to be at a fixed distance apart. If they are in relative motion both the difference of their vibrational frequencies, $\Delta \nu$, and their relative velocity are important. The larger the relative velocity the sharper is the interaction, and the smaller the distance of closest approach the stronger becomes the interaction. The probability of a vibrational energy exchange by dipole–dipole interaction depends inversely upon the difference in the vibration frequencies of the two dipoles. Thus as one molecule passes another the significant parameters are the relative velocity, the distance of closest approach and the difference in vibrational frequency, and the interaction depends on all three.

Various authors have considered models for resonant dipole–dipole interaction and have derived the probability of vibrational energy transfer (Mahan, 1967; Stephenson, Wood and Moore, 1968; Yardley, 1969). In the resonant case the probability is constant at small velocities and inversely proportional to the velocity for large velocities. In the off-resonance condition, the probability of the exchange diminishes at low velocities and displays a maximum at a velocity which is larger the greater is Δv. Fig. 1 shows the general interrelationship of the variables. For molecules moving with thermal velocities the net interaction averaged over the velocity distribution is a rapidly decreasing function of Δv, as shown in Fig. 1.

LIMITING OSMIC FREQUENCIES

It was shown above that the lower limit of vibrational frequency likely to be osmically significant probably lies somewhat below 100 cm^{-1}. An upper limit in the vicinity of 500 cm^{-1} was originally deduced by using the Planck formula to calculate the mean thermal energy of a molecule in its various vibrational modes (Wright, 1954). The calculation ignored the zero-point energy and when this term is included the mean energy *increases* with increasing frequency so that there is no "high-frequency cut-off" to be deduced from the total vibrational energy. Yardley's calculation of the resonant interaction between relatively moving dipoles depends upon the molecules being vibrationally excited to different degrees. In the simplest case this would call for one to have one quantum of vibrational energy and the other to have only the zero-point energy. The probability of this is similar at high frequency ($hv > kT$) to that given by the Planck formula used originally by Wright (1954), so that his upper cut-off frequency appears to be substantially correct though not for the reason originally assumed (see Appendix, p. 336). A study of the far infrared spectra of several families of odorous substances has so far revealed no substantial correlations higher than about 400 cm^{-1}, which is in reasonable accord with the theoretical prediction based on a high-frequency limit given by $hv = 2kT$.

THERMODYNAMIC CONSIDERATIONS

A frequently heard objection to the vibrational hypothesis is that the energy in a low-frequency oscillation is very small and of no more than thermal magnitude, and therefore it is incapable of generating any specific response or of being registered selectively by any kind of receptor system. Colloquially, this is equivalent to the proverbial saying that "in the dark all cats are black". This matter requires close scrutiny.

It may always be argued with validity that for an equilibrium system of molecules and receptors, energy transfer to the receptor necessarily occurs as often as energy transfer from the receptor. Also, the distribution of the quantum numbers for the receptor is unaffected by interaction with the odorous molecules and is determined solely by the equilibrium statistics.

Now although the occupation statistics are fixed in this way, the rate at which the transitions occur between the states is not. In particular an odorous molecule could make its presence felt by providing resonant energy transfer to (and also from) a fairly large number of receptor modes in a very short time-interval. This close coincidence in the times of the transitions between states could furnish the necessary signal of the presence of certain primary osmic stimuli. The probability of a large number of modes becoming excited (or de-excited) simultaneously by random excitation would be extremely small.

The receptor surface consists of a mat of interwoven thread-like "olfactory hairs" or cilia. A molecule with a diffusivity of 10^{-5} cm^2 s^{-1} would traverse a root mean square distance of about 1 μm along this mat in 5×10^{-4} s. If, as is probable, the interaction time for energy transfer with a receptor is much shorter, of the order of 10^{-11} s, it could cause the discharges from a considerable number of receptor cells to overlap in time closely enough to be identified by synaptic processing as a "coincident event" associated with the passage of a particular type of molecule.

Thermodynamic considerations are over-riding for the macro-system and the long run, but have no relevance to the micro-system in the short run. Thus the frequency and distribution in time of departures from the long-run average are outside its scope but are nevertheless highly relevant to any processes involving individual molecules and receptors.

It is interesting at this point to consider in what respects a non-equilibrium model would exhibit essentially different features. Suppose, for example, that the stimulus molecules are assumed to be in equilibrium at a temperature, T_m, while the receptors have a vibrational mode occupation determined by a different temperature, T_r. Note, however, that a high value of T_r does not imply that the receptor is "hot" in the usual sense, but rather that the distribution of its vibrational quantum numbers is more nearly uniform than would be the case if T_r were the same as T_m. Such a situation could arise from a biochemical "pumping" mechanism which supplies energy continuously in order to maintain a distribution different from that appropriate to an ambient temperature of 37°C. There is a reduction in the probability of energy transfer to the receptor when T_r is greater than T_m, but the reduction is comparatively small even when T_r considerably exceeds T_m.

DISCUSSION

The fact that an interaction between an odorous molecule and a receptor site can be vibrationally initiated in a specific way still leaves open the actual nature of the process of nerve excitation. This is, of course, inevitable in view of our very incomplete knowledge of the detailed structure of nerve cell membranes, and the same consideration would apply with equal force to any of the current attempts to develop a detailed model of olfactory stimulation. Some speculation is nevertheless justified because it may suggest new and possibly fruitful lines of enquiry.

The currently accepted picture of the nerve cell wall is of a gelatinous membrane of highly complex chemical composition, probably including phospholipids, proteins and carotenoid substances as well as pigments and other components of quite unknown constitution. Moreover, as part of a living structure it is the site of an unknown number of dynamic processes. It is bathed on both sides with electrolytically conducting solutions and as a consequence of its differential permeability to different ionic species there is an electrical double layer positive on the outside and negative on the inside. The first step in the initiation of a nerve impulse is generally thought to be localized discharge of the membrane potential consequent upon an increase in the overall ion permeability of some part of the cell wall so as to allow the comparatively free passage of sodium and potassium ions through it. This would imply the creation of "holes" by some kind of opening up of the membrane structure.

For such an opening to be related to the vibrational specificity of the stimulus molecules the wall of each receptor would be dotted with sites sensitive or "tuned" to respond to a certain rather narrow band of frequencies in the stimulus. The response of such a site to a passing odorous molecule might take any of several forms, alone or in combination. Until more is known about the detailed structure of the membrane it is probably idle to speculate farther. A possible mechanism for the vibrationally stimulated de-excitation of a metabolically "pumped" system was proposed some years ago and remains as one of several possible models (Wright, Reid and Evans, 1956).

Given that the osmic specificity of a particular molecule is defined by a particular pattern of vibrational frequencies, the manner in which an analogue pattern is generated within the organism is probably somewhat as follows.

We know that the primary receptor cells in the olfactory epithelium are connected to the olfactory bulb by parallel neurons without synaptic interruptions (Le Gros Clark, 1956). The structure might be compared to an array of push-buttons of several different kinds, each distinguished by a different colour. If pushing any one or more of the "red" ones turns on a "red light" at some central point, and if pressing a "blue" one turns on a "blue light" there, and so on, then with six colours we could produce 64 different colour combinations because each colour may be either on or off and 6 on/off choices can be made in 64 different ways. If instead of six differently coloured push-buttons we had twenty, the number of combinations would be 2^{20}, which is just over a million. This was Hainer's suggestion of how the nose achieves its amazing power of discrimination (Hainer, Emslie and Jacobson, 1954). The coloured push-buttons correspond to primary receptor cells in the olfactory epithelium tuned to respond to certain vibrational frequencies in a stimulus molecule. As the molecule skips over the array of push-button receptor cells it touches all kinds but only brings about discharges in those which are in rapport with it. Collisions with other kinds of primary cells have no effect. Therefore, somewhere in the olfactory centres of the brain the "colours" of the stimulus molecule are represented by a sort of "lighting up" of an equivalent pattern which is then scanned by the central nervous system and eventually passed on to the consciousness as a particular sensation, such as "musk" or "bitter almond".

With a mechanism like this the stimulus molecule does not have to penetrate the cell membrane or interact with it chemically. A momentary approach during which a vibrationally specific interchange of energy takes place is all that is required. This makes the interaction purely physical and leaves the stimulus molecule free to go on to make contact with many other receptors of all kinds. This is an important point for two reasons. The first is that it makes the amazing sensitivity of the nose much easier to understand if one stimulus molecule can stimulate many receptor cells. The second is even more important. All the primary odour qualities of a molecule must be registered if its special quality is to be recognized. As it moves over the surface it can register each of these qualities many times in a way that would not be possible if one molecule could stimulate only one receptor only once. It is well known that odours tend to retain their distinctive character even at very low intensity levels and this can best, and perhaps only, be explained in the way we have described. It is for this reason that we have come to believe that olfactory specificity is the too-long-neglected clue to a proper understanding of the phenomena of smell perception.

SUMMARY

The specificity of odour sensations and our ability to recognize subtle differences in odour quality imply a corresponding specificity in the stimulus molecules. The accumulating evidence for a correlation of distinctive odours with specific combinations of low-frequency molecular vibration opens the way for an enquiry into the way an odorous molecule might interact with a complex of receptor cells so as to impress its specificity on the organism as a distinctive odour. A detailed application of physico-chemical principles shows that the vibrational properties of a molecule are capable of imparting the requisite degree of specificity to several kinds of interaction and the upper and lower limits to the scale of "osmic frequencies" prescribed by the theory coincide with those found by experiment. A permanent or induced dipole may help to orient the stimulus molecule with respect to the polarized membrane of the receptor nerve, and the vibrational modulation of the moment could govern the transfer of energy between stimulus and receptor in several ways. The differential stimulation of selectively sensitive receptors then enables the stimulus to impress its individuality upon the organism and the essentially physical nature of the interaction makes possible the high sensitivity of the olfactory sense.

Acknowledgement

This investigation has been made possible by a grant from the Donner Canadian Foundation, and for this we offer our sincere thanks.

REFERENCES

BEETS, M. G. J. (1964). In *Molecular Pharmacology: the Mode of Action of Biologically Active Compounds*, vol. 2., pp. 3–51, ed. Ariëns, E. J. New York: Academic Press.
BLOOM, G. (1954). *Z. Zellforsch. mikrosk. Anat.*, **41**, 89–100.
DAVIES, J. T., and TAYLOR, F. H. (1957). *Proc. II Int. Congr. Surf. Activ.*, vol. 4, pp. 329–340. London: Butterworths.
DEBYE, P. (1929). *Polar Molecules.* New York: Chemical Catalog Co.
DYSON, G. M. (1937). *Perfum. essent. Oil Rec.*, **28**, 13–19.
FRENKEL, Y. A. (1954). *Kinetic Theory of Liquids.* New York: Dover.
GASSER, H. S. (1956). *J. gen. Physiol.*, **39**, 473–96.
GURNEY, R. W. (1953). *Ionic Processes in Solution*, chapters 7 and 8. New York: McGraw-Hill.
HAINER, R. M., EMSLIE, A. G., and JACOBSON, A. (1954). *Ann. N.Y. Acad. Sci.*, **58**, 158–174.

HIRSCHFELDER, J. O., CURTISS, C. F., and BIRD, R. B. (1954). *Molecular Theory of Gases and Liquids*, chapter 13. New York: Wiley.
IVANOV, E. N. (1964). *Soviet Phys. JETP*, **18**, 1041–1045.
LAFFORT, P. (1963). *C.r. hebd. Séanc. Acad. Sci., Paris*, **256**, 5618–5621.
LE GROS CLARK, W. E. (1956). *Yale J. Biol. Med.*, **29**, 83–95.
MAHAN, B. H. (1967). *J. chem. Phys.*, **46**, 98–101.
MÜLLER, A. (1936). *Perfum. essent. Oil Rec.*, **27**, 202–205.
SMITH, J. W. (1955). *Electric Dipole Moments*, chapter 5. London: Butterworths.
STEPHENSON, J. H., WOOD, R. E., and MOORE, C. B. (1968). *J. chem. Phys.*, **48**, 4790–4791.
WRIGHT, R. H. (1954). *J. appl. Chem., Lond.*, **4**, 611–615.
WRIGHT, R. H. (1964). *Ann. N.Y. Acad. Sci.*, **116**, 552–558.
WRIGHT, R. H. (1967). *Perfum. essent. Oil Rec.*, **58**, 648–650.
WRIGHT, R. H. (1968). In *Theories of Odor and Odor Measurement* (Proc. NATO Summer School, Istanbul), 1966, pp. 459–478, ed. Tanyolaç, N. Istanbul: Robert College.
WRIGHT, R. H., REID, C., and EVANS, H. G. V. (1956). *Chem. Ind.*, **37**, 973–977.
WRIGHT, R. H., and ROBSON, A. (1969). *Nature, Lond.*, **222**, 290–292.
YARDLEY, J. T. (1969). *J. chem. Phys.*, **50**, 2464–2466.

APPENDIX

PROBABILITIES OF VIBRATIONAL STATES OF HARMONIC OSCILLATORS

The probability for the quantum number n of a harmonic oscillator of frequency ν in equilibrium with a heat reservoir at absolute temperature T is

$$P(n) = e^{-nx}(1-e^{-x}) \qquad n = 0, 1, 2, \ldots$$

where $x = h\nu/kT$ (h = Planck's constant, k = Boltzmann's constant). The expectation value of n is

$$\bar{n} = \Sigma n P(n) = 1/(e^x - 1)$$

where the mean energy of the harmonic oscillator is

$$\bar{E} = (\bar{n} + \tfrac{1}{2})h\nu = \frac{h\nu}{2}\coth\frac{x}{2}$$

The zero-point contribution $\tfrac{1}{2}h\nu$ has been included and is the value of the mean energy when $h\nu$ is large compared with kT; that is, at high frequencies. At low frequencies ($h\nu \ll kT$) the classical value $\bar{E} = kT$ is obtained. This is probably the range of frequencies of major relevance for olfactory stimulation. The mean energy is a monotonically increasing function of both frequency and temperature.

The probability that n exceeds some chosen value n_0 is

$$P(n > n_0) = \exp[-(n_0+1)x]$$

Now consider two harmonic oscillators in thermal equilibrium and whose frequencies are sufficiently close that a single value can be taken for the parameter $x = h\nu/kT$. The probability that the vibrational quantum n_1 of oscillator 1 is greater than that (n_2) of oscillator 2 is

$$Q = P(n_1 > n_2) = 1/(e^x + 1)$$

This probability Q decreases steadily from the value $\frac{1}{2}$ as x increases from zero; that is, for increasing frequency.

We may now identify oscillator 1 with the odorous molecule with a vibrational frequency ν and oscillator 2 with the receptor having a vibrational band with the same or very nearly the same frequency. One prerequisite for energy transfer from 1 to 2 is that $n_1 > n_2$ and the calculated probability Q for this condition is the one factor in the net probability for energy transfer which involves temperature through the equilibrium statistics.

For body temperature ($T = 310$ K) the probability has the values 0·27 at 200 cm^{-1}, 0·05 at 650 cm^{-1} and 0·015 at 800 cm^{-1}.

This decrease of probability with increasing vibrational frequency is greatly accentuated if it be further required that more than one receptor with a vibrational frequency ν independently satisfy the condition $n_1 > n_2$ and that each such receptor is excited by interaction with a molecule within a short interval of time. If R is the number of receptors the probability that $n_1 > n_2$ for all of them is Q^R.

Even if R is a small number (say 3) the probability Q^R decreases rapidly with increasing frequency especially for frequencies above 500 cm^{-1}.

This type of correlated stimulation of receptors in any one osmic band may well be required by the olfactory nerve system for providing confirmatory signals to distinguish the presence of an odorous molecule from random excitations.

Furthermore a given molecule may be characterized osmically by 3 or 4 vibrational bands and the probability of the simultaneous stimulation of R receptors for each band by random thermal excitation becomes very small.

DISCUSSION

Martin: Why does black-body radiation not give a colossal background to all the senses all the time? Actually it is not dark to infrared in your wavelength range of 3–50 μm. There is 0·5 kW m^{-2} of radiation in this range at 300 K with an energy peak at 10 μm.

Wright: This is a black-body situation. You are inside an enclosure at uniform temperature, which is traversed by radiation, and in those conditions no special pigmentation will let anything stand out against the background.

Martin: Then putting in a molecule of an odorant will not make any difference.

Wright: It doesn't make any difference to the energy, but if you have a molecule in the receptor surface which is in the ground state, it can go up into the first level of vibrational excitation and down again. Left to itself it will go up and down with a certain random frequency. But if a passing molecule can exchange one quantum of vibrational energy and kick it up a level, that molecule will move on and after picking up another quantum of vibration from contact with the water, it will kick up another receptor, and another and another. It is true that it may equally help one receptor molecule to go down an energy level, but if the operative thing is for it to go up, if this happens in a great enough number of receptor cells within a short enough time this may count as a simultaneous event for the nervous system and the fact that a lot of these changes have happened simultaneously, or nearly so, may be registered centrally by the organism as something unusual, and so you get a stimulation.

Martin: The frequency of changes of level will be as unchanged as the energy.

Lowenstein: You postulate neural convergence, then, Professor Wright?

Wright: Presumably all the "red" buttons tend to converge on one cell body in the olfactory bulb.

Martin: Have you done the experiment of irradiating the nose with particular wavelengths of light?

Wright: I would love to do it! It's a matter of considerable difficulty because the olfactory cleft is so inaccessible, but surgical procedures are occasionally used in human cancer in which one eye is removed, enabling you to see into the olfactory surface. When we hear of such a patient we want to irradiate the surface with light and see whether this weakens, strengthens or otherwise modifies the patient's ability to smell. If properly carried out this surgical procedure apparently does not interfere with the ability to recognize odours. If you could illuminate the epithelium you might in a sense invert the usual population, and you might produce something in the nature of a laser action.

Pfaffmann: The preparation of the frog olfactory epithelium is usually open to view and people recording there should be able to determine the effect of illumination on the electrophysiological response. Is there any evidence on this, Professor Beidler?

Beidler: I don't know of any. We did not observe fluorescence on opossum olfactory epithelium when odours were applied in the dark.

Wright: That was to test for the opposite effect. At that time we thought it might be a triggered de-excitation of a metabolically excited vibrational state.

Beidler: Wouldn't the long wavelengths you are using be absorbed by water?

Wright: Yes, they would, but visible radiation would also invert the usual pattern of excitation.

Martin: It would be possible to introduce a small gallium arsenide light source on the end of a pair of wires.

Beets: Dr Wright mentioned dipole moments in connexion with odour. We have determined the dipole moments for several musks of different odour strength, in order to find whether there is a correlation between the size of the dipole moment and the odour intensity, and we found none at all.

Wright: What was the solvent? It would be most interesting if the solvent was water.

Beets: These dipole moments were determined in carbon tetrachloride.

Davies: One of the obvious ways to test this theory of the vibrational spectrum is to use deuterated derivatives of compounds and see if they smell the same as the normal compound, because if you substitute deuterium for hydrogen on some part of the molecule the far infrared spectrum is shifted by between 50 and 500 cm^{-1}. This has been done with insects (Doolittle *et al.*, 1968), without very much success in correlating the spectrum with the insect's response; it would be interesting to try this in vertebrates.

Wright: This work on insects was done by R. E. Doolittle and colleagues in the U.S. Department of Agriculture with compounds that attract the Melon fruit fly (*Dacus curcurbitae*, Coq.). The compounds used were as nearly as possible completely deuterated or deuterated in various places. The maximum change in the peak frequency was only 10 cm^{-1}, but the bands I have found are considerably wider than that, so it merely changes the position of the peak within a band. I know only one substance with a characteristic odour whose vibrations are significantly changed by deuteration, namely naphthalene. Ordinary naphthalene has peaks at 183 and 363 cm^{-1}; the fully deuterated compound has peaks at 169 and 331 cm^{-1}. These shifts are enough to be noticeable. An experiment was set up as follows. A panel of six people were presented with three bottles, two with ordinary naphthalene and one with deuterated naphthalene, and asked if they could detect any differences. Of the six people, four successfully recognized the deuterated compound as different. One could find no difference and the sixth spotted one of the ordinary naphthalenes as being different from the other two. This is the only compound I know where deuteration makes enough difference to affect the odour.

Beets: I have heard that Givaudan research in the United States has made some deuterated odorants. You may write to them in order to obtain information on their odour properties. We have never done experiments on deuteration but replacement of one hydrogen atom by fluorine in dimethyl phenylethylcarbinol does not produce a considerable change (E. T. Theimer, personal communication).

Wright: Of course it would depend on where it went in.

Beets: I doubt it, but you may be right.

Davies: Fluorination can make a big change in odour; 1 : 1 dihydroperfluorobutanol, for example, doesn't smell like butanol at all; it smells more like camphor.

Beets: This is for small molecules, much smaller than the one I mentioned.

Martin: To turn to the question of the response of the receptor to the stimulus, your vibrational hypothesis has not solved the problem of the difficulty of stimulating one cell with one molecule, for if your multiple contacts were all on different cells, there would still be the problem of exciting each individual cell.

Wright: If a number of primary receptors feed into a single glomerulus or other centre in the olfactory bulb, a signal would go through when any one or perhaps some minimum number "fired" in some time-interval short enough to be processed as a simultaneous event.

Mac Leod: Dr E. Priesner of Schneider's group in Munich (personal communication) is trying to record the electrophysiological response of nerve units in the antenna of *Bombyx mori* to bombykol, and he has reached the one molecule, one receptor, once, situation; that is, he records one spike every time one molecule of bombykol reaches the right place on the antenna. So in this case of a "specialist" receptor, you can say that "one molecule once" is true.

Wright: Is there any indication of whether that molecule is then released to go on, or is it used up in the process?

Mac Leod: He cannot have any evidence about that from these techniques, but actually there is only one spike.

Wright: That is very nice. I am glad to know that it is one spike and not a whole train, because if there was a train it would be very difficult to see how intensity could be coded. If intensity is coded by the number of spikes, you get a very simple representation of the whole stimulus.

Mac Leod: In the case of the lowest thresholds in man or other vertebrates there is certainly no more than one molecule per receptor, but then many thousand molecules are present simultaneously. So it is not necessary that one molecule hits many receptors in order to make the substance effective in different receptors simultaneously. There is only one case which has been reported by Teichmann (1959) in the eel (*Anguilla anguilla* L.), where one molecule was present at one moment in the olfactory sacs, and the eel responded. It could be only an intensity response and not a quality response.

Beidler: How sure are we, Dr Mac Leod, that a one molecule–one receptor process exists in vertebrates? The notion partly derives from a calculation of the number of molecules in a sniff, divided by the total number of receptors, and since the number of receptors is large, the number of molecules per receptor naturally is small. It may be that many odour molecules are already adsorbed by the first receptors they meet.

Mac Leod: In general this is true but certain substances, such as vanilline or butyric acid, are so effective that you can think of one molecule for one receptor being enough to stimulate.

Beidler: But here again the reason that the sensitivity per receptor appears so high is because it is assumed that the molecules are equally distributed over a large number of receptors.

Andres: From our work on the glomerular structure of the vertebrate olfactory bulb there are signs that Professor Wright's speculations are correct, because the cross-sections of the olfactory fibres differ (Andres, 1965). Some fibres have one or two microtubules, and others go up to 18 or 20 microtubules. In some glomeruli most of the fibres have four microtubules; in other glomeruli most have seven or ten. In the vomeronasal organ of the rat, which conducts another spectrum of odours to the accessory olfactory bulb, one finds many more fibres containing ten or 15 microtubules than in the normal olfactory nerve. It will perhaps be possible to correlate the fibre type with the cell type. I should add that we do not find that one glomerulus receives only one fibre type; it is a statistical distribution of fibre sizes.

Wright: These are very interesting results, and it is reasonable that the correlation should be a statistical one. I keep thinking back to the paper by Hainer, Emslie and Jacobson (1954) becuse it lays down boundary conditions that any theory has to meet. The idea is that primary receptors of a certain type feed into one filter centre, and receptors of another type feed into another filter centre, so that an analogue to the external pattern can be created internally; and that is consistent with what you say.

Andres: Unfortunately my concept of fibre differentiation is not yet sufficiently proved. For example, I have to take receptor moulting (see p. 248) into consideration. I think there has to be a corresponding change in the olfactory fibres. There are other problems as well here.

Lowenstein: Nevertheless it seems likely that statistically the same sort of distribution would be recovered in the glomerulus.

Andersen: I would like to make a general comment. I have been increasingly uneasy that the impression exists that everything happens at the periphery. However, the higher levels of the central nervous system may also have something to do with the perception of taste and odours! I would like to submit that higher levels of the central nervous system not only receive and process information, but that the central nervous system also sets the bandwidth, let us say, of the secondary neurons, or even of the receptors themselves. We know for the stretch reflex, for instance, that if you relate the firing pattern of the primary neurons to the stretch of the muscle, this is a very sharp and well-defined signal. If you look at it at the level of the secondary neurons the signal is messed up terribly, but all the information is there to be relayed further on and integrated with a feedback to peripheral elements, giving a well-defined relationship between the primary neuron response and the final output on the motor side. I would like also to submit that amplification as well as inhibition may take place at higher levels, not only in peripheral enzyme systems as suggested by Dr Martin.

Lowenstein: This unsharpening of the information in the intermediate stages of transmission to the higher centres is also very characteristic of the auditory pathway (Whitfield, 1967).

REFERENCES

ANDRES, K. H. (1965). *Z. Zellforsch. mikrosk. Anat.*, **65,** 530–561.
DOOLITTLE, R. E., BEROZA, M., KEISER, I., and SCHNEIDER, E. L. (1968). *J. Insect Physiol.*, **14,** 1697.
HAINER, R. M., EMSLIE, A. G., and JACOBSON, A. (1954). *Ann. N.Y. Acad. Sci.*, **58,** 158–174.
TEICHMANN, H. (1959). *Z. vergl. Physiol.*, **42,** 206–254.
WHITFIELD, I. C. (1967). *The Auditory Pathway.* London: Arnold.

ELECTRICAL SIGNS OF OLFACTORY TRANSDUCER ACTION

DAVID OTTOSON

Department of Physiology, Kungl. Veterinärhögskolan, Stockholm

It is generally agreed today that the particles of an odorous substance have to enter into direct contact with the olfactory receptors in order to excite them. In discussing the mechanisms of olfactory transducer action we can therefore leave out the various radiation theories which not long ago were still flourishing.

A fundamental problem in relation to the excitatory process is that of the actual site of receptor activation. Since the initial contact between the odorous particles and the sensory cells undoubtedly takes place at the surface of the olfactory cilia it seems natural to assume that these structures represent the transducer elements. The cilia form a dense network at the surface of the mucous layer (cf. Reese, 1965) and are separated from the air in the nasal cavity by only a very thin film of watery material, through which the stimulating particles have to pass before contact is established. It seems most likely that most of the particles impinging upon the mucosa are trapped in the meshwork of the cilia close to the surface of the mucous layer and that only a relatively small proportion penetrates to deeper layers. If the transducer action occurred in deeper structures like the olfactory vesicles or in the microvilli of the sustentacular cells (cf. Takagi, 1967) only a small fraction of the stimulating agent would reach the actual receptor sites. The olfactory receptors are characterized by their sensitivity; this high sensitivity would seem most unlikely if the greater amount of the odorous material was captured and inactivated by the cilia and only those particles which reached the vesicles or microvilli produced excitation. There is also more direct evidence that transducer action takes place in the cilia. As will be discussed later, recordings of the electrical response of the olfactory membranes strongly suggest that the primary reaction between the stimulus and the olfactory receptors occurs in the outer layer of the mucosa.

The cilia provide for a considerable increase of the receptive surface of the olfactory organ. The dimensions of the cilia in man are not known but if we assume that they are of the same order and number as in the frog the chemosensitive membrane at each side of the nasal cavity would have an area of 30–50 cm². In most mammals the regions lined with sensory cells are considerably larger than in man and their total receptive olfactory membrane therefore is correspondingly greater. It would appear that the high olfactory sensitivity of many mammals may to a great extent be attributed to the extensive size of their active olfactory membrane.

Experimental and anatomical evidence thus leads to the conclusion that the primary process of energy reception occurs in the cilia. It would appear therefore that the solutions of the many challenging problems which we are faced with concerning the functions of the receptors have to be sought for in the structure and functional properties of the membrane of the cilia.

The first stage in the excitatory process can be assumed to involve the absorption of the molecules of the stimulating substance on the chemosensitive membrane. This stage may include an orientation of the molecules so that they fit into certain sites of the membrane; it is followed by an interaction between the molecules of the stimulating agent and the macromolecular system of the sensory membrane. At the present time there appears to be no means by which we can directly study these events. Although the primary stages of the transducer process are inaccessible to direct analysis we still have some possibilities of obtaining information about the transducer action of the receptor elements. One such possibility is to record the potential change which is the result of the action of the stimulus on the receptors.

The discovery that sense organs produce an electrical response when stimulated was made by Beck in 1899. In recording from the eye of the cephalopod *Eledone moschata* he found that the receptor layer became negative during illumination. The response as described by Beck consisted of a purely monophasic potential with an initial peak when light was turned on and a later steady phase during constant illumination of the eye. After cessation of illumination the potential returned gradually to zero. Beck attributed this response to the activity of the visual cells, a conclusion which was borne out by the experiments of Fröhlich (1914). Later studies have provided evidence of similar potentials in other kinds of sense organs.

GENERAL CHARACTERISTICS OF THE OLFACTORY RECEPTOR POTENTIAL

The action of odorous particles on the olfactory receptors manifests itself as a potential change which can be recorded with an electrode placed in contact with the surface of the mucosa (Ottoson, 1956). The response, the electro-olfactogram (EOG), exhibits the same general characteristics as the homologous potentials recorded from the simple invertebrate eye and those obtained from single sensory endings such as the muscle spindle, the decapsulated Pacinian corpuscle or the eccentric cell of the *Limulus* eye. It is a purely monophasic potential with a fast rising phase and an exponential decline (Fig. 1). With sustained constant stimulation the initial peak is

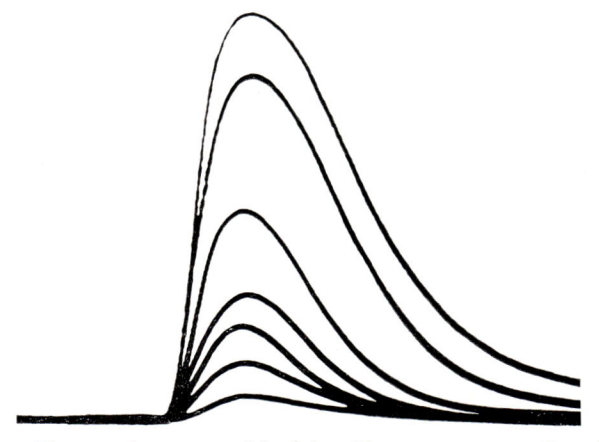

FIG. 1. The transducer potential of the olfactory receptors. Superimposed records of the EOG of the frog's olfactory organ to brief stimulations with butanol vapour of different strengths. (From Ottoson, 1956.)

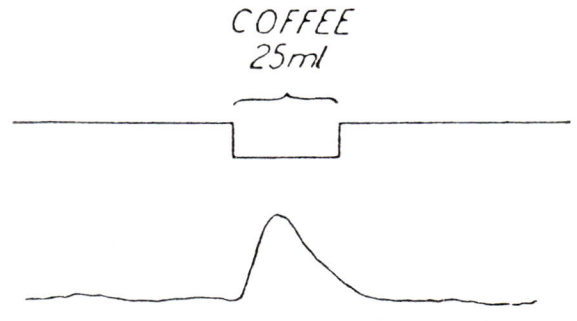

FIG. 2. The human EOG. Response of the olfactory epithelium in man to brief stimulations with coffee-saturated air. (From Osterhammel, Terkildsen and Zilstorff, 1969.)

followed by a steady phase for the duration of stimulation. The general properties of the response are essentially the same in different species. The first successful recordings in man have recently been reported by Osterhammel, Terkildsen and Zilstorff (1969). The response obtained to a brief stimulus is closely similar to the frog's EOG, as illustrated in Fig. 2.

The evidence showing that the EOG represents the activity of the receptor cells has been presented earlier (Ottoson, 1956) in detail and will therefore not be covered here. Experiments reported by Shibuya (1964) have been quoted in the literature as evidence against the interpretation of the EOG as the generator potential of the olfactory organ. The inconsistencies of Shibuya's conclusions have been subsequently pointed out (Ottoson and Shepherd, 1967; Ottoson, 1970).

Takagi and his collaborators (cf. Takagi and Shibuya, 1960a, b) have reported that an "off" response can be obtained from the olfactory mucosa with certain types of stimuli. They concluded that this response was produced by specific "off" receptors and suggested that there are two or three types of olfactory receptor cells "corresponding to the on-and-off, or on-, off-, and on-off responses" (Takagi and Shibuya, 1960b). According to this view the olfactory receptors would exhibit closely similar properties to the ganglion cells of the retina. The hypothesis that there should exist specific receptors of the kind suggested by Takagi and Shibuya (1960b) may be criticized on the ground that "off" responses are generally the result of synaptic interaction. It is well established that there are no synaptic connexions between the receptor cells in the mucosa. It seems likely that the various types of responses reported might be explained as the result of the specific effect of the stimulating substance used. To elicit the "off" responses Takagi and Shibuya used saturated vapours of ether and chloroform. It would appear obvious that responses which are only obtained under highly unphysiological conditions would have little relation to the physiological functions of the receptors.

The EOG is the first step in the excitatory chain of events which at present is amenable to direct analysis. By studying the EOG we can get some insight into the transducer processes in the receptor membrane. Some of the findings will be summarized.

LATENCY OF RESPONSE

Before the stimulating particles reach the cilia they have to pass through a layer of watery mucus. The time required for this passage is reflected in the delay of the electrical response. For a moderately strong stimulus the response appears about 200 ms after the odorous particles have reached the surface of the mucosa. This latency period includes not only the time for the passage of the stimulating agent through the mucus but also what may be defined as the "true latency"; that is, the time elapsing from the moment when contact is established to that when the depolarization of the receptors begins. The actual duration of each of these two periods cannot be directly measured. However, we can make some approximate estimations from latency data in other sense organs. In the eye of the frog the delay of the electrical response is about 50 ms but may be more than 100 ms. If we therefore assume that the "true latency" of the olfactory receptors is about 100 ms we are left with about 100 ms for the passage of the particles though the mucus. This passage probably occurs by passive diffusion and at a comparatively slow velocity. It is therefore unlikely that the stimulating particles during the short time available would be able to pass into the deeper layers of the mucus. The conclusion to be drawn from the latency measurements is that the response arises in structures which are located close to the surface of the mucosa. This conclusion is further strengthened by the changes in amplitude of the response when recordings are made at different depths within the mucosa. As shown in Fig. 3 the response decreases gradually in amplitude as the recording electrode is advanced into the sensory epithelium.

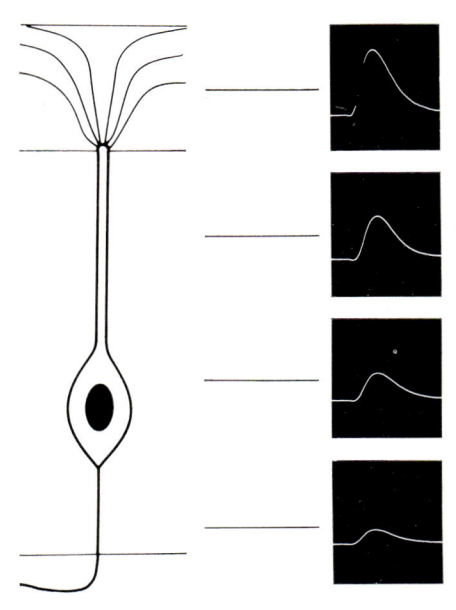

FIG. 3. Recordings of the EOG in the frog at different depths of the olfactory mucosa. (From Ottoson, 1956.)

In 1964 Shibuya reported experiments in which he applied a piece of absorbent paper to the mucosa. After removing the paper he found that the receptor potential had disappeared in the treated area. Shibuya concluded that the disappearance of the response was due to removal of the mucus and claimed that the cilia were undamaged except possibly at their tips. It would appear more reasonable to assume that the response disappeared because the cilia were broken off. Reese (1965) has shown that the cilia extend to the outer layer of the mucus. Removal of half of the mucous layer (cf. Shibuya, 1964) would therefore appear to lead to the destruction of the outer segments of the cilia. This is also clearly shown by the experiments of G. Gemne and K. B. Døving (personal communication, 1969). They applied a piece of absorbent paper to the frog's olfactory mucosa. The paper was then removed and examined electron microscopically. It was found that the mucus adhering to the paper contained a great number of cilia (see Fig. 4). This finding provides strong evidence that the cilia are the site of the olfactory transducer process. From the observation that the removed ciliary elements in the mucus mainly consisted of the distal segments it would be tempting to conclude that the transducer action primarily takes place in the outer parts of the cilia. This assumption gains support from the fact that the response disappeared in the treated area although the proximal parts of the cilia were left. However, it cannot be

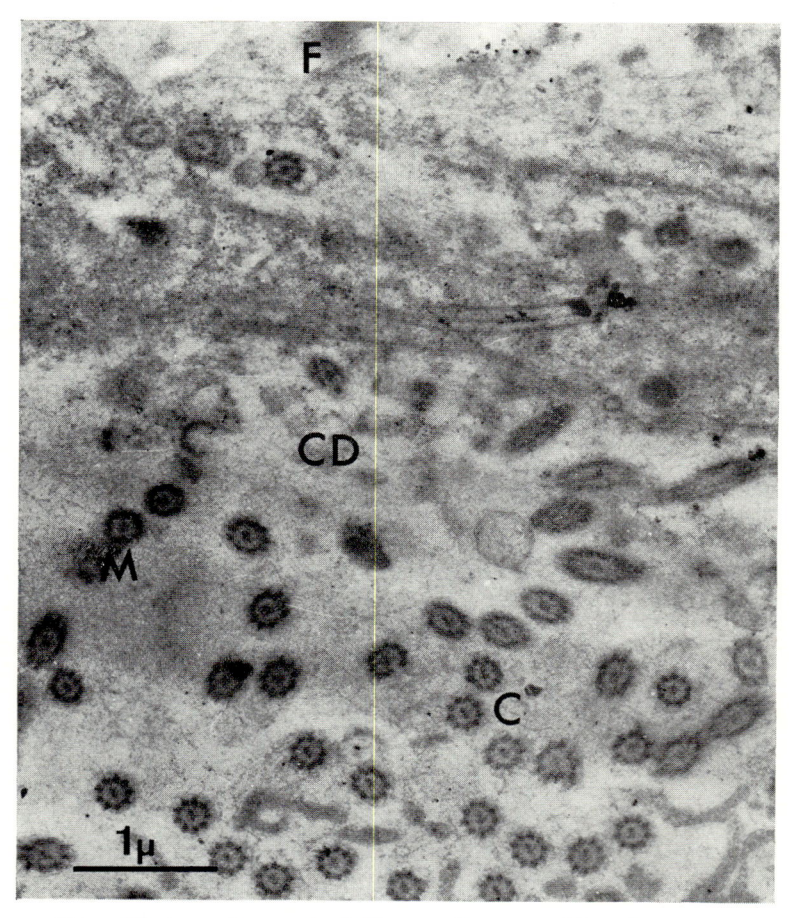

FIG. 4. Electron micrograph of olfactory cilia removed from the mucosa of a frog with a piece of filter paper. The paper with adhering mucus was fixed in glutaraldehyde and osmium tetroxide and embedded in Epon. Mucus (M), cell debris (CD) and cilia (C) adhere to the filter paper (F). (Gemne and Døving, 1969, personal communication.)

excluded that the remaining parts of the cilia were depolarized when their outer segments were broken off. It is likely that under normal conditions the entire cilium participates in the transducer process. However, since the greater portion of the stimulating molecules most likely are absorbed on the cilia lying at the surface of the mucus it may be concluded that the transduction process mainly takes place in the outer segments of the cilia.

RELATION BETWEEN STIMULUS INTENSITY AND RESPONSE AMPLITUDE

The response of the olfactory receptor is a graded potential which increases in amplitude up to a given maximum with the strength of the stimulus. This increase in amplitude may be ascribed partly to an increasing depolarization of the individual sensory elements and partly to the recruit-

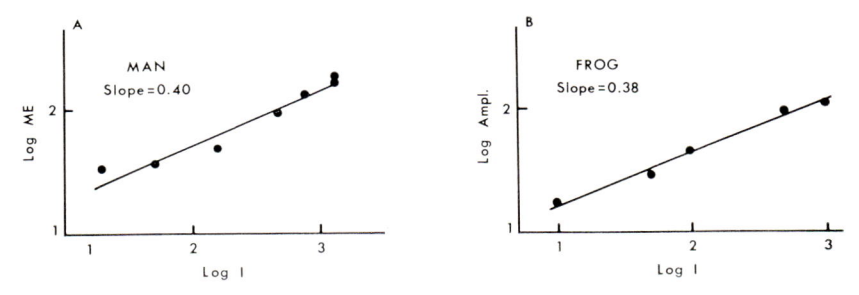

FIG. 5. EOG of frog and subjective odour intensity in man in relation to concentration of the stimulus. *Left:* magnitude estimation of *n*-butanol at different concentrations. (From Jones, 1958.) *Right:* magnitude of frog's EOG evoked with different concentrations of *n*-butanol. (From Ottoson, 1956.)

ment of receptors with higher thresholds as the intensity of the stimulus is increased. Measurements of the intensity–amplitude relation show that the magnitude of the response is a power function of the concentration of the stimulus. The exponent is 0·38 for *n*-butanol. It is interesting to note that the perceptual intensity for *n*-butanol at different concentrations as measured with psychophysical methods (Jones, 1958) is also a power function, with an exponent of 0·40 (Fig. 5).

It thus appears that there is a close similarity between the transducer action in the frog's olfactory receptors and the events underlying evaluations of odour intensity in man. This would suggest that quantitative data obtained in recordings of the EOG in the frog may be valid for man. This view gains further support from the results obtained by von Sydow (1968) and Drake and his co-workers (1969) in comparative studies of the frog's

EOG and subjective odour intensity functions in man. With methyl benzoate as test substance the EOG data were found to fit a power function with an exponent of 0·32. The psychophysical response to the same substance was also a power function, the exponent of which was 0·28 (Fig. 6) (von Sydow, 1968). Drake and co-workers (1969) compared the data obtained from similar measurements of four substances. A close correlation was found between the electrophysiological data and the psychophysical values for subjective odour intensity.

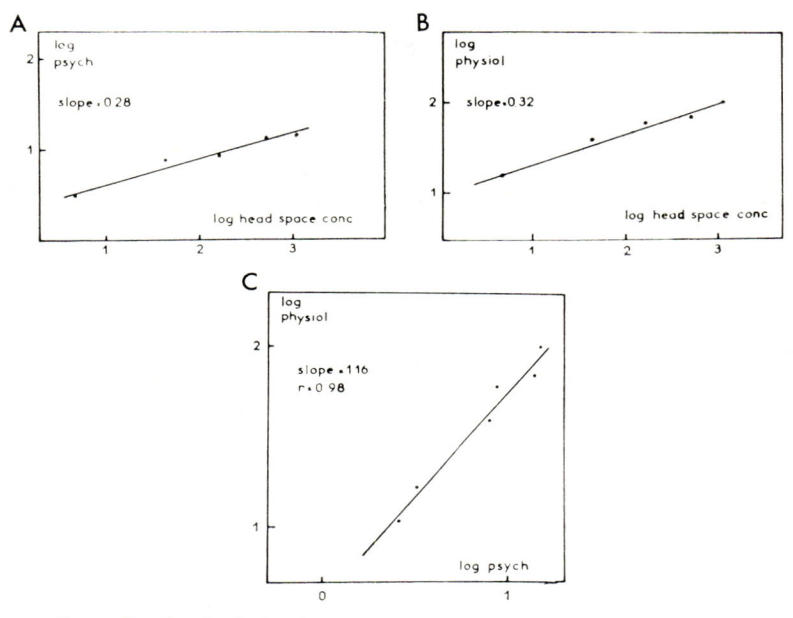

FIG. 6. Psychophysical and electrophysiological measurements of odour intensities for different concentrations of methyl benzoate. A, data from psychophysical measurements; B, measurements of EOG; C, correlation diagram between psychophysical and electrophysiological data. (From von Sydow, 1968.)

These findings are of interest in relation to the demonstration by Zotterman and his co-workers (Borg *et al.*, 1967) of the congruity between the perceptual taste intensity and the impulse activity of the chorda tympani. At present there are no corresponding data available for olfactory functions in man. There is, however, no reason to assume that the olfactory system differs significantly from the gustatory in this respect. It would appear likely therefore that the psychophysical data for olfactory intensity functions reflect the transducer process in the receptors. The agreement between the electrophysiological results in frog and the values for subjective intensity measurements further indicate that the basic mechanisms at the receptor level with respect to intensity functions are closely similar in man and frog.

RELATION BETWEEN STIMULUS QUANTITY AND RESPONSE AMPLITUDE

It is a characteristic feature of the olfactory organ that for a given concentration of an odorous substance the EOG increases in amplitude with increasing volume of the stimulating air current. This suggests that the

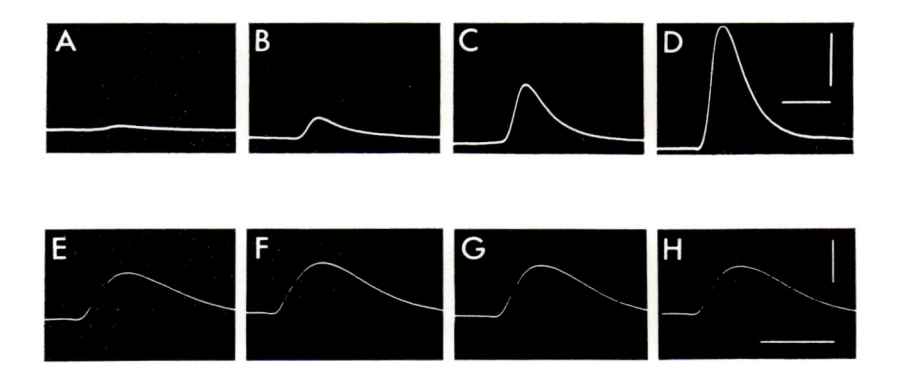

Fig. 7. Relation between stimulus quantity and response amplitude. Records A-D show the effect of increasing volumes of stimulating air of the same concentration. Records E-H show responses obtained when *volume × concentration* = k. (From Ottoson, 1956.)

olfactory membrane measures the absolute quantity of the odorous material which is brought in contact with the mucosa. This functional property of the olfactory organ can be demonstrated by stimulating the mucosa with equal amounts of odorous material distributed in different volumes of air. For a short stimulus the amplitude of the EOG is then the same, as illustrated in Fig. 7. The behaviour of the olfactory organ in this respect is closely similar to that of the eye. As demonstrated by Hartline (1928) the reaction elicited by light in the photoreceptors of the *Limulus* eye follows the Bunsen–Roscoe law so that the same number of impulses is produced when the product of time and intensity of illumination is kept constant. The olfactory receptors behave in a similar way with respect to the volume and concentration of odorous stimulation.

12*

TRANSDUCER ACTION AND TRANSMISSION OF INFORMATION TO HIGHER LEVELS

The information embodied in the receptor potential is transmitted to the olfactory brain as a frequency-coded impulse message in the olfactory nerves. In order that no essential information is lost in the encoding process the impulse frequency ideally should closely follow the transducer action. The general relationship between the receptor potential and the frequency of the afferent impulse discharge has been established in single end organs (Katz, 1950; Terzuolo and Washizu, 1962; Ottoson and Shepherd, 1965).

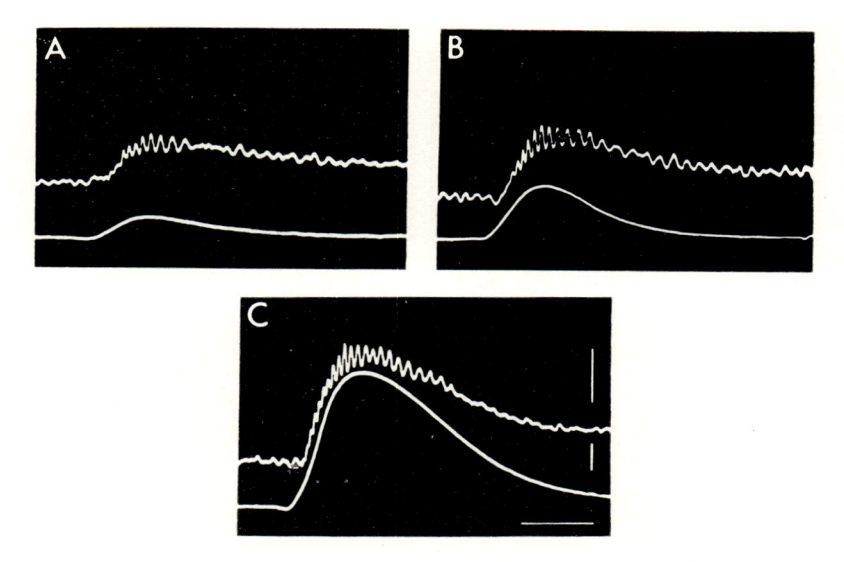

FIG. 8. Congruity between EOG (lower trace) and slow bulbar potential (upper trace). Recordings of responses to stimulation with butanol at increasing concentration (A–C). (From Ottoson, 1959.)

It has also been shown in studies on isolated spindles that the impulse discharge reproduces the features of the receptor potential (Ottoson and Shepherd, 1969). A corresponding analysis of the activity of the olfactory receptors would require recording from single units. The data available from such recordings are not yet sufficient to permit any definite conclusions. However, studies on the relation between the impulse activity in the olfactory nerve and the EOG (Kimura, 1961) suggest that the properties of the EOG are faithfully passed on to higher levels by the impulse activity of the olfactory nerves.

There is also evidence that this information with respect to intensity functions is well preserved in the following stage of the transmission process.

The afferent inflow to the bulb gives rise to a slow potential (Fig. 8) which most likely reflects the synaptic activity within the glomeruli. This potential closely follows the potential produced by the receptors. At the level of the second-order neurons the situation becomes more complicated as the interaction between the bulbar neurons enters as a dominant factor (cf. Døving, 1964, 1966; Rall and Shepherd, 1968). It is of particular interest to note that there appears to be a kind of lateral inhibition between the mitral cells similar to that in the *Limulus* eye. There is also evidence that the bulbar neurons are organized on similar functional principles to the retina of the vertebrate eye (Shepherd, 1970). Thus direct activation of the bulbar output neurons is mediated by "vertical" pathways, while "horizontal" pathways connecting the vertical ones are responsible for the integrative action. The horizontal pathways appear to have two principal functions: the first of these is related to the control of the olfactory receptor input to individual bulbar units and the second to the control of the output from the bulb to higher levels. The characteristics of the final impulse message to the olfactory cortex are the outcome of the neuronal interactions at these two levels.

SUMMARY

Anatomical and experimental evidence suggests that the olfactory transducer action takes place in the cilia of the sensory cells. The odorous particles impinging upon the mucosa are trapped in the dense network of the cilia, the membrane of which provides for a considerable receptive area. The first step in the excitatory chain of events which at present is amenable to direct study is the production of an electrical response, the electro-olfactogram. The analysis of the relationship between stimulus strength and magnitude of response shows that the amplitude of the potential is a power function of the stimulus concentration. Comparative electrophysiological and psychophysical studies have provided evidence that there is a close correlation between the response of the olfactory organ and the values for subjective odour intensity. An interesting functional feature of the olfactory organ as revealed by the EOG is that the olfactory membrane measures the absolute amount of odorous material; that is, it behaves like the eye to variations in intensity and duration of the stimulus. The properties of the receptor potential in the peripheral sense organ appear to be faithfully passed to the olfactory bulb. At this level neuronal interaction enters as an important control mechanism by which the input to individual units and their output to higher levels may be modified.

REFERENCES

BECK, A. (1899). *Pflügers Arch. ges. Physiol.*, **78**, 129–162.
BORG, G., DIAMANT, H., STRÖM, L., and ZOTTERMAN, Y. (1967). *J. Physiol., Lond.*, **192**, 13–20.
DRAKE, B., JOHANSSON, B., SYDOW, E. VON, and DØVING, K. B. (1969). *Scand. J. Psychol.*, **10**, 89–96.
DØVING, K. B. (1964). *Acta physiol. scand.*, **60**, 150–163.
DØVING, K. B. (1966). *J. Physiol., Lond.*, **186**, 97–109.
FRÖHLICH, F. W. (1914). *Z. Sinnesphysiol.*, **48**, 28–164.
HARTLINE, H. K. (1928). *Am. J. Physiol.*, **83**, 466–483.
JONES, F. N. (1958). *Am. J. Psychol.*, **71**, 305–310.
KATZ, B. (1950). *J. Physiol., Lond.*, **111**, 261–282.
KIMURA, K. (1961). *Kumamoto med. J.*, **14**, 37–46.
OSTERHAMMEL, P., TERKILDSEN, K., and ZILSTORFF, K. (1969). *J. Lar. Otol.*, **83**, 731–733.
OTTOSON, D. (1956). *Acta physiol. scand.*, **35**, suppl. 122, 1–83.
OTTOSON, D. (1959). *Acta physiol. scand.*, **47**, 136–148.
OTTOSON, D. (1970). In *Handbook of Sensory Physiology*, vol. *IV, Chemical Senses*, ed. Beidler, L. M. Berlin: Springer-Verlag. In press.
OTTOSON, D., and SHEPHERD, G. M. (1965). *Cold Spring Harb. Symp. quant. Biol.*, **30**, 105–114.
OTTOSON, D., and SHEPHERD, G. M. (1967). *Prog. Brain Res.*, **23**, 83–138.
OTTOSON, D., and SHEPHERD, G. M. (1969). *Acta physiol. scand.*, **75**, 49–63.
RALL, W., and SHEPHERD, G. M. (1968). *J. Neurophysiol.*, **31**, 884–915.
REESE, T. S. (1965). *J. Cell Biol.*, **25**, 209–230.
SHEPHERD, G. M. (1970). In *The Neurosciences: Second Study Program*, ed. Quarton, G. C., Melnechuk, T., and Schmitt, F. O. New York: Rockefeller University Press. In press.
SHIBUYA, T. (1964). *Science*, **143**, 1338–1340.
SYDOW, E. VON (1968). In *Theories of Odor and Odor Measurement* (Proc. NATO Summer School, Istanbul, 1966), pp. 297–330, ed. Tanyolaç, N. Istanbul: Robert College.
TAKAGI, S. F. (1967). In *Olfaction and Taste II* (Proceedings of the Second International Symposium, Tokyo, 1965), pp. 167–179, ed. Hayashi, T. Oxford: Pergamon Press.
TAKAGI, S. F., and SHIBUYA, T. (1960a). *Jap. J. Physiol.*, **10**, 99–105.
TAKAGI, S. F., and SHIBUYA, T. (1960b). *Jap. J. Physiol.*, **10**, 385–395.
TERZUOLO, C. A., and WASHIZU, Y. (1962). *J. Neurophysiol.*, **25**, 56–66.

DISCUSSION

Beidler: I understand that you feel that the cilia are the site of transduction; what about the olfactory vesicles (knobs) from which the cilia arise?

Ottoson: I don't think that the olfactory vesicles normally are concerned with transduction since only a very small fraction of the stimulus will reach them, the main part of the molecules being absorbed on the superficial cilia.

Beidler: Shibuya (1964) used antidromic stimulation to determine the area from which he recorded and found neural activity in response to an odour but no EOG after the mucous layer had been removed. When the cilia are removed in your experiments what happens to the activity in the primary olfactory nerves?

Ottoson: It disappears.

Beidler: Dr Tucker (1967) removes the cilia *in vivo* with detergent and records from the peripheral olfactory nerve. He finds that the nerve activity in response to odour stimulation goes down but within an hour has returned to normal.

Ottoson: It is likely that when the outer parts of the cilia are broken off the proximal parts are initially depolarized as a result of the damage. Later they may, however, recover as they seal off.

Beidler: It is possible that as a result of the treatment with detergent the cilia were breaking off, to leave the stumps that Dr Reese mentioned. (Incidentally, you can spin down the removed cilia and examine them chemically.) It may be that the stub that is left is enough to produce electrical activity. If the cilia are the site of transduction, what about non-ciliated olfactory receptors? How would you expect them to differ in the EOG or the neural response?

Ottoson: There is a great deal of evidence that the transduction process takes place in the cilia. This does not imply that the ciliary structure as such is necessary for transduction. In species having villous-like extensions from the sensory cells instead of cilia transduction most likely occurs in the membrane of the villi. The important thing is not whether the receptive extensions of the cells look like cilia or like villi; it is at the molecular level that the specific transducer properties have to be sought.

Beidler: Professor Davies, how long would it take for the passage of olfactory molecules through 50 μm of mucus?

Davies: If the distance through the mucus is 50 μm, I should expect the time of diffusion to be of the order of 1 or 2 seconds. This is for a fairly small molecule; something the size of ethanol, for instance. This assumes that the mucus is stagnant and that the odorant diffuses through the whole 50 μm.

Zotterman: There is an analogy in the skin to the regeneration of broken-off cilia. After you have scratched the skin hard you cannot feel tickle, but this sensitivity returns in about 20 minutes which is the time for the very superficially placed endings of the specific non-myelinated tickle fibres to restore their excitability after mechanical damage. Such fibres regenerate at a speed of about 10 μm an hour.

Murray: As far as the growing of cilia is concerned, they presumably would have to grow out from the basal bodies. Could they do so in an hour?

de Lorenzo: A cut myelinated nerve fibre degenerates back to its first node of Ranvier and regenerates from there; it doesn't start growing again from the point where it is severed.

Zotterman: What about the regeneration of unmyelinated fibres without nodes of Ranvier, like the "tickle" fibres of the skin?

Lowenstein: I am not sure one can equate cilia and unmyelinated nerve fibres, however.

Murray: Since there are many new basal bodies in the olfactory receptor cells, replacement of whole cilia is likely. However, the tips of broken cilia might seal over, as Dr Ottoson suggested. Perhaps you don't need complete cilia for transduction. If you take away the cilia, the remaining cell surface may still be able to function as the transducer.

REFERENCES

SHIBUYA, T. (1964). *Science*, **143**, 1338.
TUCKER, D. (1967). *Fedn Proc. Fedn Am. Socs exp. Biol.*, **26**, abstr. 1609, 544.

HIGHER OLFACTORY CENTRES

W. R. Adey

*Departments of Anatomy and Physiology, and Space Biology Laboratory,
University of California, Los Angeles*

The concept of functional centres as an intrinsic aspect of cerebral organization is far older than the trinity of investigative techniques that have laid the cornerstones of modern neurology. It is true that stimulation, ablation and electrical recording have lent credence to this concept. However, support so gained has been tempered by a growing awareness that the living brain exhibits a holistic organization, far removed from the simple schemes of fibre connexions offered by anatomists and physiologists as a basis for our understanding of the brain's role in the substrates of behaviour. In part, at least, the concept of centres and the inadequacies inherent therein are attributable to the limited tools available, and to the equally limited windows on the brain so provided.

Research on higher olfactory centres in the past two decades has moved almost a full circle. Enhanced interest in the organization of central olfactory pathways in the early 1950s emphasized with almost Calvinistic zeal the restricted primary terminations of the olfactory tracts in the frontal and temporal prepyriform cortex (Brodal, 1947; Allison, 1953; Adey, 1953). Although valuable in emphasizing a non-olfactory role for much of the medial aspect of the cerebral hemisphere classically considered as part of the "rhinencephalon", and now assigned a role in the limbic lobe, this evidence may have diverted attention from important olfactory functions in arousal through olfacto-diencephalic and olfacto-limbic paths. Further studies have emphasized a role for olfaction in hormonal release mechanisms, and conversely, the role of neurohumoral influences from diencephalic and mesencephalic levels on the sensitivity of the olfactory bulb to odorous stimuli.

The following account seeks to emphasize interactions at many levels between the so-called primary cortical olfactory areas and diencephalic structures, rather than stressing hypothetical and perhaps dubiously significant aspects of a regional autonomy in the dynamics of information transaction. We may consider central olfactory organization in terms of (1) projections of the olfactory bulb to the primary olfactory cortex; (2) projection of the olfactory bulb to the diencephalon and rostral midbrain reticular formation; (3) secondary projections of the olfactory cortex to di- and mesencephalon; and (4) centrifugal projections from diencephalon and mesencephalon to the olfactory bulb, and neurohumoral influences on odour sensitivity of the bulb.

(1) PROJECTIONS OF THE OLFACTORY BULB TO PRIMARY OLFACTORY CORTEX
Experimental anatomical studies

More than 20 years ago, Brodal (1947) drew attention to the fragmentary evidence that had been the basis for the classic scheme of a rhinencephalon or olfactory brain extending around the medial rim of the cerebral hemisphere. There followed a series of experimental anatomical studies which validated the concept that in the marsupial (Adey, 1953) and in the rabbit (Le Gros Clark and Meyer, 1947) and monkey (Meyer and Allison, 1949), the major, if not the total, termination of the lateral olfactory tract lies in quite restricted cortical zones at the junction of the frontal and temporal lobes. In the frontal lobe, this region includes the substantia perforata. In the temporal lobe, it is found as part of the uncus, and includes the peri-amygdaloid cortex. These regions form respectively the frontal and temporal prepyriform cortical areas (Allison, 1953; Adey, 1960).

Brodal's critical review had served to focus and synthesize evidence slowly accumulated by the Golgi method (Ramon y Cajal, 1911), Marchi degeneration techniques (Elliot Smith, 1909; Fox and Schmitz, 1943) and electrophysiological stimulation and recording (Hasama, 1934; Adrian, 1942; Allen, 1943; Rose and Woolsey, 1943; Fox, McKinley and Magoun, 1944) that the principal termination of fibres from the bulb lies in the anterior regions of the pyriform lobe, including the so-called anterior olfactory nucleus. No evidence was found of direct projections to the entorhinal cortex, the prominent caudal subdivision of the pyriform lobe, although bulbofugal fibres were seen in the central amygdaloid nucleus, in the bed nucleus of the stria terminalis, and in commissural terminations in the contralateral olfactory bulb (Le Gros Clark and Meyer, 1947).

The early studies by the Oxford school were made with the Glees (1946) silver staining technique. The merits of other silver techniques in displaying degenerating terminals in the olfactory system have been reviewed by Heimer (1968), with emphasis on the value of the original Nauta (1950) method. Powell, Cowan and Raisman (1965) and White (1965) applied this technique to central olfactory pathways in the rat. White concluded that projections from the bulb do, indeed, reach the entorhinal cortex, contrary to the findings of Powell, Cowan and Raisman (1965). In an attempt to resolve this difference, Heimer (1968) also studied the bulbofugal paths in the rat with the Nauta method, and concluded that, in addition to generally accepted terminations in the olfactory peduncle, the olfactory tubercle and prepyriform cortical areas, bulbar fibres also terminate in the ventrolateral entorhinal area. Although it would be easy to exaggerate the possible significance of these more caudal terminations, they provide at least one pathway for activation of the hippocampal system and other limbic structures by olfactory volleys.

Electrophysiological studies

Anatomical distribution of lateral olfactory tract fibres described above has been confirmed in the cat by electrical stimulation, which has also indicated a "parahippocampal" termination in a zone adjoining the ventral entorhinal area (Dennis and Kerr, 1968).

Prepyriform cortical areas exhibit characteristic electrical rhythms that have been intensively studied in relation to behaviour, and as an archetype of rhythmic processes in other cerebral nuclei. These studies are discussed below. Attention is directed here to microphysiological evaluation of afferent bulbar volleys in prepyriform cortex.

Electrical stimulation of the lateral olfactory tract in the anaesthetized cat elicits a prepyriform evoked potential with components that undergo polarity reversal 200 μm below the surface (Biedenbach and Stevens, 1969a), and approximately 200 μm superficial to the border between the molecular and soma layers. Microelectrodes did not record unit activity in the molecular layer, but frequently recorded unit action potentials within the soma layer. The typical response was excitation during the first wave of the evoked potential, and inhibition of spontaneously active cells during the second wave. With intracellular recording (Biedenbach and Stevens, 1969b), all cells showed a depolarization–hyperpolarization sequence. Latencies of postsynaptic potentials suggest that superficial neurons excite deep neurons, and that deep neurons make inhibitory connexions with neighbouring deep neurons, and with apical dendrites of superficial neurons (see Fig. 1). Richards and Sercombe (1968) have maintained guinea pig olfactory cortex *in vitro*, with similar findings in the location of units and in the sequence of excitatory and inhibitory postsynaptic potentials accompanying initial negative and later positive phases of the surface evoked potential.

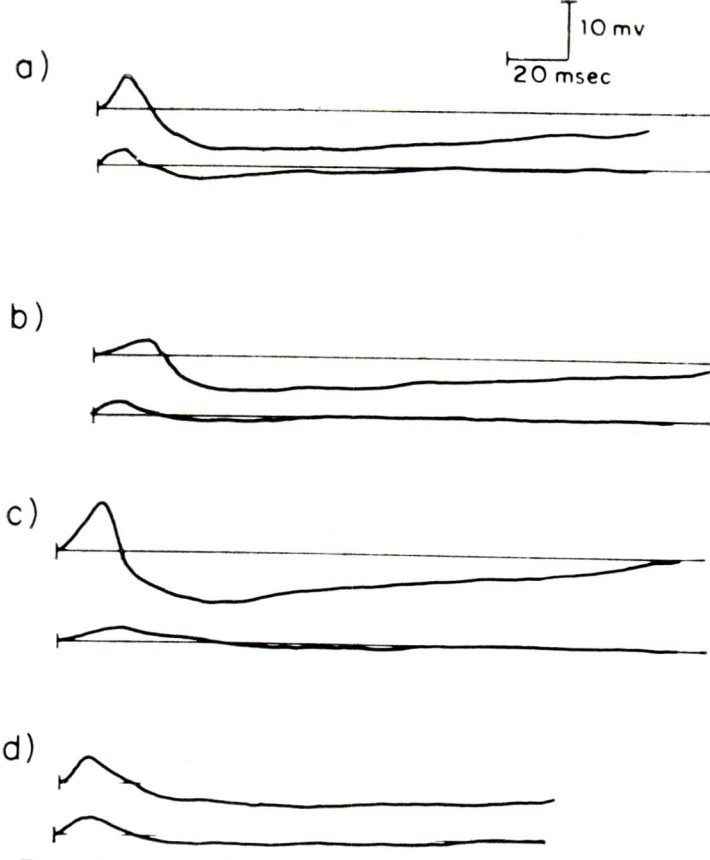

FIG. 1. Comparison of membrane potentials in intracellular records with field potentials evoked in prepyriform cortex of cat by stimulation of the lateral olfactory tract. The upper trace in each pair shows the inside-outside potential difference for a neuron, and the lower trace is the field potential recorded near that neuron. Records in (*a*), (*c*) and (*d*) are from superficial neurons and in (*b*) from a deep neuron. (From Biedenbach and Stevens, 1969*b*.)

Characteristic responses to odorous stimuli have been reported in prepyriform cortical units of the rat (Haberly, 1969). Odorants included organic compounds, such as camphor and geraniol, and also solid rat food. Units studied were spontaneously active, even under deep pentobarbitone anaesthesia, and 17 of 21 units responding unequivocally showed decreased spontaneous firing during presentation of the odour, while four were excited (Fig. 2). The inhibitory response pattern was related to stimulus intensity. "Off" responses as a post-inhibitory rebound were also noted, and occasionally reached a level many times the resting firing rate. Units excited by odours were rare and of small amplitude. The degree of excitation was related to particular odours, thus resembling the inhibitory responses of larger units, but the odour specificity of these cortical units was considered less than at the bulbar level.

These findings of predominantly inhibitory responses in apparently "spontaneously" firing neurons raise important questions about information transmission in terms of inhibition rather than excitation, as pointed

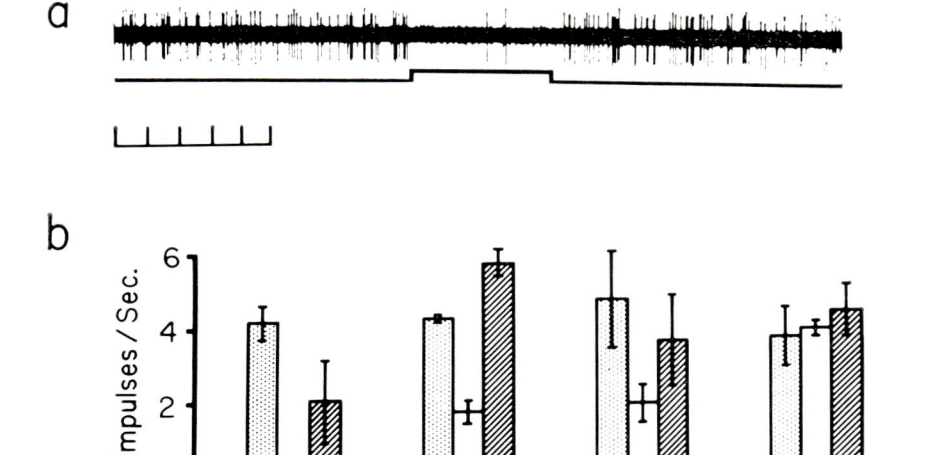

FIG. 2. Inhibitory responses to olfactory stimuli from single units in anaesthetized prepyriform cortex of rat. (*a*) Sample record with period of olfactory stimulus (rat food) shown in lower trace. Time marker, 1·0 second divisions. (*b*) Comparison of responses of same unit to three odours and pure air control. Stippled bars indicate resting firing rates (during the 10 seconds immediately preceding stimuli); open bars, firing rates during stimuli; crosshatched bars, firing rates during the 2·0 seconds immediately following stimuli. Vertical lines centred at top of each bar indicate one standard deviation. (From Haberly, 1969.)

out by Haberly. An even more fundamental issue concerns the possibility that cerebral tissue is organized as an essentially noisy processor, and that our attempts to understand its essential mechanisms may have been obfuscated by preconceived notions of the ordered activity most readily discovered in the constricting frame of deep anaesthesia (Adey, 1969, 1970; Elul, 1969). Certainly, strong trends to Gaussian amplitude distributions in resting EEG records of man and animals, and correlated trends away from a Gaussian distribution as concomitants of alerted states and task performance, indicate strong tendencies to independence in wave generators contributing to the EEG. They thus support concepts of pseudo-random behaviour. Such data have been easier to collect in relation to neuronal wave generators than to unit spike discharges, but qualitative observations of independent behaviour of units as close as 100 μm (Morrell, 1967) and quantitative correlates of firing patterns of paired cortical neurons seen with a single microelectrode (Noda, Manohar and Adey, 1969a,b) both suggest highly variable and often seemingly random relations. If such a scheme of apparently noisy elements is indeed the basis of information transactions, we might do well to confront its essential complexity, rather than to seek or impose order in ways unrelated to actual physiological processes.

(2) PROJECTIONS OF THE OLFACTORY BULB TO THE DIENCEPHALON

Early experimental anatomical studies with the Glees staining method indicated a bilateral projection to the bed nucleus of the stria terminalis, and to the central amygdaloid nucleus, in marsupials, rodents, carnivores and primates (Le Gros Clark and Meyer, 1947; Meyer and Allison, 1949; Adey, 1953; Allison, 1953). Many later studies have failed to confirm these pathways, using either the original Nauta method which does not suppress normal fibre staining (Powell, Cowan and Raisman, 1965; White, 1965; Heimer, 1968) or the suppressive Nauta-Gygax method (Cragg, 1961; Lohman, 1963; Lohman and Lammers, 1967; Powell, Cowan and Raisman, 1965; Scalia, 1966; Mascitti and Ortega, 1966; Girgis and Goldby, 1967). These anatomical uncertainties have been reviewed by Heimer (1968), including the question of pseudo-degeneration of the bed nucleus of the stria terminalis in normal tissue (Cowan and Powell, 1956), an explanation not compatible with our findings (Adey et al., 1958).

In summary, there is no consensus of anatomical opinion in favour of direct anatomical connexions between the olfactory bulb and the diencephalon. On the other hand, there is clear evidence relating olfaction to reproductive functions, and thus to hypothalamic activity, so that physiological research has sought relatively direct, oligosynaptic paths for olfacto-hypothalamic influxes, as described in the following section. Moreover, bilateral amygdalotomy in the monkey, removing essentially all the temporal prepyriform cortex (but apparently not damaging the smaller frontal prepyriform area) was not followed by defective olfactory discrimination (Schuckman, Kling and Orbach, 1969). These findings again raise questions of secondary projections from frontal olfactory cortex to the thalamus and other diencephalic regions, as well as possibly more direct diencephalic terminations of olfactory tract fibres, or of their oligosynaptic relays.

It seems inevitable that a full understanding of olfacto-diencephalic interrelations will reveal substantial parallel paths with varied internuncial patterns. In such a scheme, therefore, major limbic lesions, as in the hippocampus (Kimble and Zack, 1967), would not necessarily impair olfactory performance, even though White (1965) and Heimer (1968) have reported direct bulbo-entorhinal paths, presumably capable in turn of entorhinal-hippocampal activation through the temporo-ammonic tracts of Cajal. Such an oligosynaptic pathway for olfacto-diencephalic influences may be found in the anterior limb of the anterior commissure, with origins in the anterior olfactory nucleus, and terminating in the bed nucleus of the stria terminalis (Girgis and Goldby, 1967; Powell, Cowan and Raisman, 1965), which was classed by Dennis and Kerr (1968) as a direct bulbar contribution. Changes in adaptation to olfactory stimuli and in perception of intensity follow lesions in the anterior limb of the anterior commissure (Bennett, 1968).

(3) SECONDARY PROJECTIONS OF CENTRAL OLFACTORY STRUCTURES TO DI- AND MESENCEPHALON

The base of the mammalian forebrain typically displays a rostro-caudal sequence of nuclei subserving olfaction—the anterior olfactory nucleus, the olfactory tubercle and the cortical zones of prepyriform cortex. The medial forebrain bundle unites them as a series of association paths. It also ramifies much further, extending forward to the olfactory bulb and caudally into the lateral hypothalamus. The latter extension offers a major route for the olfactory tubercle (Heimer, 1968), and from the prepyriform cortex (Valverde-García, 1965). Electrical stimulation of the rat's olfactory bulb or lateral olfactory tract elicited unit discharges in the medial forebrain bundle region of the lateral hypothalamus, with latencies of 4 to 25 ms (Fig. 3). Many units driven by electrical stimulation could also be activated with odours (Scott and Pfaffmann, 1967). However, no responses were elicited in the dorsomedial or ventromedial nuclei of the medial hypothalamus.

Cajal noted the stria terminalis as an important path for olfactory influences on the hypothalamus (Ramon y Cajal, 1911). It originates partly in the cortico-medial amygdaloid nuclei, which receive terminals of the lateral olfactory tract (Powell, Cowan and Raisman, 1965; Valverde-García, 1965). Stria terminalis fibres terminate widely in the medial

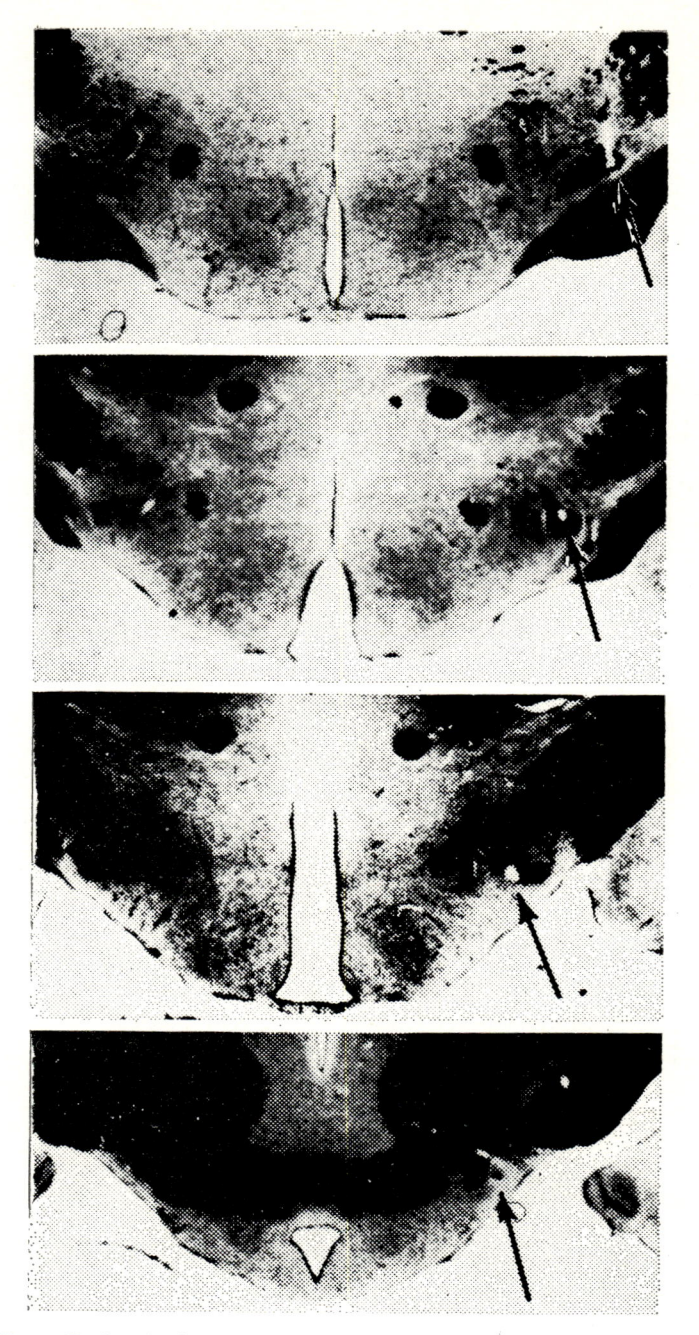

FIG. 3. Weil-stained sections of rat hypothalamus showing sites of short latency units driven by electrical stimulation of the ipsilateral olfactory bulb or olfactory tract. Arrows show lesions made in a responsive site. The two anterior (*top*) sections show tracts of bipolar electrodes. The two posterior (*lower*) sections show microelectrode recording sites. (From Scott and Pfaffmann, 1967.)

hypothalamus, including the dorsomedial and ventromedial nuclei, which exercise control over neuroendocrine mechanisms.

We have considered so far only direct projections from olfactory cortex to lateral and medial hypothalamic zones. It would be unfortunate if the impression were gained that functional interrelations between olfactory centres and medial and lateral zones of the hypothalamus could be comprehensively described by the respective contributions from the stria terminalis and the medial forebrain bundle. Such a view would overlook the broad stream of connexions, presumably polysynaptic and thus not easily accessible to current experimental anatomy, shown electrophysiologically to reach from olfactory areas to the amygdala as far caudally as the rostral midbrain tegmentum (Gloor, 1955a,b). It is through such caudal diencephalic and midbrain connexions that lowered levels of general arousal might follow bilateral destruction of olfactory bulbs in pigeons (Wenzel and Salzman, 1968). Since olfaction plays only a minor role in the discriminative behaviour necessary to avian existence, these authors also emphasize direct olfactory connexions with parts of the limbic system in birds. Similarly, evoked potentials to odorants, recorded at the vertex in man, and widely distributed over the scalp, occur with latencies of 500 ms, suggesting complex subcortical paths (Allison and Goff, 1967).

Olfactory influences on the hypothalamus play a major role in sexual behaviour and mechanisms of pregnancy (Lee and Boot, 1955; Whitten, 1956; Barraclough and Cross, 1963). Male rats prefer the odour of oestrous females to non-oestrous ones (Carr, Loeb and Dissinger, 1965) and male rats show reduced mating ability after removal of the olfactory bulbs (Heimer and Larsson, 1967). Exposure to odours from strange males leads to interruption of pregnancy in mice, and depends on connexions between the olfactory bulb and the anterior hypothalamus (Parkes and Bruce, 1961). Cyclic ovarian functions in rabbits are closely dependent on the integrity of the olfactory bulbs (Franck, 1966).

It is clear from these striking behavioural and endocrine findings that there are important functional olfacto-hypothalamic relations that are only partially described by known anatomical paths. Many of these paths may well be polysynaptic, passing through a series of short neuronal relays. They are thus not amenable to experimental anatomical methods that delineate only the synaptic limits of primary neuronal fields.

(4) CENTRIFUGAL PROJECTIONS AND NEUROHUMORAL INFLUENCES ON THE OLFACTORY BULB FROM DIENCEPHALON AND MESENCEPHALON

The central control of sensory afferent pathways has attracted much attention since direct testing by electrophysiological methods became feasible. Following demonstration of such effects in the olfactory system by Kerr and Hagbarth (1955), further studies have elucidated both morphological and functional aspects of bulbipetal fibre systems.

Anatomically, these fibres arise in many regions of the basal forebrain (Heimer, 1968), including the anterior prepyriform cortex (Cragg, 1962) a

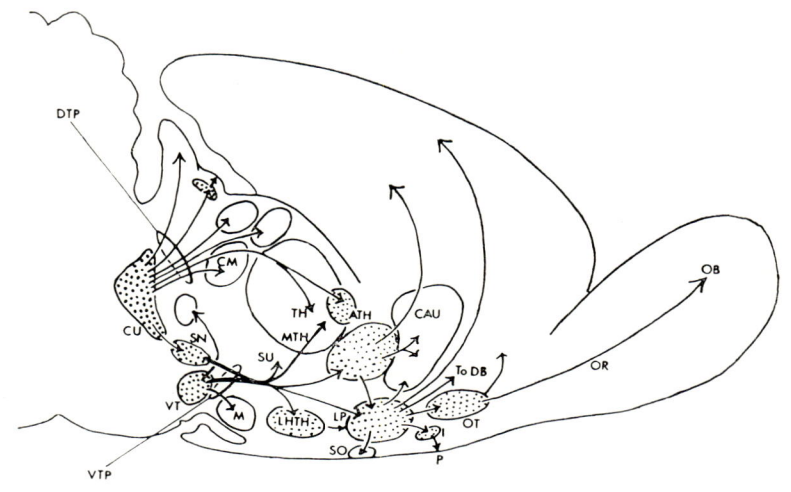

FIG. 4. Diagram of nuclei of ascending cholinergic reticular system proposed by Shute and Lewis (1967) in midbrain and forebrain, with projections to cerebellum, tectum, thalamus, hypothalamus, striatum, lateral cortex, and olfactory bulb. Abbreviations: ATH, anteroventral and anterodorsal thalamic nuclei; CAU, caudate nucleus; CM, centromedian nucleus; CU, cuneiform nucleus; DB, diagonal band of Broca; DTP, dorsal tegmental pathway; I, islets of Calleja; LHTH, lateral hypothalamic area; LP, lateral preoptic area; M, mammillary body; MTH, mammillothalamic tract; OB, olfactory bulb; OR, olfactory radiation; OT, olfactory tubercle; P, plexiform layer of olfactory tubercle; SN, substantia nigra, pars compacta; SO, supraoptic nucleus; SU, subthalamus; TH, thalamus; VT, ventral tegmental area and nucleus of basal optic root; VTP, ventral tegmental pathway. (From Shute and Lewis, 1967.)

short distance behind the olfactory tubercle. However, regions behind this anterior prepyriform zone also contribute, including the entire prepyriform and periamygdaloid cortical areas, which send some fibres to the olfactory tubercle and the olfactory peduncle (Heimer, 1968). Heimer points out that these forward-conducting association fibres and the olfactory tract fibres have contrasting terminal distributions on the dendritic fields of pyramidal cells in prepyriform cortex, and in the olfactory tubercle and peduncle. Deeply terminating association fibres on basal dendrites may modulate cell responses to olfactory volleys in terminals near the periphery of the dendritic fields.

Many centrifugal fibres pass beyond the olfactory tubercle, and in rodents ramify in the olfactory bulb in the vicinity of the glomeruli (Cragg, 1962; Powell, Cowan and Raisman, 1965; Heimer, 1968). This system of association fibres has been displayed by Shute and Lewis (1967) using a thiocholine staining method for cholinesterases (Koelle and Friedenwald, 1949). They

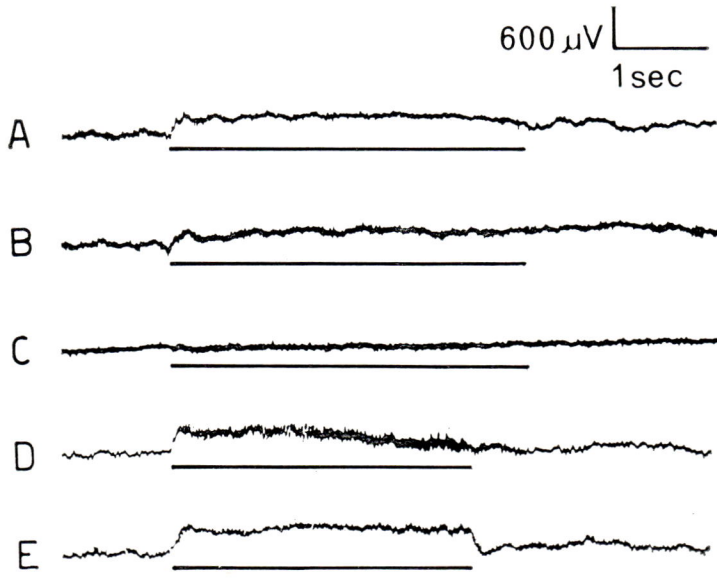

FIG. 5. D.c. surface-negative changes induced in the olfactory bulb by high-frequency stimulation (300 per second; 0·5 ms, 3 V) of midbrain reticular formation before (A) and after (B) bilateral ablation of frontal cortices. The d.c. shift disappeared after acute sectioning of the ipsilateral olfactory peduncle (C). Similar shifts followed high-frequency stimulation of the oral mucosa (D) and midbrain tegmentum (E). (From Carreras, Mancia and Mancia, 1967.)

apply the term "olfactory radiation" to fibres from the lateral preoptic area and the olfactory tubercle which travel rostrally in the lateral olfactory tract and olfactory peduncle, and supply anterior prepyriform cortex, the nucleus of the lateral olfactory tract, the olfactory tubercle and the olfactory bulb (Fig. 4). Their staining method also suggests that the lateral preoptic area and olfactory tubercle may receive activation from a "ventral tegmental pathway" arising far caudally in the central midbrain tegmentum, and running rostrally through subthalamus and hypothalamus to reach the basal forebrain.

There is thus much anatomical evidence for widespread centrifugal forebrain influences on the olfactory bulb. This evidence is corroborated by sectioning the olfactory tract, which augments the bulbar electrical response to chemical and electrical stimulation of the olfactory epithelium in the goldfish (Hara and Gorbman, 1967). In *encéphale isolé* and curarized cats, surface-negative d.c. changes were found in one bulb after stimulation of the contralateral one, the ipsilateral prepyriform cortex, midline thalamic nuclei and midbrain reticular formation (Fig. 5) (Carreras, Mancia and Mancia, 1967). Similar changes followed electrical activation of oral cavity mucosa and natural stimuli, such as whistling. The changes induced from the contralateral bulb were abolished by section of the anterior commissure, but cutting the lateral olfactory tract did not abolish changes elicited from prepyriform cortex (Dennis and Kerr, 1968). Bilateral removal of primary olfactory cortex abolished responses to thalamic stimulation, but enhanced the slow shifts to stimulation of the reticular formation. In addition to these "non-specific" influences on the bulb, Stone, Williams and Carregal (1968) found that blocking the trigeminal ganglia increased olfactory bulb excitability in rabbits with chronically implanted electrodes, but was without effect on respiration or heart rate during presentation of odours, which also failed to induce cortical desynchronization. The trigeminal nerve thus appears to exercise a central regulatory role over olfactory afferent volleys.

Control of the olfactory bulb by other brain regions is not limited to neural connexions. It is also susceptible to neurohumours circulating in its environment. These may be set free elsewhere in the brain, or may cross the blood–brain barrier after being released peripherally. For example, nerve terminals in the olfactory bulb contain noradrenaline and serotonin, and will incorporate these monoamines when they are injected into the cisterna magna, and release them in response to odorous stimuli (Chase and Kopin, 1968).

Similarly, sex hormones released under hypothalamic control change the excitability of the olfactory bulb. Intraperitoneal administration of oestradiol or testosterone did not change the spontaneous EEG activity of the olfactory bulb of the goldfish. In both sexes, oestradiol markedly increased the bulbar response to infusion of salt solution into the nasal cavity. This effect was much greater in male fish than in females (Hara, 1967). Testosterone was slightly excitatory in male fish, but depressive in females. In further studies, Oshima and Gorbman (1968) noted that oestrogen-treated goldfish exhibited a spontaneous slow olfactory bulbar EEG, which changed to a desynchronized fast pattern after section at the midbrain–hindbrain level (*cerveau isolé*) (Fig. 6). By contrast, the spontaneous "normal" desynchronized bulbar EEG of progesterone-treated fish was converted to a synchronized slow pattern after similar brainstem transection. The authors interpret these findings as indicating that these hormonal effects on olfaction are mediated through other regions of the central nervous system, but it appears that direct bulbar effects are not precluded.

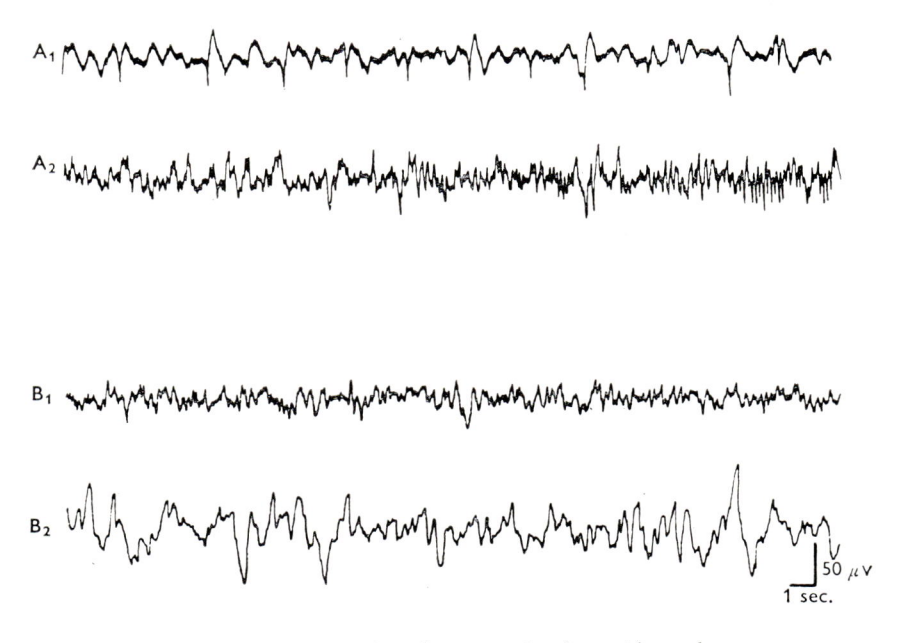

FIG. 6. Effect of oestrogenic and progestational steroids on the spontaneous olfactory bulbar EEG and on the EEG after *cerveau isolé* preparation in goldfish. A_1, goldfish treated with 17α-ethynyl-5(10)-oestraneolone in repeated doses of 40 μg. Records were similar after repeated doses of 40 μg oestradiol. A_2, 20 minutes after transection at the midbrain-hindbrain border, showing EEG change from slow to desynchronized fast. B_1, olfactory bulbar EEG in fish after repeated doses of 40 μg progesterone. B_2, 20 minutes after transection similar to A_2. (From Oshima and Gorbman, 1968.)

In summary, the spontaneous and evoked electrical activity of the olfactory bulb appears to be under strong central control. These central influences are in turn susceptible to neurohumoral and hormonal substances, and may also be modified by psychopharmacological agents (Khazen, Kandalaft and Sulman, 1967).

(5) ON THE NATURE OF ELECTRICAL RHYTHMS IN CENTRAL OLFACTORY STRUCTURES

Regular rhythmic oscillations, often at frequencies considerably higher than the typical EEG of the cerebral neocortex, characterize the olfactory bulb and higher levels of the olfactory system, including the prepyriform cortex (Adrian, 1942, 1950; Agalides, 1967). It differs from many EEG records in its regularity and tendency to occur in bursts or "spindles". It has been recorded in the region of the uncus, which includes the amygdaloid

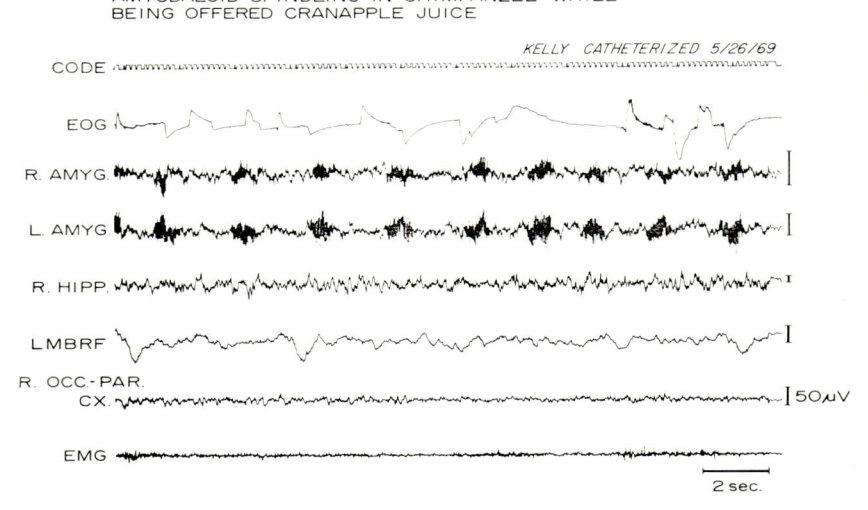

FIG. 7. Regularly recurring bursts of high-amplitude waves in left (L. AMYG) and right amygdalae (R. AMYG) of chimpanzee alerted by an offering of fruit juice. Other channels: CODE, timing code; EOG, electro-oculogram; R. HIPP, right hippocampus; LMBRF, left midbrain reticular formation; R. OCC.-PAR. CX., right occipito-parietal cortex; EMG, cervical electromyogram. (From NcNew and Adey, in preparation.)

nuclei receiving fibres of the lateral olfactory stria. The frequencies in anaesthetized preparations are 15 to 25 Hz, and are faster in the awake state. Bulbar rhythms also become faster during olfactory stimulation. However, the relationship of the bursts of waves to respiration has been carefully investigated, and the conclusion drawn in the cat that bursts appear in the alerted state, particularly during sniffing, but that air flow through the nose is not a required condition for bursting (Peneloza-Rojas and Alcocer-Cuarón, 1967; Gault and Leaton, 1963).

These bursts in the cat have been related by Freeman (1959) to alerted states with a high level of behavioural drive, as in aggression, fear, hunger and sexual gratification. Satiation leads to a decline and disappearance of the bursting. More recent studies in the chimpanzee have confirmed the occurrence of uncal spindling at 20 to 23 Hz in both sleeping and waking states (Fig. 7) (Rhodes *et al.*, 1963; Reite, Stephens and Pegram, 1967; J. J. McNew and W. R. Adey, in preparation). Reite, Stephens and Pegram point out that there is a distinct temporal relation between the spindle bursts and expiration, although olfaction *per se* does not play a role.

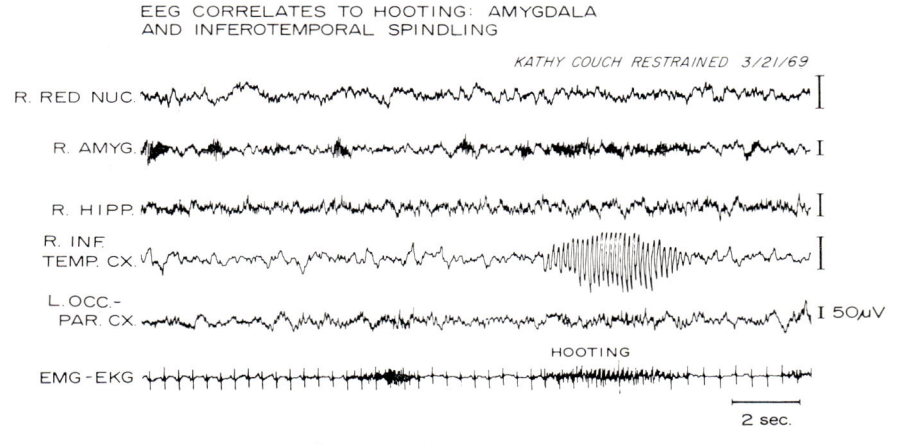

Fig. 8. Two types of "spindling" wave trains in chimpanzee olfactory and adjacent temporal lobe cortex, associated with hooting in angry behaviour. Amygdaloid records (R. AMYG) showed irregularly recurring bursts of waves 0·5 to 2·0 seconds in duration, with frequencies of 20 to 23 Hz. Slower spindles at 5 to 6 Hz occurred further posteriorly in the hippocampal gyrus (R. INF. TEMP. CX.), and were less numerous than amygdaloid spindles. Other channels: R. RED NUC., right red nucleus; R. HIPP., right hippocampus; L. OCC.-PAR. CX., left occipito-parietal cortex; EMG-EKG, electromyogram and electrocardiogram. (From McNew and Adey, in preparation.)

Rather, it appears that only when the respiratory activity took on biological (or emotional) significance were the spindles seen.

We have noted spindling in the uncus of the chimpanzee in alerted states associated with hooting or with impending food rewards (Fig. 8). The bursts at 20 to 23 Hz last 0·5 to 2·0 seconds. Additionally, they are seen in drowsy states, and in all stages of sleep (Fig. 9). They are particularly well developed in REM sleep, associated with dreaming, but their occurrence in other sleep stages makes an interpretation of an exclusive relationship to emotional experience unlikely.

The geometry of the electric field of these waves within the prepyriform cortex has been extensively studied by Freeman (1959), who has modelled its distribution as a "prepyriform dipole". The model envisages a cortical dipole organized in depth, with its outer pole in the molecular layer and its inner pole deep to the somata of the pyramidal cells. Stimulation of the lateral olfactory tract causes an outer negativity and an inner positivity. The outer pole is focal in location but the inner is more diffusely distributed. Reversal of polarity was also elicited. Freeman interpreted his data in terms

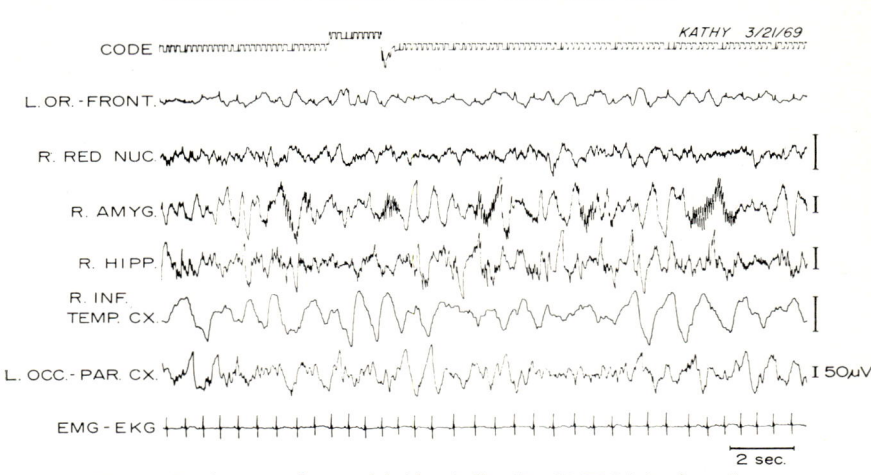

FIG. 9. Persistence of amygdaloid spindles (R. AMYG.) in deep slow wave sleep in the chimpanzee. Abbreviations: L. OR-FRONT., left orbito-frontal cortex; EMG-EKG, electromyogram and electrocardiogram; other abbreviations as in Figs. 7 and 8. (From McNew and Adey, in preparation.)

of inward negative current flow in pyramidal dendrites and outward flow in proximal axons.

Although such a model does not offer a detailed picture of the relations between the gross EEG waves of the burst and the intraneuronal processes of the contributing generators, we may infer that these relations are essentially similar to those observed by intracellular recording in neocortical and hippocampal neurons (Elul, 1967, 1968). Here, a large intraneuronal wave, up to 20 mV in amplitude, was found to have a spectral density contour similar to that of the gross EEG in the same tissue domain. However, coherence between the gross and the cellular wave process remained low, suggesting that the gross EEG represented the volume-conducted sum of a large number of independent and presumably non-linear generators. This view of the essential independence of the contributing generators has been confirmed by further studies using EEG amplitude distributions, rather than determination of frequency characteristics (Elul, 1969).

There remains a major challenge in the basis for such regionally focal high-frequency electrical rhythms. It is surely not obvious why in all mammals from rat to man the prepyriform cortex exhibits fast rhythmic oscillations in the range 20 to 45 Hz, whereas adjacent hippocampal tissue shows equally well-developed dominant rhythms at much slower frequencies. Clearly, the answer will lie in structural arrangements, but we may well ask whether it will be found on the basis of synaptic connexions, presumably in local neuronal pathways, or whether a more intimate factor may be causal. For example, hydrated networks of proteinaceous macromolecules that form glycocalyces at the neuronal surface and in cerebral intercellular spaces are probably involved in neuronal excitability, through their reversible interaction with divalent cations, particularly calcium (Adey et al., 1969; Bass and Moore, 1968; Wang and Adey, 1969). These surface coats may play an essential role in the propagation of neuronal electrotonic waves, determining their decay times and thus the inherent frequency characteristics of these perturbations in membrane potential. There is now direct evidence that alterations in divalent cations or enzymic digestion of these surface substances in the mammalian cortex changes neuronal excitability and the pattern of so-called "spontaneous" wave activity (Wang and Adey, 1969).

In such a scheme, it is also possible that these surface mucoproteins and mucopolysaccharides would endow local tissue domains with considerable regional structural and functional specificity. This is suggested by electrical impedance measurements in tissues of the uncus of the cat (Adey, Kado and Walter, 1965) and man (Porter, Kado and Adey, 1964) which both show only resistive shifts to such manipulations as hypo- and hypercapnoea. Typical cortical and subcortical responses show both resistive and reactive changes. Production of these surface macromolecules may be under the control of juxtamembranal ribosomes, as in antibody production by plasma cells (Nossal, Williams and Austin, 1967), and may reflect, by its regional character as a mosaic on the surface of each neuron, the previous experience of that cell (Adey et al., 1969). Here, also, evidence from the olfactory bulb indicates an extraordinary level of specificity in responsiveness based on previous experience, akin to an immune response. When the nasal sac of the spawning salmon is irrigated with water from the spawning site, a high-amplitude EEG of characteristic pattern occurs in the olfactory bulb. It does not occur with water from spawning sites of other groups of breeding salmon. Weaker responses are evoked by water from sites adjoining the spawning region (Ueda, Hara and Gorbman, 1967).

We have thus come to a fascinating crossroads in the elucidation of higher olfactory centres. We may expect that future studies will add substantially to our knowledge of the role of olfaction in relation to more general processes of alerting and arousal. It is also conceivable that ultrastructural studies in this most ancient of forebrain mechanisms will contribute to our knowledge of that most pressing problem, the physical substrates of information storage in brain tissue.

SUMMARY

Evidence concerning the terminal distribution of the olfactory tract in the prepyriform cortex has been reviewed. Direct projections from olfactory bulb to hypothalamic areas have not been confirmed in recent anatomical studies. On the other hand, physiological research has disclosed clear relations between olfaction and reproductive functions, suggesting widespread oligosynaptic connexions. Centrifugal projections influence bulbar excitability and arise in olfactory cortex and from diencephalic and mesencephalic regions located much further caudally. Sex hormones change the excitability of the olfactory bulb, at least partly through other central structures. The behavioural correlates of rhythmic electrical activity in olfactory nuclei are discussed.

Acknowledgements

It is a pleasure to acknowledge the assistance of the Brain Information Service of the University of California, Los Angeles in the preparation of this review. Studies from our laboratory were assisted by Contract NSR 05-007-158 with the National Aeronautics and Space Administration, and by Contract AF 49(638)-1387 with the United States Air Force Office of Scientific Research, and by Grant MH-03708 with the National Institute of Mental Health.

REFERENCES

ADEY, W. R. (1953). *Brain*, **76**, 311–336.
ADEY, W. R. (1960). In *Handbook of Physiology*, sect. 1, pp. 535–548, ed. Field, J., Magoun, H. W., and Hall, V. E. American Physiological Society. Baltimore: Williams and Wilkins.
ADEY, W. R. (1969). In *Biocybernetics of the Central Nervous System*, pp. 1–27, ed. Proctor, L. D. Boston: Little, Brown.
ADEY, W. R. (1970). In *The Neurosciences: Second Study Program*, ed. Quarton, G. C., Melnechuk, T., and Schmitt, F. O. New York: Rockefeller University Press. In press.
ADEY, W. R., BYSTROM, B. G., COSTIN, A., KADO, R. T., and TARBY, T. J. (1969). *Expl Neurol.*, **23**, 29–50.
ADEY, W. R., KADO, R. T., and WALTER, D. O. (1965). *Expl Neurol.*, **11**, 190–216.
ADEY, W. R., RUDOLPH, A. F., HINE, I. F., and HARRITT, N. J. (1958). *J. Anat.*, **92**, 219–235.

Adrian, E. D. (1942). *J. Physiol., Lond.*, **100**, 459–473.
Adrian, E. D. (1950). *Br. med. Bull.*, **6**, 330–333.
Agalides, E. (1967). *Trans. N.Y. Acad. Sci.*, **29**, 378–389.
Allen, W. F. (1943). *Am. J. Physiol.*, **139**, 553–555.
Allison, A. C. (1953). *J. comp. Neurol.*, **98**, 309–352.
Allison, T., and Goff, W. R. (1967). *Electroenceph. clin. Neurophysiol.*, **23**, 78.
Barraclough, C. A., and Cross, B. A. (1963). *J. Endocr.*, **26**, 339–359.
Bass, L., and Moore, W. J. (1968). In *Structural Chemistry and Molecular Biology*, pp. 356–369, ed. Rich, A., and Davidson, N. San Francisco: Freeman.
Bennett, M. H. (1968). *Physiol. Behav.*, **3**, 507–516.
Biedenbach, M. A., and Stevens, C. F. (1969a). *J. Neurophysiol.*, **32**, 193–203.
Biedenbach, M. A., and Stevens, C. F. (1969b). *J. Neurophysiol.*, **32**, 204–214.
Brodal, A. (1947). *Brain*, **70**, 179–222.
Carr, W. J., Loeb, L. S., and Dissinger, M. L. (1965). *J. comp. physiol. Psychol.*, **59**, 370–377.
Carreras, M., Mancia, O., and Mancia, M. (1967). *Brain Res., Amst.*, **6**, 548–560.
Chase, T. N., and Kopin, I. J. (1968). *Nature, Lond.*, **217**, 466–467.
Cowan, W. M., and Powell, T. P. S. (1956). *J. Anat.*, **90**, 188–192.
Cragg, B. G. (1961). *Expl Neurol.*, **3**, 588–600.
Cragg, B. G. (1962). *Expl Neurol.*, **5**, 406–427.
Dennis, B. J., and Kerr, D. I. B. (1968). *Brain Res., Amst.*, **11**, 373–396.
Elliott Smith, G. (1909). *Anat. Anz.*, **34**, 200–206.
Elul, R. (1967). In *Progress in Biomedical Engineering*, pp. 131–150, ed. Fogel, L. J., and George, F. W. New York: Spartan Books.
Elul, R. (1968). *Data Acquis. Process. Biol. Med.*, **5**, 93–115.
Elul, R. (1969). *Science*, **164**, 328–331.
Fox, C. A., McKinley, W. A., and Magoun, H. W. (1944). *J. Neurophysiol.*, **7**, 1–16.
Fox, C. A., and Schmitz, J. T. (1943). *J. comp. Neurol.*, **79**, 297–314.
Franck, H. (1966). *C.r. Séanc. Soc. Biol.*, **160**, 863–865.
Freeman, W. J. (1959). *J. Neurophysiol.*, **22**, 644–665.
Gault, F. W., and Leaton, R. N. (1963). *Electroenceph. clin. Neurophysiol.*, **15**, 299–304.
Girgis, M., and Goldby, F. (1967). *J. Anat.*, **101**, 33–44.
Glees, P. (1946). *J. Neuropath. exp. Neurol.*, **5**, 54–59.
Gloor, P. (1955a). *Electroenceph. clin. Neurophysiol.*, **7**, 223–242.
Gloor, P. (1955b). *Electroenceph. clin. Neurophysiol.*, **7**, 243–264.
Haberly, L. B. (1969). *Brain Res., Amst.*, **12**, 481–484.
Hara, T. J. (1967). *Comp. Biochem. Physiol.*, **22**, 209–225.
Hara, T. J., and Gorbman, A. (1967). *Comp. Biochem. Physiol.*, **21**, 185–200.
Hasama, B. (1934). *Pflügers Arch. ges. Physiol.*, **234**, 748–755.
Heimer, L. (1968). *J. Anat.*, **103**, 413–432.
Heimer, L., and Larsson, K. (1967). *Physiol. Behav.*, **2**, 207–210.
Kerr, D. I. B., and Hagbarth, K.-E. (1955). *J. Neurophysiol.*, **18**, 362–374.
Khazan, N., Kandalaft, I., and Sulman, F. G. (1967). *Psychopharmacologia*, **10**, 226–236.
Kimble, D. P., and Zack, S. (1967). *Psychon. Sci.*, **8**, 211–212.
Koelle, G. B., and Friedenwald, J. S. (1949). *Proc. Soc. exp. Biol. Med.*, **70**, 617–622.
Lee, Sovan der, and Boot, L. M. (1955). *Acta physiol. pharmac. néerl.*, **4**, 442.
Le Gros Clark, W. E., and Meyer, M. (1947). *Brain*, **70**, 304–328.
Lohman, A. H. M. (1963). *The anterior olfactory lobe of the guinea pig*. Thesis, Nijmegen. (Quoted by Heimer, 1968.)
Lohman, A. H. M., and Lammers, H. J. (1967). *Prog. Brain Res.*, **23**, 65–82.
Mascitti, T. A., and Ortega, S. N. (1966). *J. comp. Neurol.*, **127**, 121–135.
Meyer, M., and Allison, A. C. (1949). *J. Neurol. Neurosurg. Psychiat.*, **12**, 274–286.

Morrell, F. (1967). In *The Neurosciences: A Study Program*, pp. 452–468, ed. Quarton, G. C., Melnechuk, T., and Schmitt, F. O. New York: Rockefeller University Press.
Nauta, W. J. H. (1950). *Schweizer Arch. Neurol. Psychiat.*, **66,** 353–376.
Noda, H., Manohar, S., and Adey, W. R. (1969a). *Expl Neurol.*, **24,** 217–231.
Noda, H., Manohar, S., and Adey, W. R. (1969b). *Expl Neurol.*, **24,** 232–247.
Nossal, G. J. V., Williams, G. M., and Austin, C. M. (1967). *Aust. J. exp. Biol. med. Sci.*, **45,** 581–594.
Oshima, K., and Gorbman, A. (1968). *J. Endocr.*, **40,** 409–420.
Parkes, A. S., and Bruce, H. M. (1961). *Science*, **134,** 1049–1054.
Peneloza-Rojas, J. H., and Alcocer-Cuarón, G. (1967). *Electroenceph. clin. Neurophysiol.*, **22,** 468–472.
Porter, R., Kado, R. T., and Adey, W. R. (1964). *Neurology, Minneap.*, **14,** 1002–1012.
Powell, T. P. S., Cowan, W. M., and Raisman, G. (1965). *J. Anat.*, **99,** 791–813.
Ramon y Cajal, S. (1911). *Histologie du système nerveux de l'homme et des vertébrés*, vol. 2, trans. Azoulay. L. Paris: Maloine.
Reite, M., Stephens, L., and Pegram, G. V. (1967). *Brain Res., Amst.*, **3,** 392–395.
Rhodes, J. M., Reite, M. L., Brown, D., and Adey, W. R. (1963). In *Neurophysiologie des états de sommeil*, ed. Jouvet, M. Paris: CNRS.
Richards, C. D., and Sercombe, R. (1968). *J. Physiol., Lond.*, **197,** 667–683.
Rose, J. E., and Woolsey, C. N. (1943). *Fedn Proc. Fedn Am. Socs exp. Biol.*, **2,** 42.
Scalia, F. (1966). *J. comp. Neurol.*, **126,** 285–310.
Schuckman, H., Kling, A., and Orbach, J. (1969). *J. comp. physiol. Psychol.*, **67,** 212–215.
Scott, J. W., and Pfaffmann, C. (1967). *Science*, **158,** 1592–1594.
Shute, C. C. D., and Lewis, P. R. (1967). *Brain*, **90,** 497–521.
Stone, H., Williams, B., and Carregal, E. J. A. (1968). *Expl Neurol.*, **21,** 11–19.
Ueda, K., Hara, T. J., and Gorbman, A. (1967). *Comp. Biochem. Physiol.*, **21,** 133–143.
Valverde-García, F. (1965). *Studies on the Pyriform Lobe*. Cambridge, Mass.: Harvard University Press.
Wang, H. H., and Adey, W. R. (1969). *Expl Neurol.*, **25,** 70–84.
Wenzel, B. M., and Salzman, A. (1968). *Expl Neurol.*, **22,** 472–479.
White, L. E. (1965). *Anat. Rec.*, **152,** 465–480.
Whitten, W. K. (1956). *J. Endocr.*, **14,** 160–163.

DISCUSSION

Pfaffmann: In our study Scott and I were concerned with the olfactory projections to the hypothalamus, because of such remarkable effects as the "pregnancy block" when the odour of a strange male is introduced after normal mating, the so-called "Bruce" effect (1959). Leonard and Scott in my laboratory (unpublished observations) have since continued to clarify the lateral olfactory-hypothalamic pathways and they have found a very circumscribed bed nucleus in the olfactory tubercle which has large cells projecting all the way back to the midbrain (giving off branches on the way) to a pair of nuclei, the nuclei Gemini. Thus anatomical evidence shows a continuous neural tract, to support those electrophysiological findings. The interesting point is that this is an excitatory system. That is, the olfactory stimulus drives these large cells, via one or two synapses. The responses look very much like the responses obtained in the olfactory bulb to stimuli such as amyl acetate, or other odorous stimuli. That surprised us because we thought that sexual odours would be more selectively reactive in that pathway than chemical odours. This excitatory system is in contrast to the inhibitory properties of Haberly's data. There appears to be an anatomical separation of predominantly excitatory effects and predominantly inhibitory effects.

Adey: I was very interested that Haberly's (1969) findings were of an inhibitory response in an anaesthetized preparation. Were these newer studies made on anaesthetized or unanaesthetized animals?

Pfaffmann: They have all been done in an anaesthetized preparation.

Adey: The presence of inhibition in an anaesthetized preparation is always interesting, although I don't think its absence would necessarily be significant.

Døving: Haberly used only four odorants and found no units specifically activated. In my view you have to create a novelty, an excitation in a cortical unit, to get through to the perceptual system, similar to what is found in the visual system. With only four odours and recordings from only 21 units it is very unlikely that you would find the one unit which is excited by an odour.

Adey: This is true, but bear in mind that there are two studies, by Biedenbach and Stevens (1969) and also by Richards and Sercombe (1968) who used isolated guinea pig olfactory cortex, both of which suggested that electrical stimulation of the lateral olfactory tract produces inhibitory responses as the *main* type of change in the olfactory cortex. They were dealing with electrical responses of very short latency and not dependent upon the natural stimulus. So how does that affect your argument?

Døving: There is a big difference between electrical and natural stimulation. I don't think it is possible to discuss this question further at present.

Gorbman and his co-workers (personal communication) have repeated their experiments with a large number of salmon and a larger number of samples from different rivers. The correlation is no longer as good as that found in their first experiments (Hara, Ueda and Gorbman, 1965).

The results of the studies by Kandel (1964) on *Carassius auratus* might be worth bearing in mind. He recorded intracellularly from the neurosecretory cells of the preoptic area. The axons of these cells go to the pituitary. You can identify the cells by stimulating the pituitary stalk antidromically. He could activate the cells orthodromically by stimulating the olfactory tract. The latencies in his recordings indicate that this influence is mediated via unmyelinated nerves in the olfactory tract. His finding raises the question to what degree unmyelinated fibres are participating in the olfactory pathways to higher olfactory centres.

Adey: My own anatomical studies gave me the strong impression of a predominantly unmyelinated set of pathways.

Døving: The unmyelinated fibres might account for the long latencies found for many of the responses in higher brain structures when the olfactory bulb is stimulated electrically. Instead of a multisynaptic pathway it might be an unmyelinated pathway.

Zotterman: We know from B. Andersson's experiments with implanted electrodes in the "thirst area" in the hypothalamus of fully awake goats that the drinking urge elicited by electrical stimulation is completely inhibited by quinine in the water. It is very plausible that you will be able to demonstrate similar inhibitory effects of odours on the hypothalamic regulatory centres for water and food intake.

Pfaffmann: Actually, we started working on the hypothalamus because we were trying to find taste inputs—without success so far. But that led us to look more widely, and we came upon this olfactory input, which is a very striking effect.

Andersen: I have worked on the hypothalamic satiety centres and the interplay between the medial centres and the lateral tracts in the hypothalamus. One of the difficulties here is that we don't know what happens in the excitatory system when something happens simultaneously in the inhibitory system. To my mind we have to be able to record from both these sites simultaneously, to see to what extent they influence each other.

Adey: You are very rightly arguing for the need to record from a sufficient number of neurons in a single hypothalamic domain, so that we can get some concept of whether there is a predominantly inhibitory or predominantly excitatory response at that place. The input–output transforms used in this cybernetic fashion certainly require that we take account of the total influx and the total efflux. And after all, the concepts of excitation and inhibition to my mind only have meaning when one comes to the final common pathway, which is whether or not an effect or function is changed by the preceding sequence of excitatory or inhibitory steps in the neuronal chain leading to that final common path.

REFERENCES

BIEDENBACH, M. A., and STEVENS, C. F. (1969). *J. Neurophysiol.*, **32**, 193–203, 204–214.
BRUCE, H. M. (1959). *Nature, Lond.*, **184**, 105.
HABERLY, L. B. (1969). *Brain Res., Amst.*, **12**, 481–484.
HARA, T. J., UEDA, K., and GORBMAN, A. (1965). *Science*, **149**, 884–885.
KANDEL, E. R. (1964). *J. gen. Physiol.*, **47**, 691–717.
RICHARDS, C. D., and SERCOMBE, R. (1968). *J. Physiol., Lond.*, **197**, 667–683.

GENERAL DISCUSSION

Lowenstein: It might be useful in this concluding discussion if speakers were to give us their first priorities for future work, experimentally or theoretically. Professor Pfaffmann, what do you think of the future in the field of taste?

Pfaffmann: First, regarding structure. Since we know that there are local differences in the taste buds on the tongue within a species, and also differences among species in the relative density of units reactive to different chemicals, it would be desirable if a histochemical analysis could be made. For example, in the rat tongue we know that salt and acid electrolyte receptors are located on the front and sugar and quinine receptors at the back. Such functional differences should be examined with histochemical methods. I think for example of Oakley's work (1967) on crossing the chorda tympani with the IXth nerve and *vice versa*. When the nerves regenerate the regrown nerve appears to adopt the characteristic sensitivity of the tongue area to which it goes. The IXth nerve regrown to the front of the tongue is more sensitive to electrolytes than when it goes to its normal focus. Clarification of what there may be histochemically in the tongue surface would be an important step towards understanding the chemical genesis of specific mechanisms of sensitivities.

In terms of electrophysiological studies, the extension of the techniques Dr Døving has described for olfaction to the analysis of the electrophysiological patterns of taste sensitivity would be very worthwhile. We might find in the objective data some indication of a closer relation or clustering among the electrolytic types of taste receptors as opposed to those responding to organic stimuli.

My former student Erickson (1963) has done some of this work, making statistical analyses of many fibres. We tried to study intensity relations, that is, to plot more thoroughly the stimulus–response functions, as well as simply studying a single-value stimulus, as in the experiments I described in my paper.

Dr Adey asked whether recordings had been made in unanaesthetized preparations. This is very important because actually the evidence of multi-quality input in the taste system is not unique to taste. Perhaps we were naïve to be surprised by it. Later evidence from the visual and the auditory system indicates wide bands of sensitivity around a best frequency, for example. With increases in intensity of stimulation, the "spill-over" that Dr Wright was arguing for is quite apparent in hearing. A wide range of frequencies stimulate a single auditory nerve fibre, but the best frequency usually gives the highest response. The central projections of the auditory system and colour vision system show evidence of "sharpening"; that is, the side bands become narrower. It appears that lateral inhibition is operating, perhaps at the several successive cell stations. Inhibition is very sensitive to anaesthetic effects and it is of highest priority to study thalamic sensory neurons to see if there is such lateral inhibition without anaesthesia. Finally, there remains the paradoxical question of why cortical units do not respond to chemical stimulation of the tongue. Recordings can be quite easily made in the thalamic nuclei which project to the cortical taste area.

Murray: I strongly support the desirability of making chemical analyses in different topographical regions of the taste organs. It would be important also to analyse differences between the areas of the foliate and circumvallate taste buds on the one hand, and those of the fungiform, and we hope to do this. But I would also like to turn that round to say that I am pleased that there is now more physiological experimentation on taste as mediated by the IXth nerve, which I hope will be expanded and correlated with fine structure. Perhaps there will be little histological or histochemical difference between various parts of the fungiform field, but I would expect significant differences to be found between the fungiform area and the circumvallate or foliate areas.

Zotterman: I should like to focus the interest of physiologists on experiments on man, since it is only with human beings that one is able to make direct comparisons between neurophysical and perceptual events. Perceptual analysis to my mind adds much more interest to our experiments. The trend among young researchers in the field is towards the physical side of sensory events, so much that they refer to pain fibres as high threshold fibres, for example. Certainly physiology is nothing else but physics and chemistry applied to the function of the body, but we must not neglect the sensory side.

Lowenstein: We now leave the area of taste and come to the olfactory system. Here I would ask Dr Reese what he feels will come next in the ultrastructural field.

Reese: We have heard several new ideas that sound promising. Dr Ottoson showed that it is possible to make preparations of isolated olfactory cilia that could be used for biochemical analysis. These preparations can be examined with the electron microscope to see what components of the olfactory surface are present.

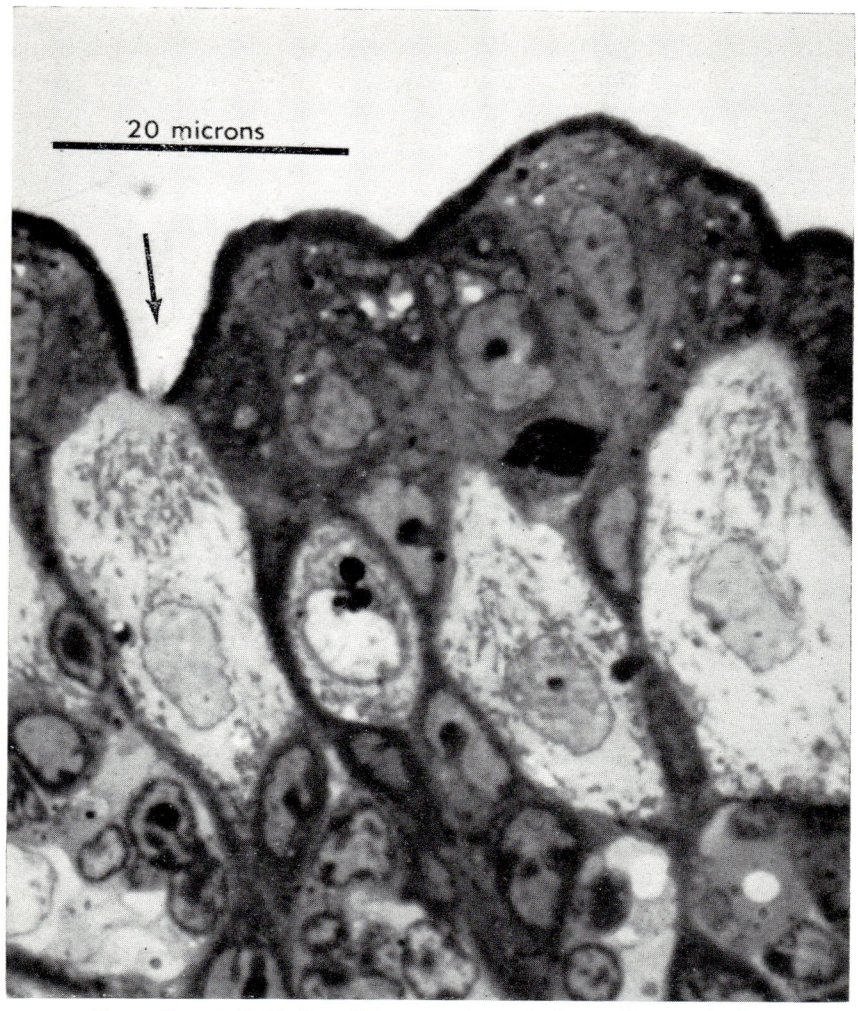

FIG. 1 (Reese). Epithelium lining troughs at the base of septae in the olfactory sac of a guitar fish (*Rhinobatus lentiginosus*; elasmobranch). Large, lightly stained cells appear to be bipolar neurons which, at their apices, extend tufts of villi (arrow) into ambient seawater. Apical surface upwards × 2000.

Dr Andres' and Dr Moulton's work promises that we will soon know exactly how rapidly olfactory cells are turning over.

Now that we have looked at the olfactory epithelium in most classes of vertebrates, we should develop a preparation with which the physiologist could record intracellularly from individual olfactory cells. Certain large cells in the elasmobranch olfactory sac look promising (Fig. 1), but amphibia known to have large cells in other locations (e.g. *Amphiuma*) should be looked at carefully for this purpose. Perhaps the olfactory epithelia that lack cilia will prove useful for evaluating the role of cilia in the electrical responses of the epithelium to odours.

To turn to the olfactory bulb, I would hope that we might eventually be able to make a wiring diagram, of the type that Dr Andres showed, which would include the sources of all the different types of synaptic connexions. Such a diagram should also indicate how many of these different types of connexions are present and give some picture of the variations in connexions between individual cells of each type. We should also know which neurotransmitters are associated with various types of synaptic ending. As physiological data of comparable detail become available we shall have a good picture of how the olfactory bulb processes the information received from the epithelium. I believe that we have already progressed far enough to predict that many of the principles of cortical anatomy and physiology will be worked out in the olfactory bulb.

Døving: I had always hoped to find one of the big receptor cells which Dr Reese has shown in the guitar fish! The question of course is what kind of odours would stimulate such a cell. It would be most interesting to know whether these cells are specific and have only one kind of receptor site on them, or if there are different sites on one cell. We could then test Dr Wright's or Dr Amoore's theories on this preparation, and also correlate the findings from the results on insect antennae.

Second, working on the bulb it is essential that we should be able to identify the cell from which we are recording and see how each particular cell responds to a large number of odours. This also applies to the higher olfactory centres. Here as a junior member of this group, I would disagree with our senior member! Work on the human olfactory system seems unnecessary as long as we don't know all the features of the olfactory system in lower vertebrates. I think we should understand these phenomena in other vertebrates before we go on to such experiments in man.

Finally, we should aim to find a group of chemicals which will be fruitful in our olfactory studies. In this connexion I would hope that physiologists and biochemists will both work on the same chemicals so that we can compare results obtained from different laboratories.

Lowenstein: Professor Davies, you presented us with a challenging hypothesis. What lines of approach would you like either to initiate yourself or to see initiated that would pursue your own ideas of what I might call a non-specific receptor theory?

Davies: There are three possible lines along which one might attack this problem. The first is that we seem to be agreed that the cilia are responsible for the transduction mechanism in olfaction, and this means that we do not have to analyse the whole olfactory epithelium with its mucus and sustentacular cells, but can now concentrate on an analysis of the lipids and the proteins in the cilia. These could be removed by the filter paper technique or by detergent. The lipids could be analysed to see if they differ in different parts of the olfactory epithelium, as Dr Døving has suggested. One could also analyse the proteins which might conceivably complex with certain odorant molecules. An analysis of cilia from those parts of the epithelium which are more or less sensitive to certain types of odorant might be revealing.

Secondly, the vibration theory of Dr Wright, because it is quantitative, is open to sharper testing than most of our theories. With the use of further deuterated or, better still, tritiated derivatives one could shift the infrared bands a little further, and so test his theory. He postulates that the peak must be in a certain band; if one took a peak which was towards the end of this band, tritiated the molecule and shifted the peak outside the band, one should lose the odour very sharply if the theory is correct.

Thirdly, Dr Mac Leod reported earlier on similarities between different compounds measured electrophysiologically, and in particular a strong similarity between isoamyl acetate and decanol. This does not seem to me to be in accord with psychological similarity, and this worried me because the previous electrophysiological work by Dr Døving seemed to be in general agreement with psychological similarity. This is something that we should look at further to see just what electrophysiological properties one is actually measuring when Dr Mac Leod says that isoamyl acetate is very similar to decanol.

Lowenstein: Professor Wright has told us something about his ideas at the transducer end; has he any observations and ideas about what olfaction looks like at the action potential end? I believe that his hypothesis rests to a certain extent on the assumption of a synchrony between input and output: is that correct?

Wright: Yes, the frequency band in the odorous molecule has to match a frequency band in the receptor, and therefore one would have to do a chemical isolation of the materials in the receptor cells and run tests on them. But this is still a long way in the future because of the quantities involved. The immediate problem I would like to see investigated, and we are now assembling the equipment for it, is the systematic investigation of olfactory thresholds. To this end we are building a booth similar to that described by the American Society for Testing Materials which will have a dozen smelling points at which accurately controlled concentrations of substances in preconditioned air will be presented and people will smell them and say whether or not they can distinguish an odour. To be really useful it would be very desirable to replicate the apparatus in a great number of places with a regular exchange of samples so that our results could be cross-checked often enough for the reliability of the measurements to be validated. In this way a large body of good data would become available for general use. A multiple facility of this sort would surely be very useful to the perfume industry, and it is possible that agencies such as the National Bureau of Standards in the United States might be persuaded to offer financial support. Similar things are done by the oil industry in relation to octane ratings.

As to the practical usefulness of these measurements, the threshold data will be interesting and useful more particularly if they could be correlated with such properties as vapour pressure or polarity or lipid solubility, but I can see many other fascinating avenues of experiment that can be quickly opened up, given this kind of facility. One example is the interesting technique developed by Dr G. H. Cheesman in Tasmania, where he tries to determine the amount by which pre-adaptation to one odour raises the threshold of another, and in this way seeks numerical evaluations of the degree of similarity. Multidimensional scaling is based on subjective estimates of similarity; Cheesman's technique offers a way of getting an objective, numerical estimate of the similarity between two odours, and this could be very useful in all sorts of ways.

In connexion with my own ideas, we have an experiment we hope to do. We believe that the bitter almond odour involves a vibrational frequency of about 175 cm^{-1}, a frequency of 225 and another frequency at 345 cm^{-1}. The cumin odour involves a frequency of 175 and two others which do not coincide with those of bitter almond. We will measure the threshold for a compound with the bitter almond odour and the threshold for a compound with a typical cumin odour, and then prepare a blend in which each of these compounds is separately below its own threshold. We hope that the quality arising from the 175 cm^{-1} frequency, which is present in both halves, will come above threshold and the subjects will be smelling a pure primary odour sensation. If that can be done, we may be able to do it again, because one of the other frequencies seems to coincide with one of the musk frequencies. You can see the possibilities. So a facility such as I envisage, which is not very expensive, opens the way to many interesting experiments and critical tests of some of the different hypotheses put forward here.

Amoore: My personal list of future objectives starts with the search for primary odours by the method of specific anosmia, which seems to be promising and should produce a working list of primary odours. Looking further afield, I feel we should encourage people working in receptor proteins to purify their proteins and look into what co-factors may be needed. With regard to the friendly rivalry between the odour theories over many years, I would suggest a "correlation coefficient" competition, where we try to work on the same set of biological data, say a set of odour similarities or a set of insect attractancies; and then each should apply his own theory and try to quantify it. One will obtain correlation graphs from which to see which theory accounts for most of the variance. I would also like to see other methods of looking for the hypothetical primary odours. When their methods are refined to the point at which they can work on single cells, electrophysiologists will have a good chance of getting at the primary odour qualities. The method of fatigue has been mentioned; such alternative methods are needed both for looking for primary odours and for checking those postulated by other approaches.

Beidler: One of the most exciting areas is certainly the identification and isolation of specific proteins or other specific materials. I would like to see experiments with the taste protein to show that it has some relation to taste buds; that is, that this protein is contained in a taste structure. With respect to neural coding, I would like to see techniques developed for multiple recording from single units, so that interaction between units can be studied over long periods of time.

Lowenstein: There seems to be a common factor emerging of multidisciplinary collaboration. Professor Zotterman, as the elder statesman here, how could one set about creating an organization in Europe or in the United States by which such multidisciplinary collaboration could be carried out, with continuous interchange of information and the application of different theories to the same problem, using mutually agreed experimental techniques, or making a search for mutually agreed parameters?

Zotterman: The main experience we have of supranational organizations is one which most of us belong to, UNESCO and its International Council of Scientific Unions. Its activities have been mainly concerned with financing meetings of research workers. The permanent congress committees for the different disciplines have however developed much wider interests. A Commission on Olfaction and Taste has just been set up within the Union of Physiological Sciences, so there will be a symposium on olfaction and taste in conjunction with the next International Congress of Physiological Sciences in Munich in 1971. But the aid obtainable financially from UNESCO or from ICSU is minimal; it primarily gives a kind of official recognition.

Lowenstein: Professor Wright, can you estimate the cost of setting up one of the chain of research facilities that you envisage?

Wright: I would estimate that the cost of the hardware for such a unit will be between 2000 and 3000 dollars, perhaps 4000. The costs of labour to build and operate it and of the people who will be tested are additional but not excessive.

Lowenstein: The order of magnitude is not as great as with much other instrumentation. I am quite sure that if the proper way of negotiating were to be found, industry would consider assisting such plans. It is mainly a question of how to approach the problem, and it might be useful if we put on record that the problem exists, and Professor Zotterman and other members of the ICSU Commission on Olfaction and Taste may be able to take it up.

Beets: Several personal talks during this symposium have given me the impression that research on chemoreception is financially in a poor position. It does not attract sufficient attention to obtain adequate support and its interests are not taken care of by an international body such as exists for chemistry, physiology and many other branches of science. Yet chemoreception is a multidisciplinary subject of which the results will probably radiate out into many other disciplines.

Would it not be possible to set up an International Society for Research on Chemoreception? Such an international body could do much better than any private research institute or individual worker in raising and distributing funds. It could approach the big foundations and the large pharmaceutical and fragrance industries for annual support with a much better chance of success than private institutes or research workers.

Lowenstein: Dr Beets' suggestion is a very good one, and it has been heard here by members of the Olfaction and Taste Commission of ICSU. If this suggestion were to come to fruition, and if it had started at this symposium, I myself would be extremely pleased and so I am quite sure would the Ciba Foundation.

Amoore: The beginning of such an organization exists in the Monell Chemical Senses Center at the University of Pennsylvania, which has promise for becoming a centre for international cooperation in chemical senses research. Dr David Moulton is the Deputy Director of the Center.

Moulton: The Center could well serve as a focus for such activities. It was founded in 1968 in the University of Pennsylvania. It is now in temporary quarters but should move into a new building late in 1970. The Director is Dr Morley R. Kare and the full-time professional staff now numbers seven. We are exploring the possibility of initiating an international journal. While we certainly need more financial support, there has been an encouraging response from industry.

Zotterman: Such a research institute would have the means of administering a society such as Dr Beets suggested. This is difficult to do from an ordinary university laboratory.

Lowenstein: Would the institute be willing to farm out research, as has been suggested?

Moulton: We would certainly consider the suggestion. Possibly there is room for a similar centre in Europe, since one centre might not be effective throughout an entire area as wide as Europe and America.

REFERENCES

ERICKSON, R. P. (1963). In *Olfaction and Taste* (Proceedings of the First International Symposium, Stockholm, 1962), pp. 205–213, ed. Zotterman, Y. Oxford: Pergamon Press.

OAKLEY, B. (1967). In *Olfaction and Taste II* (Proceedings of the Second International Symposium, Tokyo, 1965), pp. 535–547, ed. Hayashi, T. Oxford: Pergamon Press.

CHAIRMAN'S CLOSING REMARKS

O. E. LOWENSTEIN

THE Latin word *sapere* means to know, and it also means to smell or taste (because the classical Greeks and Romans did not distinguish between smell and taste as we do now). This fact alone is a pointer, although it is an anthropomorphically conceived one, to the importance of the chemical senses among all the senses. Yet we humans are behaviourally quite exceptional from the point of view of chemoreception. In this respect we are really the paupers among the animals, because I don't think it is an exaggeration to say that apart from birds and mammals that have become visually oriented behaviourally, the majority of animals, if asked, would probably name chemoreception as their basic and fundamental sense by which they find their way in the world. The pain and pleasure of the world is very often associated with chemoreception. As J. Z. Young (1968) put it once, there is a direct evolutionary pathway from feeding to memory and from memory to intelligence, and by that he meant that feeding is one of the fundamental functions, if not perhaps the fundamental function, of animals besides sex (and Freud knew that as well but probably put the emphasis on the wrong side of the picture).

I have used the term "chemoreception". This symposium is entitled "Taste and Smell"; we could also have called it "Chemoreception", and I wonder if this is a useful concept, or term, for the future. In conversation it has even been suggested to me by one or two members of this symposium that we might eventually have to separate our interests and to discuss taste and smell at separate meetings. I would be very sorry to see this, especially in the light of the experience I was privileged to have in this chair during these few days. An earlier symposium in this series was on myotatic, kinesthetic and vestibular mechanisms, quite deliberately bringing together people working on stretch receptors and those working on the labyrinth, because I felt there was a lack of communication between these two groups of research workers. In that symposium I had the rather discouraging experience of finding that these two groups had either stayed or become semantically so separated that they really did not speak the same language to any great extent. I therefore had apprehensions that the same might be the case between the workers in the fields of taste and smell, but I am very much gratified to find that you do speak the same language, and it would surely be a great shame if research on the two chemical sense modalities were to become separated. When we dealt with ultrastructure we found similar problems and puzzling "abnormalities". For instance, we found in both fields that what had been previously considered to be the sensory cells are now under suspicion of being sustentacular cells, which might, however, be involved in transmission by acting like glia cells in the handling of olfactory or taste information. We have had great emphasis placed on the kinocilium, the "motile" cilium, and yet we were faced with the fact that some obviously sensory cells, as documented by their nervous connexions, lack these cilia. This is not a new shock to me, because when Wersäll and I first pointed out the topographical importance of the kinocilium in the sensory epithelium of the vestibular organ (Lowenstein and Wersäll, 1959) we felt that the kinocilium had to play a central role in the transduction process. Then we reminded ourselves that it is absent in the cochlea, the hearing part of the ear. Although we had the strongest inducement to put the kinocilium first, we became doubtful and had to look for an alternative hypothesis of mechanoelectrical transduction even in the vestibular organ (Lowenstein, 1967). Both in the taste and the olfactory systems we find cells without cilia which seem to function perfectly well. On the other hand we have had evidence in certain animal types for the assumption that probably the most likely transduction site is not merely the cilium but perhaps only the very tip of it.

There is another point which we must not lose sight of: our efforts in physiology, biochemistry and biophysics would be useless if we did not keep the behavioural side continuously under review, and here it is a matter for regret that because of the limitation of our symposium to vertebrates we have had nobody present who could directly acquaint us with the most recent work on insects. Insects have come up only by way of reference. When one thinks of the concept of pheromones and the fantastic influence such substances have, and that they often make contact with the organism through olfactory organs and not through the whole body surface as in the lower invertebrates, and when one realizes the importance of pheromones for the guidance not only of the individual but especially of social behaviour, both in insects and perhaps on all levels of animal organization, one feels that a close collaboration between the ethologist and the physiologist as well as the biochemist is necessary to come to valid conclusions about the role played by the chemical sense in the biology of animals. With this exhortation for the complete abolition of scientific parochialism, I would close this symposium, which I think was one of the most successful which I have had the honour of chairing in this series. May I on your behalf express our thanks to the Ciba Foundation for having made possible this fifth and last of the series of symposia on vertebrate sensory systems.

REFERENCES

Lowenstein, O. (1967). *Ciba Fdn Symp. Myotatic, Kinesthetic and Vestibular Mechanisms*, pp. 121–128. London: Churchill.
Lowenstein, O., and Wersäll, J. (1959). *Nature, Lond.*, **184,** 1807–1810.
Young, J. Z. (1968). In *Biology of the Mouth* (Symp. vol. No. 89), pp. 21–35, ed. Person, P. Washington, D.C.: American Association for the Advancement of Science.

INDEX OF AUTHORS*

Entries in bold type indicate a paper; other entries are contributions to the discussions.

Adey, W. R.   50, 283, 287, 288, **357**, 377, 378
Amoore, J. E.   246, 262, **293**, 307, 309, 322, 384, 386
Andersen, H. T.   45, 46, **71**, 79, 80, 341, 378
Andres, K. H.   28, 69, 146, 175, **177**, 194, 196, 224, 248, 341
Ashton, E. H.   .   **251**, 261, 262, 263
Beets, M. G. J.   222, 224, 246, 290, 309, **313**, 321, 322, 323, 339, 340, 386
Beidler, L. M.   25, 47, **51**, 68, 69, 70, 81, 96, 113, 148, 223, 224, 250, 288, 311, 338, 341, 354, 355, 385
Borg, G. **99**
Brightman, M. W..   .   .   .   **115**
Burgess, R. E. **325**
Çelebi, G. **227**
Davies, J. T.   224, 247, **265**, 281, 282, 283, 286, 287, 288, 289, 290, 309, 311, 321, 339, 340, 355, 382
de Lorenzo, A. J. D.   25, 26, 28, 113, **151**, 173, 174, 175, 176, 355
Diamant, H. **99**
Døving, K. B.   47, 144, 149, **197**, 221, 222, 223, 224, 225, 250, 282, 283, 377, 378, 382
Duncan, C. J.   30, 69, 263, 287, 290, 310
Eayrs, J. T. 26, **251**
Fink, R. P. **227**
Hellekant, G.   68, 81, **83**, 96, 97, 113, 263
Lowenstein, O. E.   **1**, 26, 27, 68, 70, 82, 96, 112, 113, 143, 144, 148,
173, 174, 176, 261, 288, 321, 338, 341, 342, 356, 379, 380, 382, 383, 385, 386, 387, **389**
Mac Leod, P.   96, 147, 148, 194, 223, 282, 283, 340, 341
Martin, A. J. P.   79, 285, 286, 289, 290, 306, 308, 309, 310, 337, 338, 339, 340
Moulton, D. G.   146, 147, **227**, 246, 247, 248, 249, 250, 290, 323, 386, 387
Murray, Assia **3**
Murray, R. G.   **3**, 25, 26, 27, 28, 29, 30, 69, 70, 194, 248, 355, 356, 380
Ottoson, D.   69, 144, 249, 310, **343**, 354, 355
Pfaffmann, C.   25, **31**, 45, 46, 47, 50, 81, 82, 223, 338, 376, 377, 378, 379
Reese, T. S.   29, 30, 68, **115**, 143, 144, 146, 147, 148, 173, 174, 175, 176, 194, 223, 224, 249, 250, 380
Theimer, E. T. **313**
Wright, R. H.   45, 47, 50, 81, 148, 149, 174, 221, 262, 281, 282, 289, 308, 311, 321, **325**, 337, 338, 339, 340, 341, 383, 386
Zotterman, Y.   26, 29, 46, 68, 69, 70, 80, 81, 97, **99**, 112, 113, 147, 261, 262, 288, 289, 290, 310, 312, 355, 356, 378, 380, 385, 387

*Author and subject indexes prepared by William Hill.

INDEX OF SUBJECTS

Acetic acid, chorda tympani response to, 99, 100
Acrolein as test substance, 148
"Across fibre pattern" theory of taste, 72
Afferent fibres, relation to individual receptors, 37
Alcohol, perception from tongue, 68
Amphibia, olfactory bulb in, 178, 181
   olfactory receptors in, 198
   specificity of olfactory nerve, 273
Amygdaloid nucleus, 362, 365, 370
Amyl acetate, species differences in sensitivity to, 242–243, 246–247
Anaesthesia, effect on gustatory responses, 50
   effect on olfactory responses, 377
Anosmia, 296, 314, 384
   to isovaleric acid, 296–299, 302

Birds, olfactory bulb in, 180
Bitterness, affected by gymnemates, 75, 80, 81
   molecular basis, 61–62
   relationship with sweetness, 75
Blood-brain barrier, 165
   olfactory neuron and, 151–176
   relation to olfactory system, 165
Blood flow, affecting gustatory sensitivity, 84 *et seq.*
   around taste buds, 89–90, 96
Bowman's glands, cell renewal in, 228, 237
   discharge of, 148
Bruce effect, 376

Carboxypeptidase, structure, 299
Carotid occlusion, response of chorda tympani during, 84
Chorda tympani, division of, 113
   individual afferent fibres in, 35
   relation to taste buds, 55
   response to,
      acetic acid, 99, 100
      adaptation, 109–111
      carotid occlusion, 84
      citric acid, 101 *et seq.*
      quinine, 99, 100, 107, 108
      saccharin, 99, 100
      sodium chloride, 99, 100, 101 *et seq.*

Chorda tympani—*continued*
      sucrose, 99, 100, 101 *et seq.*
      sugars, 38–43, 48, 75
      taste stimuli, 99–113
   stimulation, 33
   effect on blood flow in tongue, 90, 92
Cilia, gustatory, *see Gustatory cilia*
   olfactory, *see Olfactory cilia*
Citric acid, chorda tympani response to, 101 *et seq.*
   psychophysical and neural response, 102, 103, 104, 107
Cortical olfactory areas, 357
Cortical response to taste stimuli, 113
Cyclamates, psychophysical and neural response to, 107

Dendrites, in olfactory bulb, synaptic contacts, 128–133, 182, 187, 190, 194, 195, 196
   of olfactory granule cells, 182, 194, 195
   of olfactory mitral cells, 178, 181, 182, 194, 195, 224
Desorption rate, of taste molecules, 217, 274–275
Diencephalon, 357
   centrifugal projections from, 365–369
   olfactory bulb projections to, 362–363
   secondary projections to, 363–365
Dogs, detection of hidden objects by, 251–263
   duration of concealment, 258
   mechanism, 255–257
   mode of concealment, 252, 257
   role of vision, 258, 261
   size of object, 255
   stimulus strength, 260
   structure of object, 255
   infrared receptors in nose, 261
   olfactory sensitivity, 238, 269
   tactile sense, 261
Duplexity theory of taste, 74

Efferent endings, on gustatory receptor cells, 17, 22
Electro-olfactogram, 199, 345, 349, 351
Enzymes, as amplifying mechanism in olfaction, 306

INDEX OF SUBJECTS

Enzymes—*continued*
  theory of olfactory sensitivity, 306–307, 310, 311
Ethylene glycol, psychophysical and neural response to, 107
Eye, photoreceptors, 351

Fatigue, to olfactory stimulation, 290, 307
Fatty acids, anosmia to, 296–299, 302
Fish,
  convergence in olfactory bulb, 144
  olfactory glomeruli in, 178
  olfactory pigment in, 148
Foliate taste buds, 17
  cellular lifespan, 51
Fructose, response of chorda tympani to, 38 *et seq.*
Fungiform papillae, nerve fibres to, 36
  taste buds in, 17, 36
    blood supply, 92
    cellular lifespan, 51, 53

Glossopharyngeal nerve, response of, 31
Glucose, response of chorda tympani to, 38
Granule cells, *see under* Olfactory cells
Gustatory cells, cilia, *see* Gustatory cilia
  dark, *see* Gustatory cells, supporting
  lifespan, 51–53
  light, *see* Gustatory cells, type II
  microvilli as receptor sites, 56
  mitochondria in, 26
  nerve fibre connexions, 5, 34
  receptors (=type III cells), 4, 10, 26, 28
    adaptation, 109–111
    combination with odorous molecule, 310
    cortical connexions, 112–113
    effect of anaesthesia, 50
    efferent endings on, 17, 22
    gymnemate blocking action, 49
    histochemistry, 29
    influence of circulation on response, 83–97
    in fungiform papillae, 17
    inhibition, 72
    interaction with stimuli, 74–78
    localization, 55–59
    mechanism, 60–64, 310–311
    membranes, 10, 58, 60
    microvilli, 29
    multiple sensitivity, 46, 72
    nerve endings, 9, 11, 69

Gustatory cells—*continued*
  physiological properties, 51–70
  quantitative response, 37
  reaction time, 59–60
  replacement of, 22, 25, 28, 51–53
  response during carotid artery occlusion, 84
  response to sugar, 43
  role of molecular structure, 76
  role of oxygen tension in response, 89–90, 96
  transitional, 25, 28
  turnover, 22, 25, 28, 51–53
  supporting (=type I cells; dark cells), 4, 5, 9
    nerve endings to, 26
  turnover in, 22, 25, 51–53
  types, 4, 5–10, 25, 26–28
  type I (dark cells), *see* Gustatory cells, supporting
  type II (light cells), 9–10, 26–27
  type III (sensory cells), 10, 17, 25, 27, 28
  See also Gustatory cells, receptors
  type IV (basal), 10, 27, 28
  turnover, 22
Gustatory nerve fibres, contacts, 22
  receptor connexions, 34
  relation to taste bud cells, 5 *et seq.*, 25
  relation to receptor cells, 9, 11
  relation to type III gustatory cells, 10, 11
  response to hydrochloric acid, 33 *et seq.*
    quinine, 33 *et seq.*
    sodium chloride, 33 *et seq.*
    sucrose 33 *et seq.*
    temperature, 73
    test substances, 31
  specificity, 108–109
  within taste buds, 11, 17, 28–29, 55
Gustatory primaries, 71, 74, 81, 82
  as continuum, 75, 79
  specific papillae for, 73
Gustatory stimuli, interaction with receptor, 74–78
  neural and perceptual responses, 99–113
Gymnemic acid and gymnemates, 75
  affecting bitter and sweet, 75, 80, 81
  as taste modifiers, 65
  blocking action, 48, 49

Higher olfactory centres, 357–378
Hippocampus, 363
  electrical activity, 372, 373

"Hole sharpness factor", 272, 275
Hormonal effects on olfaction, 368–369
Hydrochloric acid, nerve fibre response to, 33 *et seq.*
  by chorda tympani, 48, 50
Hydrogen-bonding hypothesis of taste, 61–62, 76–77
Hydrolysis, effect on response to sugars, 45
Hypothalamus, 378
  olfactory paths to, 362, 363–367, 376
  role in sexual behaviour, 365, 368

Insects, olfactory cilia in, 147
  response to pheromones, 307, 308
International organization, for chemoreception research, 385–386
Invertebrates, chemoreception in, 43
Isovaleric acid, anosmia to, 296–299, 302
  molecular shape, 301

Lamprey, olfactory bulb in, 177 *et seq.*
  olfactory epithelium, 228
Light, effect on olfactory bulb, 338

Maltose, response of chorda tympani to, 38 *et seq.*
Melanophores, in olfactory epithelium, 148
Merkel's touch receptor, 69
Mesencephalon, centrifugal projections from, 365–369
  secondary projections to, 363–365
Microvilli, on olfactory receptors, 56, 120, 138, 147
  on taste receptors, 9–10, 56
Mine-detection by dogs, 251 *et seq.*
Miracle fruit (*Synsepalum dulcificum*), as taste modifier, 65
Mitochondria in olfactory epithelium, 120–121, 154
Mitral cells, *see under Olfactory cells*
Molecular geometry, 217, *see also under Odorant molecules*
  in sweetness, 76
  relation to taste, 47, 61
  role of, 76
Monkeys, olfactory mucosal structure in, 117–120, 152
Mucus in olfactory epithelium, 117, 137
Musky odours, 274 *et seq.*, 309
  infrared spectra, 321–322

Nerve fibres, contacting gustatory cells, 5
  direct stimulation, 69

Nerve fibres—*continued*
  gustatory, *see Gustatory nerve fibres*
  to taste buds, 28–29
Nervous system, olfactory centres, *see Higher olfactory centres*
Nose, infrared receptors in, 261

Odorant molecules, adsorption coefficient, 270, 273, 282
  deuterated, 339–340
  diffusion in air, 326, 351
  diffusion through membrane, 271–272, 276, 277, 281, 284, 344
  dipoles, 327, 329–331, 339
  geometry of, 217, 277–279, 281, 313 *et seq.*, 321
    computer assessment, 301–303
    correlation with odour, 293–312
    importance of size, 273, 298
  interacting dipoles, 329–331
  latency of response to, 346
  numbers stimulating receptors, 268–269, 274, 340–341
  penetrating cell membrane, 267, 276, 277, 281, 284, 285–286, 289
  quantitative relation to stimulation, 268–269, 340–341
  rotational diffusion, 327
  similarity, 383
  thermodynamics, 330, 331–333
  time passing through mucus, 354
  translational diffusion, 327
  transport, 326
  vibration, 328, 333–334, 336–337
Odour(s), *see also Odorant molecules*
  absolute sensitivity to, 238
  biological effects, 197
  cell contact with, 146
  classes, 197
  concentration in air, 242–243, 274
  correlation with molecular shape, 293–312
  Davies' theory, 217, 218, 265–290
  hierarchical clustering of, 211–213, 218, 223
  molecular geometry, 217
  see also Odorant molecules
  "penetration and puncturing" theory, 217, 218, 265–290
  physicochemical parameters, 214, 217
  primary, 295–296
  quality of, 217, 271 *et seq.*

Odours—*continued*
  quality of—*continued*
    desorption rate, 217, 274–276
    key-words describing, 318–320, 321
    molecular size, 217, 272–273
  receptor cell response to, 198, 340
  receptors corresponding to, 186, 341
  response to, correlation between psychological and biological data, 204, 213–215, 216
    of glomeruli, 199, 340
  similarity between structurally related odorants, 222, 313–323
    correlation of data, 204, 213–215, 216
    multidimensional scaling, 205–211, 216, 221, 384
    "stress", 205
  specificity, 325
  statistical analysis of response to, 202–204
  structurally unrelated, similarities between, 313–323
  structure, response and, 217, 224
  type, according to "penetration and puncturing" theory, 271–279
Odour spectrum, 279
Olfactory bulb, 124–137
  accessory, 181, 187
  anatomy and ultrastructure, 124–137, 177–196
  behaviour or colloidal gold particles in, 166–168
  inhibition in, 140–141, 143, 190, 193, 194–196, 199–200, 201
  centrifugal projections to, 190, 196, 365–369
  electrical rhythms, 370
  hormonal effects, 368–369
  in amphibia, 178, 181
  in birds, 180
  in mammals, 124–137
  in reptiles, 179
  layers, 124, 177
  neurohumoral influences, 365–369
  projections, to diencephalon, 362–363
    to primary olfactory cortex, 358–362
  response to stimuli, 373
  species differences, 177
  synaptic connexions, 139–141, 182–186, 201, 353, 382
Olfactory cells, areas in contact with odour, 120, 146
  axons, 164

Olfactory cells—*continued*
  basal, 151, 153
  cilia in, 228
  clusters, 232, 237
  location of, 231, 232, 236
  mitosis in, 228, 229, 248
  nerve contact, 163
  blood-brain barrier and, 151–176
  cilia, *see Olfactory cilia*
  contact with nerves, 163
  dendrite depolarization, 147
  distribution, 276, 279
  electrophysiology, 198–202
  granule, dendrites on, 190, 195
    inhibitory action on mitral cells, 141, 143, 190
    synaptic contacts, 143, 178 *et seq.*, 182, 194, 195, 196
  inhibition patterns, scheme of, 200, 201, 223
    in olfactory bulb, 199–200
    response to stimulation, 199
    in olfactory cortex, 200
    in olfactory epithelium, 119, 120, 124, 137, 138, 140
  junctions, 139
  membranes, 133, 134, 138
    healing, 273, 286
    lipid layers, 266–267, 282, 284
    odorant molecules penetrating, 267, 271–272, 276, 277, 281, 284, 285–286, 289
    structure, 283, 284
  migration in olfactory epithelium, 227–238
  mitral, activity, 241
    branching, 224
    dendrites of, 178, 181, 182, 194, 195, 224
    inhibitory action from granule cells, 141, 143, 190
    intensity of stimulus, 223
    synaptic contacts, 143, 166, 177 *et seq.*, 182, 194, 195, 196
  periglomerular, 124–125, 134, 141
  proliferation, 227–238
  receptors, 151, 153, 154, 187, 194, 283, 382
  cilia, *see Olfactory cilia*
  contacts, 162–165
  contact with other receptors, 163
  contact with supporting cells, 163
  contiguity, 154, 163

Olfactory cells—*continued*
  receptors—*continued*
    correlation with psychological data, 222
    corresponding to different odours, 186
    effect of stimuli concentrations, 223
    electrophysiology, 198, 241
    latency of response, 346–349
    membrane, 154–159
    microvilli on, 56, 147
    mitosis, 228
    non-ciliated, 355
    number of odorant molecules stimulating, 340–341
    number per unit area, 250
    potential, 344–346, 352
    ratio to olfactory tract neurons, 144
    relation to blood-brain barrier, 166 *et seq.*
    renewal, 237–238
    response to odorant, 273, 326, 328, 332, 333, 337, 340
    response to specific odour, 198, 200, 202–204
    ribosomes in, 159
    site of activation, 343
    statistical analysis of response, 202–204
    specificity, 186, 198, 273, 341
    stimulation according to penetration theory, 266–271
    synaptic contacts, 182
  single-cell recording from, 281
  supporting (sustentacular), 148, 151, 153
    contact with receptors, 119–120, 163–164
    microvilli, 119, 343
    renewal, 238
    mitosis, 228, 248
    synaptic contacts, 163, 166, 174–175, 201
  tight junctions, 119–120, 124
  types, 119–120, 151, 153
Olfactory centres, higher, *see* Higher olfactory centres
Olfactory cilia, 56, 115, 117, 120, 137, 138, 159–162, 267
  analysis, 380
  as site of transducer process, 343, 348–349, 354, 383
  conducting properties, 147
  contact with supporting cell microvilli, 163, 164

Olfactory cilia—*continued*
  electrical properties, 147
  energy reception in, 344
  extent of, 347–348
  growth of, 355
  in basal cells, in lamprey, 228
  membrane, 162
  modifications, 117, 160–162
  movement, 138–139, 173
  orientation, 154, 173
  renewal, 144
  role of, 137–138, 146
  vacuoles in, 146, 162
Olfactory cortex, inhibition in, 200, 202, 359–361, 377, 378,
  olfactory bulb projections to, 358–362
Olfactory epithelium, 137–139, 381
  air turbulence at, 174
  behaviour of colloidal gold particles in, 166–168
  cell distribution, 279
  cell migration in, 227–238
  cells in, 119, 120, 124, 137, 138, 140
  cell proliferation in, 227–238
  cell types in, 117–124, 151, 153
  cell zones, 231, 236, 237
  ciliary movement in, 138–139, 174
  effect of light, 338
  melanophores in, 148
  mid-zone, 231, 236, 249
  mitosis in, 284
  mucus on, 117, 137
  of elasmobranch, 120–124, 139
  of monkey, 117–120
  perimeter zone, 231, 236
  peripheral zone, 231, 237
  regeneration, 227, 248
  stimulation of, 148, 221
  thickness, 249–250, 287, 327
  villi in, 119
Olfactory glomerulus, axons, 125, 128, 133, 139, 144
  dendrites, 129–137, 178–187
    synaptic contacts of, 131–133, 139–142, 180–187, 190, 194–196
  electrophysiology, 198–199
  fibre types in, 190, 341
  in fish, 178
  in lamprey, 177
  in rat, 124–137, 139–142, 144
  nodules, 128, 129
  periglomerular cells, 124–125, 134, 141

INDEX OF SUBJECTS

Olfactory glomerulus—*continued*
  relation to blood-brain barrier, 166 *et seq.*
  response of, 223
    to specific odours, 199
  synaptic junctions, 133, 139–140, 353
Olfactory hair, role of, 137
Olfactory membranes, ions penetrating, 266, 281, 282, 285, 286, 288, 289
Olfactory mucosa, of squirrel monkey, 117–120, 152 *et seq.*
  relation to blood-brain barrier, 165
  structure, 115 *et seq.*, 151
Olfactory nerve fibres, activity in, 355
  impulses, origin of, 137–138, 147, 267
  progression of virus infection along, 166
  specificity, 200–202, 273
  synaptic contacts in olfactory bulb, 128–134, 143–144, 182–186, 194–196
  synaptic contacts in mucosa, 174–175, 194
Olfactory paths to hypothalamus, 362, 363–367, 376
Olfactory pigment, 148
Olfactory rod, 115, 154, 157–159
  contact with supporting cell microvilli, 163
  cytoplasm, 159
  mitochondria in, 154
  vacuoles in, 157–159
Olfactory sensitivity, absolute, 238–243, 249, 250
  adaptation, 290, 307
  amplification in, 306, 311
  biochemical aspects, 299–300
  correlation between psychological and biological data, 283
  differential, 243–244
  enzyme theory, 306–307, 310, 311
  factors, concentration of odour, 274
    limiting osmic frequency, 331
    molecular dipoles, 327, 329–331, 339
    molecular geometry, 217, 277–279, 281–282, 293–305, 313 *et seq.*, 321
    molecular transport, 326
    molecular vibration, 328, 333–334, 336–337
    numbers of odorant molecules, 268, 269
    rate of diffusion through cell membrane, 272
  in mammals, 238–245
  mechanism, 255–257, 306–307

Olfactory sensitivity—*continued*
  of dog, 251–263
    training, 252
  range of thresholds, 271, 283
  role of higher nervous centres, 342
  species difference to amyl acetate, 242–243, 246–247
  stimuli-response relation, 217, 224, 241
  thermodynamic considerations, 331–333
  vibrational theory, 325 *et seq.*, 336 *et seq.*, 383
Olfactory stimuli, *see also Odorous molecules*
  physicochemical mechanisms, 325–342
  response of pyriform cortex, 360
Olfactory surface, 115 *et seq.*, 151 *et seq.*
Olfactory tract, electrical stimulation, 359
  fibres, distribution, 359
  integrative processes, 201
  mechanism, 143–144
  stimulation, 377
    statistical analysis of response to, 202–204
  terminations of, 357
Olfactory transducer action, cilia as site of, 146–147, 343, 348–349, 354–355, 383
  electrical signs, 343–357
  transmission to higher levels, 352–353
Olfactory tubercle, 363, 367
Olfactory vesicles, 146, 354
Operant conditioning procedures, testing sensory functions with, 40
Opossum, olfactory sensitivity, 242, 243
Osmic stimulus, 325
Oxygen tension and gustatory sensitivity, 84, 85, 96

Parosmia, 314, 318, 319, 322, 323
"Penetration and puncturing" theory of odour, 265–290
  cell membranes in, 266 *et seq.*
  "hole sharpness factor", 272, 275
  mathematical model, 270–271
  molecular basis, 267, 268, 269
  odour type according to, 271–279
  stimulation according to, 266–271
Peripheral interaction, 73
Pheromones, 295, 307
  insect response to, 307, 308
Pregnancy block, 376
Proteins, electrogenic, 285, 287
  in retinal rods, 53
  receptor, *see Receptor proteins*

INDEX OF SUBJECTS

Pyriform cortex, electrical rhythms, 358, 359, 370, 372, 373
  projections to, 363, 366, 368
  response to olfactory stimuli, 360

Quinine,
  response to, 49, 50
    by chorda tympani, 99, 100, 107, 108
    by nerve fibre, 33 et seq.

Receptor cells, gustatory, 4, 5, 9, 26, 28
  adaptation, 109–111
  antidromic response, 70
  cortical connexions, 112–113
  cytoplasm, 9
  difference in response, 37
  effect of anaesthesia, 50
  efferent endings on, 17, 22
  for intravenous stimulation, 56
  gymnemate blocking action, 49
  histochemistry, 29
  influence of circulation on response, 83–97
  in fungiform papillae, 17
  inhibition, 72
  in invertebrates, 43
  interaction with stimuli, 74–78
  intravenous response, 68
  localization of, 55–59
  mathematical description, 63, 64–65
  mechanism of stimulation, 60–64, 310–311
  membranes, 58, 60
  microvilli, 29
  multiple sensitivity, 46, 72
  nerve endings, 69
  physiological properties, 51–70
  reaction time, 59–60
  relation to fibres, 9, 11
  replacement, 53
  response during carotid artery occlusion, 84
  response to hydrochloric acid, 48, 50
  response to quinine, 49, 50
  response to sugars, 43
  role of molecular structure, 76; see also Odorant molecules
  role of oxygen tension in response, 89–90, 96
  sequence of events, 310
  transitional, 25
  turnover, 22

Receptor cells, gustatory—*continued*
  Merkel's touch receptor, 69
  olfactory, 139, 151, 153, 154, 187, 194, 283, 382
    cilia, see *Olfactory cilia*
    contacts, 162–165
    contact to supporting cell, 119–120, 124, 154, 163
    correlation with psychological data, 222
    corresponding to odours, 186
    effect of stimuli concentration, 223
    electrophysiology, 198
    in contact with odour, 146
    latency of response, 346–349
    membrane, 154–159
    microvilli on, 56, 147
    mitosis, 228
    non-ciliated, 120, 138, 146, 355
    number of odorant molecules stimulating, 340–341
    number per unit area, 250
    potential, 344–346, 352
    ratio to olfactory tract neurons, 144
    relation between stimulus and response, 349–351
    relation to blood-brain barrier, 166 et seq.
    renewal, 227, 237–238, 247–248
    response to odorant, 326, 328, 332, 333, 337, 340
    response to specific odour, 198, 200, 202–204
    ribosomes in, 159
    site of activation, 343
    statistical analysis of response, 202–204
    stimulation according to penetration theory, 266–271
    synaptic contacts, 124 et seq., 182
Receptor proteins, 385
  in cortex, 373
  olfactory, 293, 294–295
    conformational change in, 284–285, 299–301
    mutations, 297
  taste, 61, 65, 76, 269, 294, 385
  visual, 53
Reptiles, olfactory bulb in, 179
Reticular formation, 368
Retina, 283, 286
Retinene, 288
Rhodopsin, 53, 287

INDEX OF SUBJECTS

Ribonucleoprotein particles in olfactory glomerulus, 131, 134

Saccharin, aversion to, 59
  chorda tympani response to, 99, 100
  psychophysical and neural response, 107, 108
Salivary glands, 56, 68
Salty taste, molecular basis, 62
Sex hormones, 368
Sexual behaviour, 371, 376–377
  olfactory influences, 365
Sexual odours, 377
Skin, bimodality fibres in, 46
Smell, *see under headings beginning Olfactory* etc.
Sodium chloride, affecting response to sugars, 80
  chorda tympani response to, 99, 100, 101 *et seq.*
  nerve fibre response to, 33 *et seq.*
  psychophysical and neural response, 103, 104, 107
    adaptation, 109–111
    response to, 50
Sourness, molecular basis, 62
Steroid ketones, 316
Sucrose, chorda tympani response to, 99, 100, 101 *et seq.*
  nerve fibre response to, 33, 37, 38 *et seq.*
  psychophysical and neural response to, 103, 107, 108
Sugar(s), chorda tympani response to, 48
  discrimination by monkeys, 38–42
  psychophysical and neural response, 107
  response to, by chorda tympani, 38 *et seq.*, 49, 50
    effect of hydrolysis, 45
    effect of solubility, 75
    role of molecular structure, 61–62
Supporting (sustentacular) cells,
  of taste buds *see under Gustatory cells*
  olfactory *see under Olfactory cells*
Sweaty odour of isovaleric acid, 298, 302, 309
Sweetness, affected by gymnemates, 75, 80
  appreciation of, 38, 50
  molecular basis, 61–62, 76
  relationship to water fibres, 77, 80
  relationship with bitterness, 75
  sodium chloride affecting response, 80

Synaptic activity, in olfactory glomeruli 353
Synaptic contacts,
  in olfactory bulb, 182, 201
    bidirectional synapses, 134, 182, 187, 190, 193–195

Taste, *see also under headings beginning Gustatory*
  "across fibre pattern" hypothesis, 72
  animal preferences, 38
  behavioural studies, 38–43
  continuous spectrum, 75, 79
  discrimination, 34
  duplexity theory, 74
  hydrogen-bonding theory, 61–62, 76–77
  inhibition, 72
  intensity in relation to chorda tympani activity, 350
  intravenous, 56, 68
  modifiers, 47–49, 65, 80, 81
  physiology of, 31–50
  primary qualities, *see Gustatory primaries*
  psychophysical and electrophysiological approach, 71, 72–74
  relation of molecular structure to, 47, 61
  response, mathematical description, 63, 64–65
  sensitivity, 379
  sensory coding, 37
  specificity, 71–82
  terminology, 81
Taste buds, anatomy and ultrastructure, 3–30
  blood supply, 26, 89–90, 94, 96
  cell turnover in, 22, 51–54
  cell types, 4, 5–10
  chemical analysis, 380
  comparison between regions, 17
  effect of X-irradiation, 52
  foliate, 17
  fungiform papillae, 17
  histochemistry, 379
  innervation, 55
  interactions, 55
  lifespan of cells, 51–53
  light microscopy, 3
  mapping, 35–36
  multiple response, 46
  nerve supply, 11, 17, 28–29
  protein in, 269

Taste buds—*continued*
  sites of, 3
  vallate papillae, 17
Taste pits, 10–11, 29–30
Taste proteins, 61, 65, 76, 269, 294, 385
Temperature, gustatory fibres response to, 73
  of test solutions, 45
Tight junctions,
  in taste buds, 10–11, 29–30
  olfactory, 116, 119–120, 124, 139
Tongue, alcohol perception from, 68
  blood flow, effect of chorda tympani stimulation, 90, 92
  stimulation,
    latency, 55
    of single papillae, 34
    of whole surface, 34

Tongue—*continued*
  taste bud mapping, 35
Tracking, 260, 262
Transduction, olfactory, 146–147, 343, 348–349, 352, 354–355
Two-bottle preference test, 38

Vallate papillae taste buds, 17
  cellular lifespan, 51
Vibrational theory of smell, 325 *et seq.*, 336 *et seq.*, 383
Villi, in olfactory cells, *see Microvilli*
Virus infection, progress along nerves, 166, 169
Vision, comparison with smell, 287–288, 351
Vomeronasal organ, 56, 138, 147, 187, 236

X-irradiation, effect on taste buds, 52

Printed by Spottiswoode, Ballantyne & Co. Ltd., London and Colchester